Industrial and Applied Mathematics

The **Industrial and Applied Mathematics** series publishes high-quality research-level monographs, lecture notes, textbooks, contributed volumes, focusing on areas where mathematics is used in a fundamental way, such as industrial mathematics, bio-mathematics, financial mathematics, applied statistics, operations research and computer science.

Bipan Hazarika · Santanu Acharjee ·
Dragan S. Djordjević
Editors

Advances in Functional Analysis and Fixed-Point Theory

An Interdisciplinary Approach

Springer

Editors
Bipan Hazarika
Department of Mathematics
Gauhati University
Guwahati, Assam, India

Santanu Acharjee
Department of Mathematics
Gauhati University
Guwahati, Assam, India

Dragan S. Djordjević
Department of Mathematics
University of Niš
Niš, Serbia

ISSN 2364-6837 ISSN 2364-6845 (electronic)
Industrial and Applied Mathematics
ISBN 978-981-99-9206-5 ISBN 978-981-99-9207-2 (eBook)
https://doi.org/10.1007/978-981-99-9207-2

Mathematics Subject Classification: 11B05, 26A06, 26A33, 26B12, 35B05, 39B42, 40A05, 40C05, 40G15, 44A45, 46A45, 47A53, 47H10, 47S10, 47A55, 54H25, 65T60, 74Q20, 90C26, 90C30, 90C34

This Springer imprint is published by the registered company Springer Nature Singapore Pte Ltd.
The registered company address is: 152 Beach Road, #21-01/04 Gateway East, Singapore 189721, Singapore

Paper in this product is recyclable.

Preface

In mathematics, functional analysis and fixed-point theory are two promising areas. Until the end of the nineteenth century, mathematical research trends were confined to only areas like differential equations, mechanics, fluid dynamics, etc. Most of these areas are applied in nature. But, at the beginning of the twentieth century, the scenario was changed, due to Banach, Brouwer, and Schauder. They initiated fundamental ideas of functional analysis and fixed-point theory. Banach's fixed-point theorem is one of the milestones in functional analysis, and it has encouraged several mathematicians to develop various fixed-point theorems in metric spaces, Banach spaces, convex sets, etc. Similarly, functional analysis is used to study various properties of many applied areas of mathematics viz. differential equations, integral equations, dynamical equations, fractional calculus, etc. On the other hand, fixed-point theory is one of the useful tools to solve different real-life problems in mathematical economics, theoretical computer science, mathematical biology, etc. It is well known that John Nash used Kakutani's fixed-point theorem to develop non-cooperative game theory. Recently, functional analysis and fixed-point theory have been found to be interconnected in various aspects. Moreover, their interconnections have helped researchers to solve many problems of differential equations, integral equations, dynamical equations, fractional calculus, etc. But, it is very difficult to find any suitable book on cutting-edge research trends connecting functional analysis, fixed-point theory, and their interdisciplinary applications. Hence, we are presenting this book entitled "Advances in Functional Analysis and Fixed-Point Theory: An Interdisciplinary Approach". There are seventeen chapters in this book. All of these chapters are written by the experts of their respective fields.

In Chap. 1, some results on functions of several variables in fractional calculus are introduced. In addition to defining the non-conformable vector α-operators, α-gradient, α-divergence, and α-rotational, the concept of non-conformable line α-integral is introduced, and its properties are established. The relationship of this definition of fractional order with double integrals is also established. Chapter 2 comprises some results in infinite programming and application in the best proximity point theory. In Chap. 3, the authors study the J-iteration process in Kohlenbach hyperbolic spaces. They prove the weak w^2-stability and data dependence theorems of this iteration process for contraction mappings. They also establish some strong and Δ-convergence theorems for generalized α-nonexpansive mappings and the finite families of total asymptotically nonexpansive mappings by using the J-iteration process. In Chap. 4, the authors construct standard fixed-point theorems for integral type contractions involving rational terms and analyse their ramifications. In addition, they provide an example to support the outcome. Finally, an application of the Fredhlom integral equation is discussed. In Chap. 5, the concepts of $(H, 1)_\lambda$-summability, λ-strongly harmonical summability, λ-statistical logarithmic convergence, and λ-statistically logarithmic boundedness of sequences of real numbers are introduced and relations between these concepts are investigated. In addition, (f, λ)-harmonic summability and (f, λ)-statistical logarithmic convergence of sequences of real numbers are defined and some inclusion relations between these spaces are given. Chapter 6 is concerned with the deferred weighted summability mean in various aspects defined in modular spaces associated with generalized double sequence of functions based on (p, q)-integers. Here, the authors introduce the concept of relative modular deferred weighted statistical convergence and statistical relative modular deferred weighted summability for a double sequence of functions, and an inclusion relation between them is established.

Haar wavelet is the only wavelet family with an explicit expression among all Daubechies wavelet families. This beneficial capability is exploited in Chap. 7 to compute some bounds for Haar wavelet coefficients corresponding to real-valued continuous functions on [0, 1]. The author hypothesizes that when there are any abrupt changes in the size of detail coefficients, the underlying function has discontinuity and this discontinuity could be located in a small interval. Location of this kind of interval using Haar wavelet provides the smallest interval among all the Daubechies wavelet families across the given resolution. In Chap. 8, the authors discuss gradual normed linear spaces and some related algebraic and topological properties of the space. Different convergence concepts of the sequences in the said space and the implication relations between them are discussed. Chapter 9 studies the spectral properties of class of sequences with respect to matrix operators. In Chap. 10, some new condensing operators are introduced in Banach spaces and with the help of these condensing operators, a number of new fixed point theorems are established. Using these new results, the existence of solutions of integral equations is studied. In Chap. 11, the authors extend Wardowski's idea of F-contraction by introducing the dual F-Suzuki type contraction and the triple F-Suzuki type contraction. They prove fixed-point theorems related to these contractions. Further, they find the existence of a solution to an integral equation. The author, in Chap. 12, has given some new properties of fractional Elzaki transformation by using its relationship with Laplace transformation. As an application of it, he also finds out fractional Elzaki transformation of some special functions along with applications to conformable fractional differential equations and relationship with other fractional integral transformations. The purpose of Chap. 13 is to investigate the basic results of the tempered fractional integrodifferential equation via the concept of Caputo fractional derivative. The existence and uniqueness results of a solution for BVP-FDEs are discussed by utilizing the Banach fixed-point theorem and Schaefer's fixed-point theorem. Moreover, the Ulam-Hyers stability results under the Lipschitz condition are studied. In Chap. 14, the authors introduce multivalued generalized ϕ-F-contraction mappings. They establish the existence of the best proximity point for such types of mappings in complete metric space. Moreover, they define multivalued generalized ϕ-F-contraction pair of mappings and obtain the best proximity point results for these. A stability result is obtained for best proximity point sets and an analogous study is done for fixed-point sets via fixed-point formulation for such mappings. A new α-$\lambda-$ rational contractive condition is defined in Chap. 15. Moreover, some fixed-points and periodic-point results, based on aforesaid contractive conditions for a mapping in extended Branciari b-distance spaces, are established. The authors obtain sufficient conditions for the existence of solutions for the two-point boundary-value differential equation as an application of the main result. Chapter 16 considers a single server vacation queuing model that requires some additional time to start a busy period under N-policy subject to some unpredictable server breakdown. In Chap. 17, the authors introduce alterning JS-metric space and study some fundamental properties connecting b-metric space, dislocated metric space, and JS-metric space.

Finally, we would like to thank Mr. Shamim Ahmad, Executive Editor, Springer, India, for his keen interest on our proposal to edit this book. Many thanks to all the members of the Springer family for their various support. Gauhati University and University of Niš deserve special thanks for providing us with suitable facilities to complete the task of editing this book. We are highly thankful to our family members and well wishers for their support, encouragement, and constructive suggestions. At the end, we are highly thankful to the Almighty for blessing us.

Assam, India — Bipan Hazarika
Assam, India — Santanu Acharjee
Niš, Serbia — Dragan S. Djordjević

Contents

Chapter 1
Some Results Related with n—variables Non Conformable Fractional Derivatives

Miguel Vivas-Cortez, Janneth Velasco Velasco, and Jorge E. Hernández Hernández

Abstract The present work introduces some results on functions of several variables in the fractional calculus setting. In addition to defining the non-conformable vector α—operators: $\alpha-$gradient, $\alpha-$divergence and $\alpha-$rotational, the concept of non-conformable line α—integral is introduced; also its properties are established. The relationship of this definition of fractional order with double integrals is also established in the same study scenario.

Keywords Non conformable α-gradient · Non conformable α-divergence · Non conformable α-rotational · Non-conformable line α-integral

1.1 Introduction

In the history of the development of Calculus, there is an important moment in which the first signs of what we now call Fractional Calculus appear. Starting with some dissertations made by Ross [22] regarding the meaning of a derivative of non-integer order, and later developments of these ideas led by Laplace, Fourier, Abel, Liouville and Riemaan, Fractional Calculus has had a great development in recent decades. From 1975 to the present, about 400 papers have been published in the subject.

M. Vivas-Cortez (✉)
Facultad de Ciencias Exactas y Naturales, Escuela de Ciencias Físicas y Matemáticas , Pontificia Universidad Católica del Ecuador, Sede Quito, Ecuador
e-mail: mjvivas@puce.edu.ec

J. Velasco Velasco
Departamento de Ciencias Exactas, Universidad de las Fuerzas Armadas ESPE, Quito, Ecuador
e-mail: javelasco3@espe.edu.ec

J. E. Hernández Hernández
Decanato de Ciencias Económicas y Empresariales, Departamento de Técnicas Cuantitativas, Universidad Centroccidental Lisandro Alvarado, Barquisimeto, Venezuela
e-mail: jorgehernandez@ucla.edu.ve

B. Hazarika et al. (eds.), *Advances in Functional Analysis and Fixed-Point Theory*, Industrial and Applied Mathematics,
https://doi.org/10.1007/978-981-99-9207-2_1

In the period 1975 to the present, about 400 articles have been published in connection with the Fractional Calculus [18, 23]. In many fields of science and engineering, including fluid flow, rheology, diffusive transport similar to diffusion, electrical networks, electromagnetic theory and probability, Fractional Calculus has found an important role. Also, Phillips [19, 20] have used the fractional method calculation in statistics [16].

After the appearance of the Riemann–Liouville fractional derivative, other authors generalized it, and on some occasions, they constructed new definitions based on the study of specific problems. Hilfer [10] makes a chronological study about this topic.

We find several definitions for fractional integrals and fractional derivatives. Just to mention a few, for fractional integrals, given an integrable function $f : [a, b] \to R$ and a positive real number α, we have the following:

Riemann–Liouville fractional integral:

$$ {}^{RL}I^{\alpha}_{a+}f(x) = \frac{1}{\Gamma(\alpha)} \int_a^x (x-t)^{\alpha-1} f(t)dt $$

Hadamard fractional integral

$$ {}^{H}I^{\alpha}_{a+}f(x) = \frac{1}{\Gamma(\alpha)} \int_a^x \left(\ln \frac{x}{t}\right)^{\alpha-1} \frac{f(t)}{t} dt $$

Erdélyi–Kober fractional integral

$$ {}^{EK}I^{\alpha}_{a+,\sigma,\eta}f(x) = \frac{\sigma x^{-\sigma(\alpha+\eta)}}{\Gamma(\alpha)} \int_a^x t^{\sigma\eta+\sigma-1}(x^{\sigma}-t^{\sigma})^{\alpha-1} f(t)dt, \quad \sigma > 0, \eta \in \mathbb{R} $$

Then, from these fractional integrals, using the classical derivative, the following fractional derivatives were obtained [12]

Riemann–Liouville fractional derivative:

$$ {}^{RL}D^{\alpha}_{a+}f(x) = \left(\frac{d}{dx}\right)^n ({}^{RL}I^{n-\alpha}_{a+}f)(x) $$

Hadamard fractional derivative

$$ {}^{H}D^{\alpha}_{a+}f(x) = \left(x\frac{d}{dx}\right)^n ({}^{H}I^{\alpha}_{a+}f)(x) $$

Erdélyi–Kober fractional derivative

$$ {}^{EK}D^{\alpha}_{a+,\sigma,\eta}f(x) = x^{-\sigma\eta} \left(\frac{1}{\sigma x^{\sigma-1}}\frac{d}{dx}\right)^n \left(x^{\sigma(n+\eta)EK}I^{n-\alpha}_{a+,\sigma,\eta+\sigma}f\right)(x) $$

In 1967, the italian Mathematician Caputo introduced a modification of the Riemann–Liouville fractional derivative of order $n-1<\alpha\leq n$ as follows:

$$ {}^{C}D_{a}^{\alpha}f(x)=\begin{cases}\frac{1}{\Gamma(n-\alpha)}\int_{a}^{x}\frac{f^{(n)}(t)}{(x-t)^{\alpha+1-n}}dt, & n-1<\alpha<n\\ \frac{d^{n}}{dx^{n}}f(x), & \alpha=n\end{cases} $$

There is a lot of useful literature for the study of fractional calculus, its advances and applications [5, 12, 15, 16, 21].

Other definitions have appeared with the intention of improving and satisfying other properties of the Riemann derivative other than linearity. By example:

1. The Riemann–Liouville fractional derivative does not satisfy $D_{\alpha}^{a}(1)=0$
2. The fractional derivatives proposed until now lose some of the basic properties that usual derivatives have, such as the product rule, the quotient rule, or the chain rule.
3. In general, the fractional derivatives proposed until now do not satisfy $D^{\alpha}D^{\beta}=D^{\alpha+\beta}$
4. The Caputo definition implies that function f must be differentiable.

Recently, Khalil in [11] introduced a new definition of fractional derivative namely "conformable fractional derivative" and Abdeljawad in [1] introduced a slight modification centering the definition in a fixed point. Based on these works, Guzman et al. [7–9, 13, 17] have developed an extensive investigation of the usefulness of these concepts treating subjects as stability of solutions of fractional non conformable differential equations, generalizations of fractional integral operators, fractional calculus of n—variable and others.

In [14], Mhailan et al., from the work of Khalil [11], gave a continuity toward a fractional vector analysis.

Motivated by the aforementioned articles, we propose to define the non-conformable vector α—operators: α—gradient, α—divergence and α—rotational, the concept of non conformable line α—integral, and establish some properties.

1.2 Preliminaries

We begin this section with a compilation of definitions and properties that will be useful in the development of this work.

The Riemaan–Liouville αth−derivative of a function f, where $\alpha\in[n-1,n)$ for some positive integer n, is given by

$$ T_{a}^{\alpha}f(t)=\frac{1}{\Gamma(n-\alpha)}\frac{d^{n}}{dt^{n}}\int_{a}^{t}\frac{f(x)}{(t-x)^{\alpha-n+1}}dx, $$

and the Caputo fractional αth−derivative of a function f is given by

$$T_a^\alpha f(t) = \frac{1}{\Gamma(n-\alpha)} \int_a^t \frac{f^{(n)}(x)}{(t-x)^{\alpha-n+1}} dx.$$

Both definitions do not satisfy all the properties of the classical ordinary derivative, however, in [11], Khalil et al. give a new definition called α—fractional conformable derivative for which some similar properties of the ordinary derivative have been proved (See [4, 6, 11, 14]). This definition is established as follows: for a fucntion $f : [0, \infty) \to R$ the α—conformable fractional derivative of order $\alpha \in (0, 1]$ of f in $t > 0$ is given by

$$T_\alpha f(t) = \lim_{\varepsilon \to 0} \frac{f(t + \varepsilon t^{1-\alpha}) - f(t)}{\varepsilon}. \tag{1.1}$$

The authors specify that they use the notation $f^{(\alpha)}$ instead $T_\alpha f$ and a function f is α—differentiable on an arbitrary interval (a, b) if the limit (1.1) exists for all $t \in (a, b)$. Also, if f is α—differentiable on $(0, a)$ for some $a > 0$ and

$$\lim_{t \to 0^+} f^{(\alpha)}(t)$$

exists then

$$f^{(\alpha)}(0) = \lim_{t \to 0^+} f^{(\alpha)}(t).$$

In [14], Mailan et al., using (1.1), introduced some results concerning fractional vectorial analysis. The authors considered scalar functions and vectorial functions of several variables and defined the fractional gradient and the fractional divergence respectively, as follows. For a function $f : R^3 \to R$, the fractional conformable partial derivative with respect to the first variable is given by

$$D_x^\alpha f(x, y, z) = \lim_{\varepsilon \to 0} \frac{f(x + \varepsilon x^{1-\alpha}, y, z) - f(x, y, z)}{\varepsilon}$$

(similarly are defined the fractional conformable partial derivative with respect the other variable) and then the gradient of f is defined as

$$\nabla^\alpha f(x, y, z) = D_x^\alpha f(x, y, z) \overrightarrow{i} + D_y^\alpha f(x, y, z) \overrightarrow{j} + D_z^\alpha f(x, y, z) \overrightarrow{k}$$

and if $\mathbf{F} : R^3 \to R^3$ is defined by $\mathbf{F}(x, y, z) = P(x, y, z) \overrightarrow{i} + Q(x, y, z) \overrightarrow{j} + S(x, y, z) \overrightarrow{k}$ then the fractional divergence of f is given by

$$\nabla^\alpha \mathbf{F}(x, y, z) = D_x^\alpha P(x, y, z) + D_y^\alpha Q(x, y, z) + D_z^\alpha S(x, y, z),$$

also it is given the fractional the fractional curl of a function $\mathbf{F} : R^3 \to R^3$ such that $\mathbf{F}(x, y, z) = (P(x, y, z), Q(x, y, z), S(x, y, z))$ as follows:

$$
\begin{aligned}
(\nabla^{\alpha} \wedge \mathbf{F})(x, y, z) = &\left(D_y^{\alpha} S(x, y, z) - D_z^{\alpha} Q(x, y, z)\right) \vec{i} \\
&- \left(D_x^{\alpha} S(x, y, z) - D_z^{\alpha} P(x, y, z)\right) \vec{j} \\
&+ \left(D_x^{\alpha} Q\ (x, y, z) - D_y^{\alpha} P(x, y, z)\right) \vec{k}.
\end{aligned}
$$

From these definitions, some properties are obtained and some theorems concerning with fractional line integrals and a fractional Green functions are established.

In [7], Guzman et al. gave a new definition of fractional derivative of local type as follows.

Definition 1.1 Given a function $f : [0, +\infty) \to \mathbb{R}$. Then the N-derivative of f of order α is defined by

$$
N_1^{\alpha} f(t) = \lim_{\varepsilon \to 0} \frac{f(t + \varepsilon e^{t^{-\alpha}}) - f(t)}{\varepsilon}
$$

for all $t > 0$, $\alpha \in (0, 1)$. If f is α—differentiable on $(0, a)$ for some $a > 0$ and $\lim_{t\to 0^+} N_1^{\alpha} f(t)$ exists then $N_1^{\alpha} f(0) = \lim_{t\to 0^+} N_1^{\alpha} f(t)$.

This new definition satisfies the following properties [7, Page 90]:

1. $N_1^{\alpha}(af + bg)(t) = aN_1^{\alpha} f(t) + bN_1^{\alpha} g(t), a, b \in R$
2. $N_1^{\alpha} f(t) = 0$ if f is a constant function
3. $N_1^{\alpha}(fg)(t) = g(t)N_1^{\alpha} f(t) + f(t)N_1^{\alpha} g(t)$
4. $N_1^{\alpha}(f/g)(t) = \dfrac{g(t)N_1^{\alpha} f(t) - f(t)N_1^{\alpha} g(t)}{(g(t))^2}$, with $g(t) \neq 0$
5. $N_1^{\alpha}(t^p) = e^{t^{-\alpha}} p t^{p-1}, p \in R$
6. If f is differentiable then $N_1^{\alpha} f(t) = e^{t^{-\alpha}} f'(t)$
7. Being f differentiable and $\alpha = n$ then $N_1^{\alpha} f(t) = e^{t^{-n}} f'(t)$

There exist other definitions of non conformable fractional derivatives [8, 13, 17].

This type of non conformable fractional derivative follows the corresponding fractional integral operator.

Now, we give the definition of non conformable fractional integral:

Definition 1.2 Let $\alpha \in (0, 1]$ and $0 \le u \le v$. We say that a function $h : [u, v] \to \mathbb{R}$ is α-fractional integrable on $[u, v]$, if the integral

$$
{}_{N_1}J_u^{\alpha} h(x) = \int_u^x e^{-t^{-\alpha}} h(t) dt
$$

exists and is finite.

The following statement is analogous to the one known from the Ordinary Calculus (see [3]).

Theorem 1.1 *Let f be N-differentiable function in (t_0, ∞) with $\alpha \in (0, 1]$. Then for all $t > t_0$, we have:*

(a) If f is differentiable ${}_{N_1}J^{\alpha}_{t_0}\left(N^{\alpha}_1 f(t)\right) = f(t) - f(t_0)$.
(b) $N^{\alpha}_1\left({}_{N_1}J^{\alpha}_{t_0} f(t)\right) = f(t)$.

Proof

(a) From definition, we have

$$\begin{aligned} {}_{N_1}J^{\alpha}_{t_0}\left(N^{\alpha}_1 f(t)\right) &= \int_{t_0}^{t} e^{-s^{-\alpha}} N^{\alpha}_1 f(s)ds \\ &= \int_{t_0}^{t} e^{-s^{-\alpha}} e^{s^{-\alpha}} f'(s)ds \\ &= f(t) - f(t_0). \end{aligned}$$

(b) Analogously we have

$$N^{\alpha}_1\left({}_{N_1}J^{\alpha}_{t_0} f(t)\right) = e^{t^{-\alpha}} \frac{d}{dt}\left[\int_{t_0}^{t} e^{-s^{-\alpha}} f(s)ds\right] = f(t).$$

The proof is complete. □

Also, in [13], Martinez et al. introduced the fractional non conformable partial derivative as follows. Given a real-valued function $f : R^n \rightarrow R$ and $\mathbf{a} = (a_1, ..., a_n) \in R^n$ with positive components. Then the non conformable partial $N-$derivative of f of order α in the point $\mathbf{a}$ is defined by

$$N^{\alpha}_{1,x_i} f(a_1, ..., a_n) = \lim_{\varepsilon \rightarrow 0} \frac{f(a_1, ..., a_i + \varepsilon e^{a_i^{-\alpha}}, ..., a_n) - f(a_1, ..., a_n)}{\varepsilon} \quad (1.2)$$

if it exists, is denoted by $N^{\alpha}_{1,x_i} f(\mathbf{a})$, and called the ith non conformable partial derivative of f of the order $\alpha \in (0, 1]$ at $\mathbf{a}$. Using the last definition, also is defined the non conformable N_1-gradient for a real-valued function of several variable

$$N^{\alpha}_1 - grad(f)(\mathbf{a}) = \nabla^{\alpha}_{N_1} f\,(\mathbf{a}) = \left(N^{\alpha}_{1,x_1} f(\mathbf{a}), ..., N^{\alpha}_{1,x_n} f(\mathbf{a})\right). \quad (1.3)$$

For this non conformable fractional derivative, the mean value, chain rule and implicit function theorems for real-valued functions of several variables are proved.

Measure of noncompactness is an important tool in Banach spaces. It can be used in fixed point theory, differential equations, integral equations, integro-differential, functional equations, etc.

1.3 Main Results

The following describes the setting where this work is developed. From now on, we denote with SV the space of real-valued functions of several variables and with VF the space of vector-valued functions of several variables. We will continue using the notations and definitions given in (1.2) and (1.3). If a function $\mathbf{F} \in VF$ is the non conformable α—gradient of a function $f \in SV$, i.e. $\mathbf{F} = \nabla^{\alpha}_{N_1} f$, then f will be called the non conformable α—potential function for $\mathbf{F}$.

Let $f : U \subset R^n \to R$ be a function in SV and $\mathbf{a} \in R^n$ with positive components, if $N^{\alpha}_{1,x_i} f(\mathbf{a})$ exists for all $\mathbf{a} \in U$, then $N^{\alpha}_{1,x_i} f$ is a real- valued function to which we can consider to find its non conformable N−partial derivatives. So, if $1 \leq j \leq n$ the expression $N^{\alpha}_{1,x_j} \left(N^{\alpha}_{1,x_i} f\right)(\mathbf{a})$ represent the non conformable N−partial derivative of $N^{\alpha}_{1,x_i} f$ with respect to x_j in $\mathbf{a}$. This last will be called the second non conformable partial α—derivatives, and from now on, the following notations will be used

$$N^{\alpha}_{1,x_j} \left(N^{\alpha}_{1,x_i} f\right) = N^{(2)\alpha}_{1,x_j x_j} f, \quad (i \neq j),$$
$$N^{\alpha}_{1,x_i} \left(N^{\alpha}_{1,x_i} f\right) = N^{(2)\alpha}_{1,x_i x_i} f, \quad (i = j).$$

Definition 1.3 Let $\mathbf{F} : R^3 \to R^3$ defined by $\mathbf{F}(x, y, z) = (P(x, y, z), Q(x, y, z), S(x, y, z))$ where $P, Q, S : R^3 \to R$ are real-valued functions for which the partial α—derivatives $N^{\alpha}_{1,x_1} P$, $N^{\alpha}_{1,x_2} Q$ and $N^{\alpha}_{1,x_3} S$ exist. Then

$$\nabla^{\alpha}_{N_1} \cdot F = N^{\alpha}_{1,x_1} P + N^{\alpha}_{1,x_2} Q + N^{\alpha}_{1,x_3} S$$

is a scalar field called the non-conformable α—divergence of $\mathbf{F}$.

Theorem 1.2 *Let $f : U \subset R^2 \to R$ be a real-valued function defined over an open set $U \subset R^2$, and a point $(a, b) \in U$. If $N^{\alpha}_{1,x_1} f$, $N^{\alpha}_{1,x_2} f$, $N^{(2)\alpha}_{1,x_1,x_2} f$ and $N^{(2)\alpha}_{1,x_2,x_1} f$ are continuous in (a, b) then*

$$N^{(2)\alpha}_{1,x_1,x_2} f(a, b) = N^{(2)\alpha}_{1,x_2,x_1} f(a, b). \tag{1.4}$$

Proof Let $\varepsilon > 0$, such that the points (a, b), $(a + \varepsilon e^{a^{-\alpha}}, b)$, $(a, b + \varepsilon e^{b^{-\alpha}})$ and $(a + \varepsilon e^{a^{-\alpha}}, b + \varepsilon e^{b^{-\alpha}})$ are vertices of a rectangle $[a, a + \varepsilon e^{a^{-\alpha}}] \times [b, b + \varepsilon e^{b^{-\alpha}}] \subset U$. Let Ω be the expression defined by means of

$$\Omega = f(a + \varepsilon e^{a^{-\alpha}}, b + \varepsilon e^{b^{-\alpha}}) - f(a + \varepsilon e^{a^{-\alpha}}, b) - f(a, b + \varepsilon e^{b^{-\alpha}}) + f(a, b). \tag{1.5}$$

By an application of the unidimensional non conformable Mean value Theorem [17, Theorem 2.7] to the real-valued function defined by

$$F(x) = f(x, b + \varepsilon e^{b^{-\alpha}}) - f(x, b) \tag{1.6}$$

we have

$$\begin{aligned}\Omega &= F(a+\varepsilon e^{a^{-\alpha}}) - F(a)\\ &= \varepsilon(e^{a-c_1})^{-\alpha}\left[N^{\alpha}_{1,x_1} f(c_1, b+\varepsilon e^{b^{-\alpha}}) - N^{\alpha}_{1,x_1} f(c_1, b)\right]\end{aligned}$$

for some $c_1 \in [a, a+\varepsilon e^{a^{-\alpha}}]$. If we consider the function $g(y) = N_{1,x_1} f(c_1, y)$, then

$$\Omega = \varepsilon(e^{a-c_1})^{-\alpha}\left[g(b+\varepsilon e^{b^{-\alpha}}) - g(b)\right].$$

Since $N^{(2)\alpha}_{1,x_1,x_2} f$ exists and it is continuous then $g'(y) = N^{(2)\alpha}_{1,x_1,x_2} f(c_1, y)$ exists and is continuous provided $y \in [b, b+\varepsilon e^{b^{-\alpha}}]$; again, by an application of the non conformable Mean Value Theorem we have

$$\Omega = \varepsilon^2(e^{a+b-(c_1+d_1)})^{-\alpha} N^{(2)\alpha}_{1,x_1,x_2} f(c_1, d_1).$$

for some $d_1 \in [b, b+\varepsilon e^{b^{-\alpha}}]$.

Now, let's define the function H by means of

$$H(y) = f(a+\varepsilon e^{a^{-\alpha}}, y) - f(a, y)$$

so we have

$$\Omega = H(b+\varepsilon e^{b^{-\alpha}}) - H(b)$$

and with the same reasoning as above, we obtain

$$\Omega = \varepsilon^2(e^{a+b-(c_2+d_2)})^{-\alpha} N^{(2)\alpha}_{1,x_2,x_1} f(c_2, d_2).$$

for some $c_2 \in [a, a+\varepsilon e^{a^{-\alpha}}]$ and $d_2 \in [b, b+\varepsilon e^{b^{-\alpha}}]$. In this way, we have found that

$$\varepsilon^2(e^{a+b-(c_1+d_1)})^{-\alpha} N^{(2)\alpha}_{1,x_1,x_2} f(c_1, d_1) = \varepsilon^2(e^{a+b-(c_2+d_2)})^{-\alpha} N^{(2)\alpha}_{1,x_2,x_1} f(c_2, d_2),$$

and since $\varepsilon > 0$, then

$$(e^{a+b-(c_1+d_1)})^{-\alpha} N^{(2)\alpha}_{1,x_1,x_2} f(c_1, d_1) = (e^{a+b-(c_2+d_2)})^{-\alpha} N^{(2)\alpha}_{1,x_2,x_1} f(c_2, d_2). \tag{1.7}$$

Since $c_1, c_2 \in [a, a+\varepsilon e^{a^{-\alpha}}]$ then there exist $\theta_1, \theta_2 \in (0, 1)$ such that

$$c_1 = a + \theta_1 \varepsilon e^{a^{-\alpha}} \text{ and } c_2 = a + \theta_2 \varepsilon e^{a^{-\alpha}},$$

and since $d_1, d_2 \in [b, b+\varepsilon e^{b^{-\alpha}}]$ then there exist $\vartheta_1, \vartheta_2 \in (0, 1)$ such that

$$d_1 = b + \vartheta_1 \varepsilon e^{b^{-\alpha}} \text{ and } d_2 = b + \vartheta_2 \varepsilon e^{b^{-\alpha}}.$$

So,

$$(e^{a+b-(c_1+d_1)})^{-\alpha} = (e^{-\varepsilon(\theta_1 e^{a^{-\alpha}}+\theta_2 e^{b^{-\alpha}})})^{-\alpha},$$
$$(e^{a+b-(c_2+d_2)})^{-\alpha} = (e^{-\varepsilon(\vartheta_1 e^{a^{-\alpha}}+\vartheta_2 e^{b^{-\alpha}})})^{-\alpha},$$
$$N^{(2)\alpha}_{1,x_1,x_2} f(c_1, d_1) = N^{(2)\alpha}_{1,x_1,x_2} f(a + \theta_1 \varepsilon e^{a^{-\alpha}}, b + \vartheta_1 \varepsilon e^{b^{-\alpha}}) \text{ and}$$
$$N^{(2)\alpha}_{1,x_2,x_1} f(c_2, d_2) = N^{(2)\alpha}_{1,x_2,x_1} f(a + \vartheta_1 \varepsilon e^{a^{-\alpha}}, b + \vartheta_2 \varepsilon e^{b^{-\alpha}}).$$

By the continuity of the involved functions and the second non conformable partial α—derivatives, taking limit in (1.7) when $\varepsilon \to 0$, we have

$$N^{(2)\alpha}_{1,x_1,x_2} f(a, b) = N^{(2)\alpha}_{1,x_2,x_1} f(a, b).$$

The proof is complete. □

Definition 1.4 Let $\mathbf{F} : R^3 \to R^3$ be a vector- valued function with non conformable partial α—derivatives of order $\alpha \in (0, 1]$ in $\mathbf{a} = (x, y, z)$, then the expression

$$\alpha - div(\mathbf{F})(\mathbf{a}) = \nabla^{\alpha}_{N_1} \cdot \mathbf{F}(\mathbf{a}) = N^{\alpha}_{1,x_1} P(\mathbf{a}) + N^{\alpha}_{1,x_2} Q(\mathbf{a}) + N^{\alpha}_{1,x_3} S(\mathbf{a})$$

is called the non conformable α—divergence of $\mathbf{F}$ in $\mathbf{a}$. Also, the expression

$$\begin{aligned}\alpha - rot(\mathbf{F})(\mathbf{a}) = (\nabla^{\alpha}_{N_1} \wedge \mathbf{F})(\mathbf{a}) = \left(N^{\alpha}_{1,x_2} R(\mathbf{a}) - N^{\alpha}_{1,x_3} \; Q(\mathbf{a})\right) \overrightarrow{i} \\ - \left(N^{\alpha}_{1,x_1} R(\mathbf{a}) - N^{\alpha}_{1,x_3} P(\mathbf{a})\right) \overrightarrow{j} \\ + \left(N^{\alpha}_{1,x_1} Q(\mathbf{a}) - N^{\alpha}_{1,x_2} P(\mathbf{a})\right) \overrightarrow{k}.\end{aligned}$$

is called the non conformable α—rotational of $\mathbf{F}$.

Theorem 1.3 *let $f : U \subset R^3 \to R$ and $\mathbf{F} : U \subset R^3 \to R^3$ be a real an vector-valued functions, respectively, where $\mathbf{F} = (F_1, F_2, F_3)$. Then*

1. *The non conformable $N^{\alpha}_1 - grad$ operator $\nabla^{\alpha}_{N_1}$ is linear*
2. *$\nabla^{\alpha}_{N_1} f = 0$ if f is a constant function*
3. *The non conformable α—divergence is a linear operator*
4. *The non conformable α—rotational is a linear operator.*

Proof If $a, b \in R$ and f, g are real functions defined on an open set $U \in R^n$, then we have

$$\begin{aligned}\nabla^{\alpha}_{N_1}(af + bg)(\mathbf{a}) &= \left(N^{\alpha}_{1,x_1}(af + bg)(\mathbf{a}), ..., N^{\alpha}_{1,x_n}(af + bg)(\mathbf{a})\right) \\ &= (aN^{\alpha}_{1,x_1} f(\mathbf{a}) + bN^{\alpha}_{1,x_1} g(\mathbf{a}), ..., aN^{\alpha}_{1,x_n} f(\mathbf{a}) + bN^{\alpha}_{1,x_n} g(\mathbf{a})) \\ &= a\nabla^{\alpha}_{N_1} f(\mathbf{a}) + b\nabla^{\alpha}_{N_1} g(\mathbf{a}),\end{aligned}$$

so 1. is proved.

Suppose that $f(\mathbf{a}) = k$ for some $k \in R$ and for all $\mathbf{a} \in U$, then the N_1^α partial derivatives are 0, therefore, $\nabla_{N_1}^\alpha (f)(\mathbf{a}) = 0$ for all $\mathbf{a} \in U$, so item 2 is proved.

If $a, b \in R$ and $\mathbf{G} : U \in R^3 \to R^3$ where $\mathbf{G} = (G_1, G_2, G_3)$are vector-valued functions defined on an open set $U \in R^n$ then

$$\begin{aligned}\nabla_{N_1}^\alpha \cdot (a\mathbf{F} + b\mathbf{G})((\mathbf{a}) &= \nabla_{N_1}^\alpha \cdot (af_1 + bg_1)(\mathbf{a}), ..., af_n + bg_n)(\mathbf{a})) \\ &= N_{1,x_1}(af_1 + bg_1)(\mathbf{a}) + ... + N_{1,x_n}(af_n + bg_n)(\mathbf{a}) \\ &= a\nabla_{N_1}^\alpha \cdot \mathbf{F}(\mathbf{a}) + b\nabla_{N_1}^\alpha \cdot \mathbf{G}(\mathbf{a}),\end{aligned} \tag{1.8}$$

also

$$\begin{aligned}\nabla_{N_1}^\alpha \wedge (a\mathbf{F} + b\mathbf{G})(\mathbf{a}) &= \begin{vmatrix} \vec{i} & \vec{j} & \vec{k} \\ N_{1,x_1}^\alpha & N_{1,x_2}^\alpha & N_{1,x_3}^\alpha \\ (aF_1 + bG_1) & (aF_2 + bG_2) & (aF_3 + bG_3) \end{vmatrix} \\ &= \left(N_{1,x_2}^\alpha (aF_3 + bG_3)(\mathbf{a}) - N_{1,x_3}^\alpha (aF_2 + bG_2)(\mathbf{a})\right)\vec{i} \\ &\quad - \left(N_{1,x_1}^\alpha (aF_3 + bG_3)(\mathbf{a}) - N_{1,x_3}^\alpha (aF_1 + bG_1)(\mathbf{a})\right)\vec{j} \\ &\quad + \left(N_{1,x_1}^\alpha (aF_2 + bG_2)(\mathbf{a}) - N_{1,x_2}^\alpha (aF_1 + bG_1)(\mathbf{a})\right), \vec{k}.\end{aligned}$$

and using the linearity of the non conformable N_{1,x_i}^α partial α—derivative we have

$$\nabla_{N_1}^\alpha \wedge (a\mathbf{F} + b\mathbf{G})(\mathbf{a}) = a\nabla_{N_1}^\alpha \wedge \mathbf{F}(\mathbf{a}) + b\nabla_{N_1}^\alpha \wedge \mathbf{G}(\mathbf{a}) \tag{1.9}$$

By the results (1.8) and (1.9) we obtain items 3. and 4., respectively. □

Definition 1.5 Let $f : R^3 \to R$ be a real-valued function. Then

$$\nabla_{N_1}^{(2)\alpha} f = \nabla_{N_1}^\alpha \cdot (\nabla_{N_1}^\alpha f) = N_{1,x_1}^{(2)\alpha} f + N_{1,x_2}^{(2)\alpha} f + N_{1,x_1}^{(3)\alpha} f$$

is called the non conformable N_1^α—Laplacian of f.

Functions that satisfy the condition $\nabla_{N_1}^{(2)\alpha} f = 0$ are called non conformable N_1^α—harmonics.

Theorem 1.4 *Let $f : U \subset \mathbb{R}^3 \to \mathbb{R}$ and $\mathbf{F} : U \subset \mathbb{R}^3 \to \mathbb{R}^3$ be a real-valued and vector-valued functions, respectively. If the second non conformables partial N_1^α—derivatives exist and are continuous, then*

1. $\nabla_{N_1}^\alpha \wedge (\nabla_{N_1}^\alpha f) = \mathbf{0}$
2. $\nabla_{N_1}^\alpha \cdot (\nabla_{N_1}^\alpha \wedge f) = 0$
3. $\nabla_{N_1}^\alpha \cdot (f\mathbf{F}) = f \cdot (\nabla_{N_1}^\alpha \mathbf{F}) + \mathbf{F} \cdot (\nabla_{N_1}^\alpha f)$.

Proof From the definition of the wedge product, we have

$$\nabla^\alpha_{N_1} \wedge (\nabla^\alpha_{N_1} f) = \begin{vmatrix} \overrightarrow{i} & \overrightarrow{j} & \overrightarrow{k} \\ N^\alpha_{1,x_1} & N^\alpha_{1,x_2} & N^\alpha_{1,x_3} \\ N^\alpha_{1,x_1} f & N^\alpha_{1,x_2} f & N^\alpha_{1,x_3} f \end{vmatrix}$$

$$= \left(N^\alpha_{1,x_2}(N^\alpha_{1,x_3} f) - N^\alpha_{1,x_3}(N^\alpha_{1,x_2} f)\right) \overrightarrow{i}$$
$$- \left(N^\alpha_{1,x_1}(N^\alpha_{1,x_3} f) - N^\alpha_{1,x_3}(N^\alpha_{1,x_1} f)\right) \overrightarrow{j}$$
$$+ \left(N^\alpha_{1,x_1}(N^\alpha_{1,x_2} f) - N^\alpha_{1,x_2}(N^\alpha_{1,x_1} f)\right), \overrightarrow{k}$$
$$= \mathbf{0}.$$

Also, from definition of the scalar product and the wedge product, we have

$$\nabla^\alpha_{N_1} \cdot (\nabla^\alpha_{N_1} \wedge \mathbf{F}) = \nabla^\alpha_{N_1} \cdot \begin{vmatrix} \overrightarrow{i} & \overrightarrow{j} & \overrightarrow{k} \\ N^\alpha_{1,x_1} & N^\alpha_{1,x_2} & N^\alpha_{1,x_3} \\ F_1 & F_2 & F_3 \end{vmatrix}$$

$$= \nabla^\alpha_{N_1} \cdot \Big[\left(N^\alpha_{1,x_2} F_3 - N^\alpha_{1,x_3} F_2\right) \overrightarrow{i}$$
$$- \left(N^\alpha_{1,x_1} F_3 - N^\alpha_{1,x_3} F_1\right) \overrightarrow{j}$$
$$+ \left(N^\alpha_{1,x_1} F_2 - N^\alpha_{1,x_2} F_1\right) \overrightarrow{k}\Big]$$
$$= N^\alpha_{1,x_1}\left(N^\alpha_{1,x_2} F_3 - N^\alpha_{1,x_3} F_2\right) + N^\alpha_{1,x_2}\left(-N^\alpha_{1,x_1} F_3 + N^\alpha_{1,x_3} F_1\right)$$
$$+ N^\alpha_{1,x_3}\left(N^\alpha_{1,x_1} F_2 - N^\alpha_{1,x_2} F_1\right)$$
$$= 0.$$

So, item 1 and 2 are proved.

Following the definition of scalar product, we get the proof of item 3. □

1.4 Non Conformable N_1^α—line Integral

Let $\mathbf{F} = (F_1, \cdots, F_n)$ be a vector-valued function with $F_i : \mathbb{R}^n \to \mathbb{R}$ for $i = 1, \cdots, n$. Let

$$r(\alpha) = \left(\int_0^{x_1} e^{-t^{-\alpha}} dt, \int_0^{x_2} e^{-t^{-\alpha}} dt, \ldots, \int_0^{x_n} e^{-t^{-\alpha}} dt\right).$$

then

$$dr(\alpha) = \left(e^{-x_1^{-\alpha}} dx_1, e^{-x_2^{-\alpha}} dx_2, \ldots, e^{-x_n^{-\alpha}} dx_n\right).$$

Definition 1.6 Let γ be a piecewise smooth curve in the domain of F which lies in $E = \{(x, y) : x > 0, y > 0\}$. Then

$$\int_\gamma \mathbf{F}\cdot dr(\alpha) = \int_\gamma F_1 e^{-x_1^{-\alpha}} dx_1 + \cdots + F_n e^{-x_n^{-\alpha}} dx_1$$

Example 1 Let $\mathbf{F}(x, y) = (e^{x^{-1/2}}, e^{y^{-1/2}})$, $\gamma(t) = (t, t)$, $(0 \leq t \leq 1)$ and $\alpha = 1/2$. Then

$$\int_\gamma \mathbf{F}\cdot dr(\alpha) = \int_\gamma e^{x^{-1/2}} e^{-x^{-1/2}} dx + e^{y^{-1/2}} e^{-y^{-1/2}} dy$$
$$= \int_\gamma dx + dy = 2$$

Theorem 1.5 *Let* $\mathbf{F}, \mathbf{G}$ *be vector-valued functions defined on an open set* $U \subset \mathbb{R}^n$, γ *a curve defined on an interval* $[a, b]$ *and a scalar* k. *Then:*

1. $\int_\gamma (\mathbf{F} + \mathbf{G}) \cdot dr(\alpha) = \int_\gamma \mathbf{F} \cdot dr(\alpha) + \int_\gamma \mathbf{G} \cdot dr(\alpha)$
2. $\int_\gamma k\mathbf{F} \cdot dr(\alpha) = k \int_\gamma \mathbf{F} \cdot dr(\alpha)$

Proof The proof easily follows from the properties of the scalar product. □

Theorem 1.6 *Let* Ω *be a simply connected open set in* $E \subset \mathbb{R}^n$. *Assume* $F : \Omega \to \mathbb{R}^n$ *be an* N_1^α*—conservative force. Then* $\int_\gamma \mathbf{F} \cdot dr(\alpha)$ *is path independent.*

Proof Since $\mathbf{F}$ is N_1^α—conservative force then

$$\mathbf{F}(\mathbf{a}) = \nabla_{N_1}^\alpha f(\mathbf{a}) = \left(N_{1,x_1}^\alpha f(\mathbf{a}), ..., N_{1,x_n}^\alpha f(\mathbf{a})\right).$$

for some $f : \Omega \to \mathbb{R}$. Now, if $\mathbf{F} = (F_1, \cdots, F_n)$ then $F_i = N_{1,x_i}^\alpha f$, $(i = 1, \cdots, n)$. Using the properties of the non conformable N_1^α derivative, we have

$$\mathbf{F} \cdot dr(\alpha) = f_{x_1} dx_1 + \cdots + f_{x_n} dx_n = df.$$

Consequently

$$\int_\gamma \mathbf{F} \cdot dr(\alpha) = \int_\gamma df = f(\mathbf{a}_1) - f(\mathbf{a}_0),$$

whwere γ is the path joining $\mathbf{a}_0$ to $\mathbf{a}_1$. Hence the integral is independent of γ.

The proof is complete. □

Theorem 1.7 *Let* $F : \Omega \to \mathbb{R}^2$ *with* $\mathbf{F} = (F_1, F_2)$. *Assume that*

1. F_i *is a continuous function for* $i = 1, 2$.
2. F_i *have continuous* N_1^α*—partial derivative for* $i = 1, 2$.
3. Ω *is simply connected.*

Then for any closed piecewise smooth curve γ in Ω, we have

$$\int_{\gamma} \mathbf{F} \cdot dr(\alpha) = \int_{\gamma} F_1 e^{-x_1^{-\alpha}} dx_1 + F_2 e^{-x_2^{-\alpha}} dx_2$$
$$= \iint_{\Omega} \left(N^{\alpha}_{1,x_1} F_2 - N^{\alpha}_{1,x_2} F_1 \right) e^{-x_1^{-\alpha}} e^{-x_2^{-\alpha}} dx_1 dx_2.$$

Proof We fix two arbitrary points $z_1 = (x_{1,1}, x_{2,1})$ and $z_2 = (x_{1,2}, x_{2,2})$ on the path γ. Let γ_1 and γ_2 be the parts of γ that join the points from z_1 to z_2, and from z_2 to z_1, respectively. Then $\gamma = \gamma_1 \cup \gamma_2$. Now,

$$\int_{\gamma} \mathbf{F} \cdot dr(\alpha) = \int_{\gamma} F_1 e^{-x_1^{-\alpha}} dx_1 + F_2 e^{-x_2^{-\alpha}} dx_2$$
$$= \int_{\gamma_1} F_1 e^{-x_1^{-\alpha}} dx_1 + F_2 e^{-x_2^{-\alpha}} dx_2 - \int_{\gamma_2} F_1 e^{-x_1^{-\alpha}} dx_1 + F_2 e^{-x_2^{-\alpha}} dx_2$$

Suppose that γ_1 has a domain $[a, b]$ such that $x_1 = t, a \le t \le b$ and $x_2 = g_1(t)$ with $x_{1,1} = g_1(a)$ and $x_{2,1} = g_1(b)$. Similarly, for γ_2, we have $x_1 = t, a \le t \le b$ and $x_2 = g_2(t)$ with $x_{2,2} = g(a)$ and $x_{1,2} = g(b)$. Then

$$\int_{\gamma_1} F_1 e^{-x_1^{-\alpha}} dx_1 - \int_{\gamma_2} F_1 e^{-x_1^{-\alpha}} dx_1 = \int_a^b F_1(t, g_1(t)) e^{-t^{-\alpha}} dt - \int_a^b F_1(t, g_2(t)) e^{-t^{-\alpha}} dt$$
$$= \int_a^b \left(F_1(t, g_1(t)) - F_1(t, g_2(t))\right) e^{-t^{-\alpha}} dt$$
$$= \int_a^b \left(\int_a^b N^{\alpha}_{1,x_2} F_1 e^{-x_2^{-\alpha}} dx_2 \right) e^{-x_1^{-\alpha}} dx_1$$

Similarly, we use the same technique for the other piece of the line integral and we get the result. This ends the proof. □

1.5 Conclusion

In our present investigation, we introduce some results on functions of several variables in the fractional calculus setting, specifically when we use a non conformable fractional derivative. In addition to defining the non conformable vector α—operators: $\alpha-$gradient, $\alpha-$divergence and $\alpha-$rotational, the concept of non-conformable line α—integral is introduced; also, its properties are established. The relationship of this definition of fractional order with double integrals is also established in the same study scenario.

References

1. Abdeljawad, T.: On conformable fractional calculus. J. Comput. Appl. Math. **279**, 57–66 (2015)
2. Anastassiou, G.A.: Advances on Fractional Inequalities. Springer, New York (2011)
3. Apostol, T.: Calculus, vol. 2. Willey USA (1969)
4. Antangana, A., Baleanu, D.: New properties of conformable derivatives. Open Math. **13**, 889–898 (2015)
5. Diethelm, K.: The Analysis of Fractional Differential Equations: An Application-Oriented Exposition Using Differential Operators of Caputo Type. Springer, Berlin (2010)
6. Gozutok, N.Y., Gozutok, U.: Multi variable conformable fractional calculus. Filomat **32(1)**), 45–55 (2018)
7. Guzman, P.M., Langton, G., Lugo, L.M., Medina, J., Nápoles Valdés, J.E.: A new definition of a fractional derivative of local type. J. Math. Annal. **9**(2), 88–98 (2018)
8. Guzman, P.M., Lugo, L.M., Nápoles Valdés, J.E.: On the stability of solutions of fractional non conformable differential equations. Stud. Univ. Babes-Bolyai **65**(4), 495–502 (2020)
9. Guzman, P.M., Lugo Motta, L.M., Nápoles Valdés, J.E., Vivas-Cortez, M.: On a new generalized integral operator and certain operating properties. Axioms **9**, 1–10 (2020)
10. Hilfer, R.: Threefold Introduction to Fractional Derivatives. Anomalous Transport: Foundations and Applications. Wiley-Weinheim (2008)
11. Khalil, R., Horani, A., Yousef, A., Sababheh, M.: A new definition of fractional derivative. J. Comput. Appl. Math. **264**, 65–70 (2014)
12. Kilbas, A.A., Srivastava, H., Trujillo, J.J.: Theory and Applications of Fractional Differential Equations, North-Holland Mathematics Studies 204. Elsevier Science B.V., Amsterdam (2006)
13. Martínez, F., Nápoles Valdés, J.E.: Towards a non conformable fractional calculus of n-variables. Math. Appl. **43**, 87–98 (2020)
14. Mhailan, M., Hammad, M.A., Horani, M.A., Khalil, R.: On fractional vector analysis. J. Math. Comput. **6**, 2320–2326 (2020)
15. Mathai, A.M., Haubold, H.J.: Erdélyi-Kober Fractional Calculus From a Statistical Perspective. Inspired by Solar Neutrino Physics. Springer Nature Singapore (2018)
16. Miller, K., Ross, B.: An Introduction to the Fractional Calculus and Fractional Differential Equations. Wiley, New York (1993)
17. Nápoles Valdés, J.E., Guzman, P.M., Lugo, L.M.: Some new results on non conformable fractional calculus. Adv. Dyn. Syst. Appl. **13**(2), 167–175 (2018)
18. Oldham, K.B., Spanier, J.: The Fractional Calculus. Academic Press, New York (1974)
19. Phillips, P.C.: Fractional matrix calculus and the distribution of multivariate tests. In: Cowles Foundation Paper 767, Department of Economics. Yale University, New Haven, Conn (1989)
20. Phillips, P.C.: Operational calculus and regression t-tests. Cowles Foundation Paper 948, Department of Economics. Yale University, New Haven, Conn (1990)
21. Podlubny, I.: Fractional Differential Equations. Academic Press, USA (1999)
22. Ross, B.: The development of fractional calculus from 1695 to 1900. Historia Math. **4**(1), 75–89 (1977)
23. Samko, S., Marichev, 0., Kilbas, A.: Fractional Integrals and Derivatives and Some of Their Applications, Science and Technica, Minsk (1987)

Chapter 2
Infinite Programming and Application in the Best Proximity Point Theory

Mohammad Reza Haddadi, Mohammad Mursaleen, Mohammad Valaei, and Vahid Parvaneh

Abstract Various types of unconnected optimization problems in infinite space are explored. In particular, many papers have been published on the best approximation and the best proximity point. The purpose of this chapter is to start a new path to solve constrained optimization problems in the infinite space of the best approximation and the best proximity point types. Hence, we give a new extension of the KKT optimality conditions in a Hilbert space, and we solve the KKT system of a nonlinear programming problem with infinite constraints.

Keywords Best proximity point · Infinite programming · Optimality condition

2.1 Introduction

The point $\rho \in \Delta$ is a best proximity point (BPP) of the pair (Δ, Λ), if $d(\rho, \Upsilon\rho) = dist(\Delta, \Lambda)$ such that $\Upsilon : \Delta \to \Lambda$. The set of all best proximity points of Δ is indicated by $P_\Upsilon(\Delta)$, that is,

$$P_\Upsilon(\Delta, \Lambda) := \{\rho \in \Delta : d(\rho, \Upsilon\rho) = dist(\Delta, \Lambda)\}.$$

M. R. Haddadi · M. Valaei
Department of Mathematics, Ayatollah Boroujerdi University, Boroujerd, Iran
e-mail: haddadi@abru.ac.ir

M. Valaei
e-mail: mohamad.valaei@abru.ac.ir

M. Mursaleen
Department of Medical Research, China Medical University Hospital, China Medical University (Taiwan), Taichung, Taiwan

V. Parvaneh (✉)
Department of Mathematics, Gilan-E-Gharb Branch, Islamic Azad University, Gilan-E-Gharb, Iran
e-mail: zam.dalahoo@gmail.com

B. Hazarika et al. (eds.), *Advances in Functional Analysis and Fixed-Point Theory*, Industrial and Applied Mathematics,
https://doi.org/10.1007/978-981-99-9207-2_2

A function $\phi : \mathscr{S} \to R$ defined on a convex subset $\mathscr{S}$ of a real vector space is quasiconvex if for every $x, y \in \mathscr{S}$ and for every $\lambda \in [0, 1]$, we have

$$\phi(\lambda x + (1 - \lambda)y) \leq \max\{\phi(x), \phi(y)\}.$$

Let ϕ be a real-valued function, which is defined on an open set in Hilbert space $\mathscr{H}$ containing the set $\mathscr{S}$. Also, let $h(x; \vartheta)$ be some directional derivative of ϕ at the point $x \in \mathscr{S}$ in direction $\vartheta \in \mathscr{H}$. Then ϕ is said to be pseudoconvex on $\mathscr{S}$ in terms of the derivative h if the following term holds:

$$x \in \mathscr{S}; \, y \in \mathscr{S}, \phi(y) < \phi(x) \Rightarrow h(x; y - x) < 0.$$

There are numerous generalizations of this point in the literature. A BPP theorem for relatively nonexpansive mappings has been aspirated by Eldred et al. [6]. This is an additional demeanor to that found in Sankar Raj and Veeramani [17].

Recently, many authors have studied the existence of best proximity points defined in metric spaces and Banach spaces (see, for example, [1, 4, 6, 7, 13, 14, 17]).

In this chapter, we give global optimal approximate solutions of the best proximity points in the form of unconstrained optimization problems. As an application of the paper, we present an example of BPP problems in the form of a constrained optimization problem. Hence, the following prelims are necessary.

2.2 Extension of KKT Sufficient Conditions

2.2.1 *Preliminaries*

In what follows, we will write $\mathscr{H}$ for an Hilbert space.

Definition 2.1 Having $\mathscr{S}$ as a nonempty set in a Hilbert space $\mathscr{H}$ and $\bar{\rho} \in \mathscr{S}$, the cone of feasible directions of $\mathscr{S}$ at $\bar{\rho}$, denoted by $\mathscr{D}$, is given by

$$\mathscr{D} = \{\vartheta : \vartheta \neq 0, \text{ } and \text{ } \bar{\rho} + \lambda\vartheta \in S, \text{ } \forall\lambda \in (0, \delta) \text{ } for \text{ } some \text{ } \delta > 0\}.$$

Every nonzero element $\vartheta \in \mathscr{D}$ is said to be a feasible direction. Also, given a function $\phi : \mathscr{H} \to R$, the cone of improving directions at $\bar{\rho}$, denoted by $\mathscr{G}$, is given by

$$\mathscr{G} = \{\vartheta : \phi(\bar{\rho} + \lambda\vartheta) < \phi(\bar{\rho}), \text{ } \forall\lambda \in (0, \delta) \text{ } for \text{ } some \text{ } \delta > 0\}.$$

Each direction $\vartheta \in \mathscr{G}$ is said to be an improving direction, or a descent direction of ϕ at $\bar{\rho}$.

Let $\varphi : \mathscr{H} \to (-\infty, +\infty]$ be a function. Then the limit

$$\lim_{t \to 0} \frac{\varphi(\rho + tu) - \varphi(\rho)}{t}$$

is said to be the directional derivative of φ at the point $\rho \in \mathscr{H}$ in the direction $u \in \mathscr{H}$. If it exists, it is denoted by $D_u\varphi(\rho)$. φ is called Gateaux differentiable at a point $\rho \in \mathscr{H}$ if there exists a continuous linear functional z on $\mathscr{H}$ such that $\langle u, z \rangle = D_u\varphi(\rho)$ for all $u \in \mathscr{H}$. The element z is denoted by $\nabla\varphi(\rho)$ (i.e., grad $\varphi(\rho)$) and is called the Gateaux derivative of φ at ρ.

An immense quantity of investigations of this conditions appears in the literature. We can quote the works of Kortanek and Evans [12], Fiacco and McCormick [8], Zlobec [12], McShane [18], Craven [5], Trudzik [15] and Giorgi [9].

Recently, the study of extension of the optimization problem to infinite programming has been devoted to several works. For instance, Braunschweiger and Clark gave an extension of the Farkas theorem in infinite dimensional spaces [2]. A Fritz John type sufficient optimality theorem is demonstrated for nonlinear programming problems in infinite dimensional complex space by Craven [3] and Gulati [11]. Several works have been devoted to the study of extension of the KKT necessity conditions to infinite programming. For instance, Tapia [16], Guignard [10] and Oliveira [14]. This article provides a simpler form of KKT proof than others, with fewer conditions.

In the following, we give important results on nonlinear programming problems in an infinite dimensional space.

Theorem 2.1 ([2]) *If $f : \mathscr{H} \to R$ be Gâteaux differentiable at $\bar{\rho}$ and there is a vector $d \in \mathscr{H}$ such that $\langle \nabla f(\bar{\rho}), d \rangle < 0$, then there exists a $\delta > 0$ such that $f(\bar{\rho} + \lambda d) < f(\bar{\rho})$ for each $\lambda \in (0, \delta)$, where d is a descent direction of f at $\bar{\rho}$.*

Theorem 2.2 ([2] (Extension of Farkas's Theorem)) *Let $c \in \mathscr{H}$. Then, precisely, one of the following two systems is solvable:*
System 1: *for all $a_1, \ldots, a_m \in \mathscr{H}$, $\langle a_i, \rho \rangle \leq 0$ for all $i = 1, \ldots, m$ and $\langle c, \rho \rangle > 0$ and for some $\rho \in \mathscr{H}$.*
System 2: *there exists nonzero vector $\varrho = (\varrho_1, \ldots, \varrho_m) \in R^m$ such that $\varrho_i \geq 0$ and $\sum_{i=1}^m \varrho_i a_i = c$.*

Theorem 2.3 ([2] (Extension of Gordan's Theorem)) *Precisely one of the following two systems is solvable:*
System 1: *for all $a_1, \ldots, a_m \in \mathscr{H}$, $\langle a_i, x \rangle < 0$, for all $i = 1, \ldots, m$ for some $\rho \in \mathscr{H}$.*
System 2: *there exists nonzero vector $\varrho = (\varrho_1, \ldots, \varrho_m) \in R^m$ such that $\varrho_i \geq 0$ and $\sum_{i=1}^m \varrho_i a_i = 0$.*

Theorem 2.4 ([11] (Extension of Fritz John Sufficient Conditions)) *Let Π be a nonempty set in a Hilbert space $\mathscr{H}$, $\bar{\rho}$ a solution and $\mathscr{I} = \{i : \psi_i(\bar{\rho}) = 0\}$. Define S as the relaxed feasible region for problem P in which the non binding constraints are dropped.*
(i). If there exists an ε-neighborhood $\Lambda_\varepsilon(\bar{\rho})$, $\varepsilon > 0$, such that f is pseudoconvex over $\Lambda_\varepsilon(\bar{\rho}) \cap S$ and ψ_i, $i \in \mathscr{I}$ are strictly pseudoconvex over $\Lambda_\varepsilon(\bar{\rho}) \cap S$, then $\bar{\rho}$ is

a local minimum for problem P.
(ii). If f be pseudoconvex at $\bar{\rho}$, and if ψ_i, $i \in \mathscr{I}$ are both strictly pseudoconvex and quasiconvex at $\bar{\rho}$, then $\bar{\rho}$ is a global optimal solution for problem P.

Let $f : \mathscr{H} \to R$ and $\psi_i : \mathscr{H} \to R$ for $i = 1, \ldots, m$. In this paper, we consider problem P to minimize $f(\rho)$ subject to $\rho \in S = \{\rho \,:\, \psi_i(\rho) \leq 0,\ i = 1, \ldots, m\}$.

2.2.2 Main Results

In this section, we consider the problem to minimize $f(\rho)$ subject to $\rho \in S$ where $f : \mathscr{H} \to R$ and S is a nonempty set in $\mathscr{H}$. We now specify the feasible region S as fol1ows:

$$S = \{\rho \in \Pi : \psi_i(\rho) \leq 0,\ i = 1, \ldots, m\},$$

where $\psi_i : \mathscr{H} \to R$, for $i = 1, \ldots, m$, and Π is a nonempty open set in S. This gives us the following constrained nonlinear programming problem (NPP):

$$\begin{aligned} P: \qquad & Minimize f(\rho) \\ & s.t.\ \psi_i(\rho) \leq 0, \quad i = 1, \ldots, m \\ & \rho \in \Pi. \end{aligned}$$

Theorem 2.5 *Let f be a function which is Gâteaux differentiable at a point $\bar{\rho} \in S$ and let $\bar{\rho}$ be a optimal solution. Then $\mathscr{G}_0 \cap D = \emptyset$, where*

$$\mathscr{G}_0 = \{\vartheta : \langle \nabla f(\bar{\rho}), \vartheta \rangle < 0\}$$

and D is the cone of feasible directions of S at $\bar{\rho}$. Conversely, suppose that $\mathscr{G}_0 \cap D = \emptyset$, f is pseudoconvex at $\bar{\rho}$ and there is an ε-neighborhood $\Lambda_\varepsilon(\bar{\rho})$, $\varepsilon > 0$, such that $\vartheta = (\rho - \bar{\rho}) \in D$ for any $\rho \in S \cap \Lambda_\varepsilon(\bar{\rho})$. Then $\bar{\rho}$ is a local minimum of f.

Proof Let there exist a vector $\vartheta \in \mathscr{G}_0 \cap D$. By Theorem 2.1, there exists a $\delta_1 > 0$ such that

$$f(\bar{\rho} + \lambda\vartheta) < f(\bar{\rho}), \quad for\ all\ \lambda \in (0, \delta_1). \tag{2.1}$$

Furthermore, by Definition 2.1, there exists a $\delta_2 > 0$ such that

$$\bar{\rho} + \lambda\vartheta \in S, \ \ for\ all\ \lambda \in (0, \delta_2). \tag{2.2}$$

The hypothesis that $\bar{\rho}$ is a local minimum to the problem is not matchable with (2.1) and (2.2). Thus, $\mathscr{G}_0 \cap D = \emptyset$. Conversely, suppose that $\mathscr{G}_0 \cap D = \emptyset$ and let the given conditions in the converse statement of the theorem hold true. Then we must have $f(\rho) \geq f(\bar{\rho})$ for all $\rho \in S \cap \Lambda_\varepsilon(\bar{\rho})$. To understand this, suppose that $f(\hat{\rho}) < f(\bar{\rho})$ for some $\hat{\rho} \in S \cap \Lambda_\varepsilon(\bar{\rho})$. By the assumption on $S \cap \Lambda_\varepsilon(\bar{\rho})$, we have $\vartheta = (\hat{\rho} - \bar{\rho}) \in$

D. Moreover, by the pseudoconvexity of f in $\bar{\rho}$, we have $\langle \nabla f(\bar{\rho}), \vartheta \rangle < 0$; or else, if $\langle \nabla f(\bar{\rho}), \vartheta \rangle \geq 0$, we would obtain $f(\hat{\rho}) = f(\bar{\rho} + \vartheta) \geq f(\bar{\rho})$. Therefore, we have shown that ρ is not a local minimum over $S \cap \Lambda_\varepsilon(\bar{\rho})$, there exists a direction $\vartheta \in \mathscr{G}_0 \cap D$, which is a contradiction. □

Lemma 2.1 *Let $f : \mathscr{H} \to R$ be a Gâteaux differentiable mapping, and $\mathscr{G}$, $\mathscr{G}_0$, $\mathscr{G}_0'$ be defined in the following way:*

$$\mathscr{G}_0 = \{\vartheta : \langle \nabla f(\bar{\rho}), \vartheta \rangle < 0\},\ \mathscr{G}_0' = \{\vartheta \neq 0 : \langle \nabla f(\bar{\rho}), \vartheta \rangle \leq 0\},$$

$$\mathscr{G} = \{\vartheta : f(\bar{\rho} + \lambda\vartheta) < f(\bar{\rho}),\ \forall \lambda \in (0, \delta)\ for\ some\ \delta > 0\}.$$

Then we have $\mathscr{G}_0 \subseteq \mathscr{G} \subseteq \mathscr{G}_0'$.

Proof Let $\vartheta \in \mathscr{G}_0$. By Theorem 2.1, $\vartheta \in \mathscr{G}$. Now, let $\vartheta \notin \mathscr{G}_0'$. So, $\langle \nabla f(\bar{\rho}), \vartheta \rangle > 0$. Then ϑ is a increscent direction for f. Hence, $\vartheta \notin \mathscr{G}$. □

Theorem 2.6 (Extension of Fritz John Necessary Conditions) *Let Π be a nonempty open set in $\mathscr{H}$ and $\bar{\rho}$ be a feasible solution, and let $\mathscr{I} = \{i : \psi_i(\rho) = 0\}$. Furthermore, suppose that f and ψ_i $(i \in \mathscr{I})$ are Gâteaux differentiable in $\bar{\rho}$ and ψ_i $(i \notin \mathscr{I})$ are continuous in $\bar{\rho}$. If $\bar{\rho}$ solves Problem P locally, there exist scalars ϱ_0 and ϱ_i where $i \in \mathscr{I}$ such that*

$$\begin{aligned} \varrho_0 \nabla f(\bar{\rho}) + \sum_{i \in \mathscr{I}} \varrho_i \nabla \psi_i(\bar{\rho}) &= 0 \\ \varrho_0, \varrho_i &\geq 0,\ i \in \mathscr{I} \\ (\varrho_0, \varrho_{\mathscr{I}}) &\neq (0, 0), \end{aligned}$$

Also, if ψ_i $(i \notin \mathscr{I})$ are Gâteaux differentiable in $\bar{\rho}$, then the foregoing conditions can be written in the following form:

$$\begin{aligned} \varrho_0 \nabla f(\bar{\rho}) + \sum_{i=1}^{m} \varrho_i \nabla \psi_i(\bar{\rho}) &= 0 \\ \varrho_i \psi_i(\bar{\rho}) &= 0,\ i = 1, \ldots, m \\ \varrho_0, \varrho_i &\geq 0,\ i = 1, \ldots, m \\ (\varrho_0, \varrho) &\neq (\mathbf{0}, \mathbf{0}), \end{aligned}$$

that ϱ is a vector whose components are ϱ_i for $i = 1, \ldots, m$.

Proof Since $\bar{\rho}$ solves problem P locally, by Theorem 2.2, there exists no d such that $\langle \nabla f(\bar{\rho}), \vartheta \rangle < 0$ and $\langle \nabla \psi_i(\bar{\rho}), \vartheta \rangle < 0$ for each $i \in \mathscr{I}$. By Gordan's Theorem 2.3, there exists a nonzero $\varrho = (\varrho_0, \varrho_1, \ldots, \varrho_m) \in R^{m+1}$ such that $\varrho_i \geq 0$ and $\varrho_0 \nabla f(\bar{\rho}) + \sum_{i=1}^{m} \varrho_i \nabla \psi_i(\bar{\rho}) = 0$. □

Theorem 2.7 (Extension of KKT Necessary Conditions) *Having Π as a nonempty open set in $\mathscr{H}$ and $\bar{\rho}$ as a Fritz John solution, $\mathscr{I} = \{i : \psi_i(\bar{\rho}) = 0\}$, f and ψ_i $(i \in \mathscr{I})$ are Gâteaux differentiable in $\bar{\rho}$, ψ_i $(i \notin I)$ are continuous in $\bar{\rho}$, $\nabla\psi_i(\bar{\rho})$ $(i \in \mathscr{I})$ are linearly independent, if $\bar{\rho}$ solves problem P locally, then there exist scalars ϱ_i for all $i \in \mathscr{I}$ such that*

$$\nabla f(\bar{\rho}) + \sum_{i \in \mathscr{I}} \varrho_i \nabla\psi_i(\bar{\rho}) = 0$$

$$\varrho_i \geq 0 \quad \forall\, i \in \mathscr{I}.$$

as well as, if ψ_i for each $i \notin I$ is also Gâteaux differentiable in $\bar{\rho}$, then the foregoing conditions can be written in the following equivalent form:

$$\nabla f(\bar{\rho}) + \sum_{i=1}^{m} \varrho_i \nabla\psi_i(\bar{\rho}) = 0$$

$$\varrho_i \psi_i(\bar{\rho}) = 0 \quad \forall\, i = 1, \ldots, m$$

$$\varrho_i \geq 0 \quad \forall\, i = 1, \ldots, m.$$

Proof By Theorem 2.6, there exist scalars ϱ_0 and $\hat{\varrho}_i$ where $i \in \mathscr{I}$, not all equal to zero, such that

$$\varrho_0 \nabla f(\bar{\rho}) + \sum_{i \in \mathscr{I}} \hat{\varrho}_i \nabla\psi_i(\bar{\rho}) = 0 \tag{2.3}$$

$$\varrho_0, \hat{\varrho}_i \geq 0\, \forall\, i \in \mathscr{I}. \tag{2.4}$$

Note that $\varrho_0 > 0$, because (2.4) is contradict by the assumption of linearly independence of $\nabla\psi_i(\bar{\rho})$ for all $i \in \mathscr{I}$ if $\varrho_0 = 0$. The first part of the theorem follows if $\varrho_i = \frac{\hat{\varrho}_i}{\varrho_0}$ for each $i \in \mathscr{I}$. The equivalent form follows by letting $\varrho_i = 0$ for all $i \notin \mathscr{I}$. □

Lemma 2.2 *Let Π be a nonempty open set in $\mathscr{H}$ and $\bar{\rho}$ be a feasible solution and let $\mathscr{I} = \{i : \psi_i(\bar{\rho}) = 0\}$. Define $\mathscr{G}_0 = \{\vartheta : \langle \nabla f(\bar{\rho}), \vartheta \rangle < 0\}$ and $G_0' = \{\vartheta \neq 0 : \langle \nabla\psi_i(\bar{\rho}), \vartheta \rangle \leq 0,\ \forall i \in \mathscr{I}\}$ as before. Let $G' = \{\vartheta : \langle \nabla\psi_i(\bar{\rho}), \vartheta \rangle \leq 0,\ \forall i \in \mathscr{I}\} = G_0 \cup \{0\}$. Then $\bar{\rho}$ is a KKT solution if and only if $\mathscr{G}_0 \cap G' = \emptyset$ which is equivalent to $\mathscr{G}_0 \cap G_0' = \emptyset$.*

Proof The feasible solution $\bar{\rho}$ is a KKT point if and only if there exists a solution u_i $(i \in \mathscr{I})$ to the system $\sum_{i \in \mathscr{I}} u_i \nabla\psi_i(\bar{\rho}) = -\nabla f(\bar{\rho})$ and $u_i \geq 0$ for all $i \in \mathscr{I}$. By Farkas's Theorem 2.2, this holds true if and only if there does not exist a solution to the system $\langle \nabla\psi_i(\bar{\rho}), \vartheta \rangle \leq 0$, for each $i \in \mathscr{I}$ and $\langle \nabla f(\bar{\rho}), \vartheta \rangle > 0$. Hence, $\bar{\rho}$ is a KKT point if and only if $\mathscr{G}_0 \cap G' = \emptyset$. Clearly, this also holds true if and only if $\mathscr{G}_0 \cap G_0 = \emptyset$. □

Theorem 2.8 *(Extension of KKT Sufficient Conditions) Let Π be a nonempty open set in $\mathscr{H}$, $\bar{\rho}$ a KKT solution and $\mathscr{I} = \{i : \psi_i(\bar{\rho}) = 0\}$. Define S as the relaxed*

feasible region for problem P in which the constraints which are not binding at $\bar{\rho}$ have been dropped. Then:
a. If there exists an ε-neighborhood $\Lambda_\varepsilon(\bar{\rho})$ of $\bar{\rho}$, $\varepsilon > 0$, such that f is pseudoconvex in $\Lambda_\varepsilon(\bar{\rho}) \cap S$ and ψ_i, $i \in \mathscr{I}$, are differentiable at $\bar{\rho}$ and are quasiconvex on $\Lambda_\varepsilon(\bar{\rho}) \cap S$, then $\bar{\rho}$ is a local minimum for problem P.
b. If f is pseudoconvex in $\bar{\rho}$, and if ψ_i, $i \in \mathscr{I}$, are differentiable and quasiconvex in $\bar{\rho}$, then $\bar{\rho}$ is a global optimal solution for problem P. Also, if this assumption holds true with the domain of the feasible restriction to $\Lambda_\varepsilon(\bar{\rho})$, for some $\varepsilon > 0$, then $\bar{\rho}$ is a local minimum for P.

Proof First, consider Part a. Since $\bar{\rho}$ is a KKT point, we have, equivalently, by Lemma 2.2 that $\mathscr{G}_0 \cap G' = \emptyset$. Since $G_0 \subseteq D \subseteq G'_0$, this means that $\mathscr{G}_0 \cap D = \emptyset$. Since ψ_i $(i \in \mathscr{I})$ are quasiconvex over $\Lambda_\varepsilon(\bar{\rho}) \cap S$, we have $\Lambda_\varepsilon(\bar{\rho}) \cap S$ is a convex set. By restricting attention to $\Lambda_\varepsilon(\bar{\rho}) \cap S$, the condition of the converse statement of Theorem 2.5 hold true. So, $\bar{\rho}$ is a minimum over $\Lambda_\varepsilon(\bar{\rho}) \cap S$. Hence, $\bar{\rho}$ is a local minimum for the more restricted original problem P. This proves Part a.

Next, consider Part b. Let ρ be any feasible solution to problem P. Then, $\psi_i(\rho) \leq \psi_i(\bar{\rho})$, for all $i \in \mathscr{I}$, since $\psi_i(\rho) \leq 0$ and $\psi_i(\bar{\rho})$. By the quasiconvexity of ψ_i in $\bar{\rho}$, it follows that

$$\psi_i[\bar{\rho} + \lambda(\rho - \bar{\rho})] = \psi_i[\lambda\rho + (1-\lambda)\bar{\rho}] \leq \max\{\psi_i(\rho), \psi_i(\bar{\rho})\} = \psi_i(\bar{\rho})$$

for all $\lambda \in (0, 1)$. This implies that ψ_i does not increase by moving from $\bar{\rho}$ along the direction $\rho - \bar{\rho}$. Thus, by Theorem 2.1

$$\langle \nabla\psi_i(\bar{\rho}), (\rho - \bar{\rho})\rangle \leq 0.$$

Multiplying this by the Lagrange multiplier u_i corresponding to the KKT point $\bar{\rho}$ and summing over f, we get

$$\langle [\sum_{i \in \mathscr{I}} u_i \nabla\psi_i(\bar{\rho})^t], (\rho - \bar{\rho})\rangle \leq 0.$$

But, since $\nabla f(\bar{\rho}) + \sum_{i \in \mathscr{I}} u_i \nabla\psi_i(\bar{\rho}) = 0$, it follows that $\langle \nabla f(\bar{\rho}), (\rho - \bar{\rho})\rangle \geq 0$. Then, by the pseudoconvexity of f in $\bar{\rho}$, we must have $f(\rho) \geq f(\bar{\rho})$. □

2.3 Application

Various types of unconnected optimization problems in infinite dimension spaces are explored. In particular, many papers have been published on the best proximity point. The purpose of this chapter is to start a new path to solve constrained optimization problems in the infinite dimension space of the best proximity point. Hence, in this section, we give some examples of this problem. We start with a remark.

Remark 2.1 The norm of $\mathscr{H}$ is Gâteaux differentiable with $\nabla\|\rho\| = \frac{\rho}{\|\rho\|}$, $\rho \neq 0$. Indeed, for each $\rho \in H$ where $\rho \neq 0$, we have

$$\begin{aligned}\lim_{t\to 0}\frac{\|\rho+t\varrho\|-\|\rho\|}{t} &= \lim_{t\to 0}\frac{\|\rho+t\varrho\|^2-\|\rho\|^2}{t(\|\rho+t\varrho\|+\|\rho\|)}\\ &= \lim_{t\to 0}\frac{2t\langle\varrho,\rho\rangle+t^2\|\varrho\|^2}{t(\|\rho+t\varrho\|+\|\rho\|)}\\ &= \langle\varrho, \frac{\rho}{\|\rho\|}\rangle.\end{aligned}$$

Therefore, the norm of $\mathscr{H}$ is Gâteaux differentiable with $\nabla\|\rho\| = \frac{\rho}{\|\rho\|}$.

Remark 2.2 $\varphi(\rho) = \|\rho\|$ is not differentiable in $\rho = 0$, but it is subdifferentiable. Indeed,

$$\begin{aligned}\partial\varphi(0) = \partial\|0\| &= \{z \in \mathscr{H} : \langle\rho, z\rangle = \|\rho\|,\ \forall\rho \in H\}\\ &= \{z \in \mathscr{H} : \|z\| \leq 1\}.\end{aligned}$$

Lemma 2.3 *The norm* $\|.\|$ *has pseudoconvex property.*

Proof Since $\nabla\|\rho\| = \frac{\rho}{\|\rho\|}$, hence for all $\rho, \varrho \in \mathscr{H}$ we have

$$\begin{aligned}0 &\leq \frac{\rho}{\|\rho\|}.(\varrho-\rho)\\ &= \frac{<\rho,\varrho>}{\|\rho\|} - \frac{<\rho,\rho>}{\|\rho\|}\\ &= \frac{<\rho,\varrho>}{\|\rho\|} - \frac{\|\rho\|^2}{\|\rho\|}\\ &= \frac{<\rho,\varrho>}{\|\rho\|} - \|\rho\|\\ &\Rightarrow \|\rho\|^2 \leq \langle\rho,\varrho\rangle \leq \|\rho\|\|\varrho\|\\ &\Rightarrow \|\rho\| \leq \|\varrho\|.\end{aligned}$$

□

Example 2.1 Let $C = 3\mathscr{B}_H$ and $\rho_0 = 4e_1 \in \mathscr{H}$. Consider the following NPP:

$$\begin{aligned}&\min_{\rho\in C}\ \|\rho-\rho_0\|\\ &\qquad g_1(\rho) = \rho_1^2+\rho_2^2 \leq 5\\ &\qquad g_2(\rho) = \rho_1+\rho_3 \leq 3\\ &\qquad g_3(\rho) = -\rho_1 \leq 0.\end{aligned}$$

One can observes that $\rho^* = \sqrt{5}e_1$ is a KKT point, because

$$\nabla\|\rho_0-\rho\| + \lambda_1\nabla g_1(\rho) + \lambda_2\nabla g_2(\rho) + \lambda_3\nabla g_3 = \mathbf{0}.$$

It is clear that $D_u g_1(\rho) = 2u_1\rho_1 + 2u_2\rho_2$. If $z = 2\rho_1 e_1 + 2\rho_2 e_2$, then $\langle z, u\rangle = 2u_1\rho_1 + 2u_2\rho_2$, and so $\nabla g_1(\rho) = z$. Similarly, $\nabla g_2(\rho) = e_1 + e_3$ and $\nabla g_3(\rho) = -e_1$, and we have

$$\frac{\rho - \rho_0}{\|\rho - \rho_0\|} + 2\rho_1\lambda_1 e_1 + 2\rho_2\lambda_1 e_2 + \lambda_2 e_1 + \lambda_2 e_3 - \lambda_3 e_1 = \mathbf{0}.$$

Hence,

$$\frac{\rho_1 - 4}{\|\rho - \rho_0\|} + 2\rho_1\lambda_1 + \lambda_2 - \lambda_3 = 0,$$
$$\frac{\rho_2}{\|\rho - \rho_0\|} + 2\rho_2\lambda_1 = 0,$$
$$\frac{\rho_3}{\|\rho - \rho_0\|} + \lambda_2 = 0,$$

and so $\lambda_1 > 0$, $\lambda_2 = \lambda_3 = 0$ and $\rho^* = (\sqrt{5}, 0, 0, ..)$. Since norm is a convex function and feasible region is a convex set, according to Theorems 2.7 and 2.4 it consequences that ρ^* is a global optimal solution.

In the following, we give an example of a BPP problem in the form of a constrained optimization problem.

Example 2.2 Let $\Delta = -2e_1 + \mathscr{B}_H$, $\Lambda = 2e_2 + \mathscr{B}_H$ and $\Upsilon : \Delta \to \Lambda$ be such that $\Upsilon(\{\rho_n\}) = (\frac{1-\rho_1}{2}, \rho_2, ..., \rho_n, ...)$. It is clear that for $\rho = -e_1$ we have

$$\|\Upsilon\rho - \rho\| = d(\Delta, \Lambda).$$

Consider the following NPP:

$$\min_{\rho\in\Delta} \ \|\Upsilon\rho - \rho\|$$
$$g(\rho) = \rho_1 + \frac{3}{2} \leq 0.$$

By Theorem 2.7, KKT conditions are as follows:

$$\nabla\|\Upsilon\rho - \rho\| + \lambda\nabla g(\rho) = \mathbf{0}$$

$$\lambda g(\rho) = \mathbf{0},\ \lambda \geq 0,\ g(\rho) \leq \mathbf{0}.$$

It is clear that $D_u g(\rho) = u_1$. If $z = e_1$, then $\langle z, u\rangle = u_1$, and so $\nabla g(\rho) = z$. On the other hand, by Remark 2.1, we have

$$\frac{\Upsilon\rho - \rho}{\|\Upsilon\rho - \rho\|} + \lambda e_1 = \mathbf{0}$$

and so

$$\lambda = \frac{\frac{1-\rho_1}{2} + \rho_1}{\|\Upsilon\rho - \rho\|} \; s.t. \; \rho = \rho_1 e_1. \tag{2.5}$$

$$\lambda g(\rho) = \mathbf{0} \Rightarrow \lambda(\rho_1 + \frac{3}{2}) = 0. \tag{2.6}$$

If $\lambda = 0$, by (2.5) $\rho_1 = -1$ is not a feasible point and so, $\lambda \neq 0$. By (2.6), we have $\rho_1 = -\frac{3}{2}$, i.e., $\rho^* = -\frac{3}{2}e_1$. Because norm is a convex function and feasible region is a convex set, by Theorems 2.7 and 2.4, ρ^* is a global optimal solution.

References

1. Alfuraidan, M.R., Jorquera, E.D., Khamsi, M.A.: Fixed point theorems for monotone Caristi inward mappings. Numer. Funct. Anal. Optim. **39**, 1092–1101 (2018)
2. Braunschweiger, C.C., Clark, H.E.: An extension of the farkas theorem. Am. Math. Monthly **69**(4), 272–277 (1962)
3. Craven, B.D.: Sufficient Fritz John optimality conditions. Bull. Austral. Math. Soc. **13**, 411–419 (1975)
4. Chaira, K., Kabil, M., Kamouss, A., Lazaiz, S.: Best proximity points for monotone relatively nonexpansive mappings in ordered Banach spaces. Axioms **8**, 121 (2019)
5. Craven, B.D.: Modified Kuhn-Tucker conditions when a minimum is not attained. Oper. Res. Lett. **3**, 47–52 (1984)
6. Eldred, A.A., Kirk, W.A., Veeramani, P.: Proximinal normal structure and relatively nonexpansive mappings. Studia Math. **171**(3), 283–293 (2005)
7. Eldred, A.A., Veeramani, P.: Existence and convergence of best proximity points. J. Math. Anal. Appl. **323**, 1001–1006 (2006)
8. Fiacco, A.V., McCormick, G.P.: The slacked unconstrained minimization technique for convex programming. SIAM J. Appl. Math. **15**, 505–515 (1967)
9. Giorgi, G., Jiménez, B., Novo, V.: Approximate Karush-Kuhn-Tucker condition in multiobjective optimization. J. Optim. Theory Appl. **171**, 70–89 (2016)
10. Guignard, M.: Generalized Kuhn-Tucker conditions for mathematical programming problems in a Banach space. SIAM J. Control **7**, 232–241 (1969)
11. Gulati, T.R.: A Fritz John type sufficient optimality theorem in complex space. Bull. Austral. Math. Soc. **11**, 219–224 (1974)
12. Kortanek, K.O., Evans, J.P.: Asymptotic Lagrange regularity for pseudoconcave programming with weak constraint qualification. Oper. Res. **16**, 849–857 (1968)
13. Nashine, H.K., Kumam, P., Vetro, C.: Best proximity point theorems for rational proximal contractions. Fixed Point Theory Appl. **2013**, 95 (2013)
14. Oliveira, A.V., Santos, B.L., Osuna-Gómez, R., Rojas-Medar, M.A.: Optimality conditions for nonlinear infinite programming problems. Optim. Lett. **9**(6), 1131–1147 (2015)
15. Trudzik, L.I.: Asymptotic Kuhn-Tucker conditions in abstract spaces. Numer. Funct. Anal. Optim. **4**, 355–369 (1982)
16. Tapia, R.A., Trosset, M.W.: Extension of the Karush-Kuhn-Tucker necessity conditions to infinite programming. SIAM Rev. **36**(1), 1–17 (1994)
17. Sankar, R.V., Veeramani, P.: Best proximity pair theorems for relatively nonexpansive mappings. Appl. Gen. Topol. **10**(1), 21–28 (2009)
18. Zlobec, S.: Extensions of asymptotic Kuhn-Tucker conditions in mathematical programming. SIAM J. Appl. Math. **21**, 448–460 (1971)

Chapter 3
On the Fixed Point Outcomes of a Modified Iteration Procedure on Kohlenbach Hyperbolic Space with an Application

Aynur Şahin and Metin Başarır

Abstract We modify an iteration (repetition) process and prove the weak h-w^2-stability and data dependence theorems of this modified (altered) iteration process on Kohlenbach hyperbolic space (KHS) in this chapter. Furthermore, for two different forms of generalized non-expansive map (NM) on KHS, certain $\triangle$-convergence and strong convergence theorems utilizing the altered iteration process are proved. Finally, we show how our outcomes can be applied to non-linear integral equations. The outcomes obtained in this chapter are extensions, improvements, and generalizations of several known outcomes in fixed point (FP) theory literature.

Keywords Kohlenbach hyperbolic spaces · Convex Banach spaces

3.1 Introduction

Analysis, topology, and linear algebra are all naturally combined with non-linear functional analysis and the fixed point (FP) theory is a significant and quickly expanding field of non-linear functional analysis. FP theorems are obviously conclusions that establish the existence of FP, which those are very crucial tools for demonstrating the existence and uniqueness of solutions to a lot of mathematical models.

One of the most important outcomes of FP theory is the Banach Contraction Principle. Early discoveries in FP theory revolved around the expansion of Banach Contraction Principle. For the first FP theorem of non-expansive map (NM), the mathematics world had to wait for 43 years. Following, a lot of mathematicians have defined and studied different generalizations of NM (see e.g., [1–6, 14, 22, 26, 27]).

We now give some concepts.

A. Şahin (✉) · M. Başarır
Department of Mathematics, Sakarya University, 54050 Sakarya, Türkiye
e-mail: ayuce@sakarya.edu.tr

M. Başarır
e-mail: basarir@sakarya.edu.tr

B. Hazarika et al. (eds.), *Advances in Functional Analysis and Fixed-Point Theory*,
Industrial and Applied Mathematics,
https://doi.org/10.1007/978-981-99-9207-2_3

Let (Q, ϱ) be a metric space and $S \neq \emptyset \subset Q$. A point $t^\star \in S$ is called a *fixed point* of h self map on S if $ht^\star = t^\star$ and $F(h)$ indicates the set of all fixed points of h. The map h is called

(i) *contraction* if there is $a \in [0, 1)$ with $\varrho(hu, hv) \leq a\varrho(u, v)$;

(ii) *non-expansive* if $\varrho(hu, hv) \leq \varrho(u, v)$;

(iii) *quasi-non-expansive* if $\varrho(hu, t^\star) \leq \varrho(u, t^\star)$;

(iv) *asymptotically non-expansive* if there is a sequence $\{\phi_j\} \subset [1, \infty)$ with $\phi_j \to 1$ such that $\varrho(h^j u, h^j v) \leq \phi_j \varrho(u, v)$;

(v) *uniformly L-Lipschitzian* if there is a constant $L > 0$ with $\varrho(h^j u, h^j v) \leq L\varrho(u, v)$;

for all $u, v \in S$ and $j \geq 1$ and for each $t^\star \in F(h)$.

Alber et al. [5] presented a new generalized map called total asymptotically non-expansive map (TANM) in 2006, which is defined below:

Definition 3.1 A map $h : S \to S$ is called TANM if there are non-negative real sequences $\{\theta_j\}, \{\mu_j\}$ with $\theta_j, \mu_j \to 0$ as $j \to \infty$ and a strictly increasing, continuous function $\zeta : [0, \infty) \to [0, \infty)$ with $\zeta(0) = 0$ with

$$\varrho(h^j u, h^j v) \leq \varrho(u, v) + \theta_j \zeta(\varrho(u, v)) + \mu_j$$

for all $u, v \in S$ and $j \geq 1$. Then, h is said to be $(\{\theta_j\}, \{\mu_j\}, \zeta)$-TANM.

According to the definitions, each NM is an asymptotically NM, and each asymptotically NM is a TANM and is a uniformly L-Lipschitzian map. However, as shown in Example 1.1 (iv) of [16], the inverses are not true in general.

Every continuous and TANM defined on a non-empty, bounded, closed, convex subset of a complete uniformly convex hyperbolic space (UCHS) has an FP, it has been demonstrated in Theorem 3.1 of [16].

In 2008, Suzuki [26] defined the group of Suzuki generalized NM, satisfying a condition

$$\frac{1}{2}\varrho(u, hu) \leq \varrho(u, v) \Rightarrow \varrho(hu, hv) \leq \varrho(u, v),$$

which is weaker than non-expansive but stronger than quasi-non-expansive.

Aoyama and Kohsaka [6] gave a new group of NM, σ-non-expansive map (σ-NM) in 2011 and obtained a FP outcome for such maps. A map h is called σ-NM if there is a real number $\sigma < 1$ with each $u, v \in S$,

$$\varrho(hu, hv)^2 \leq \sigma\varrho(hu, v)^2 + \sigma\varrho(u, hv)^2 + (1 - 2\sigma)\varrho(u, v)^2.$$

Following this, Pant and Shukla [22] presented an intriguing group of generalized NM known as generalized σ-non-expansive map, which includes the Suzuki generalized NM.

Definition 3.2 (see [22]) A map $h : S \to S$ is called generalized σ-NM if there is an $\sigma \in [0, 1)$ with

$$\frac{1}{2}\varrho(u, hu) \leq \varrho(u, v) \Rightarrow \varrho(hu, hv) \leq \sigma\varrho(hu, v) + \sigma\varrho(u, hv) + (1 - 2\sigma)\varrho(u, v)$$

for all $u, v \in S$.

In the following proposition, which can be found in [22], we list some basic facts regarding generalized σ-NM.

Proposition 3.1 *(i) If h is a generalized σ-NM with a FP, then h is quasi-non-expansive.*

(ii) If h is a generalized σ-NM, then

$$\varrho(u, hv) \leq \left(\frac{3+\sigma}{1-\sigma}\right)\varrho(u, hu) + \varrho(u, v)$$

for all $u, v \in S$.

(iii) If h is a generalized σ-NM, then $F(h)$ is a closed set.

In 2019, Bhutia and Tiwary [8] defined a new repetition (iteration) procedure in Banach space, which is as follows:

$$\begin{cases} u_1 \in S, \\ w_j = h[(1-\tau_j)u_j + \tau_j hu_j], \\ v_j = h[(1-\sigma_j)w_j + \sigma_j hw_j], \\ u_{j+1} = hv_j, \quad j \geq 1. \end{cases} \tag{3.1}$$

They demonstrated that this repetition procedure is faster than the previous repetitions such as K [13], M^* [31], K^* [32], and M [33] repetitions for contraction maps. Two years later, Izhar-ud-din et al. [15] changed the repetition procedure (3.1) and established several strong convergence theorems and Δ-convergence theorems for TANM on CAT(0) space in [9] using the altered repetition procedure.

In this chapter, we give the weak h-w^2-stability theorems and data dependence theorems and some convergence properties of the repetition procedure (3.1) in KHS. There are six sections in this study. Section 3.2 recollects basic facts and provides a summary of the fundamental outcomes. In Sect. 3.3, we alter the repetition procedure (3.1) into KHS and study weak h-w^2-stability theorems and data dependence outcomes of the altered (modified) repetition procedure for contraction maps. In Sect. 3.4, we give some strong convergence theorems and Δ-convergence theorems of this altered repetition procedure for a generalized σ-NM. Sect. 3.5 modifies the repetition procedure (3.1) for a finite class of TANM into KHS and proves the strong convergence theorems and Δ-convergence theorems for this finite class by using the altered repetition procedure. At the end, we give an application to a non-linear integral equation in Sect. 3.6.

3.2 Preliminaries

Kohlenbach [17] defined the notion of hyperbolic space, which is explained below and is important in many areas of mathematics.

A *Kohlenbach hyperbolic space (KHS)* $(Q, \varrho, \mathcal{K})$ is a metric space (Q, ϱ) together with a map $\mathcal{K} : Q \times Q \times [0, 1] \to Q$ satisfying

$(\mathcal{K}1)$ $\varrho(w, \mathcal{K}(u, v, \sigma)) \leq (1 - \sigma)\varrho(w, u) + \sigma\varrho(w, v)$,

$(\mathcal{K}2)$ $\varrho(\mathcal{K}(u, v, \sigma), \mathcal{K}(u, v, \tau)) = |\sigma - \tau|\, \varrho(u, v)$,

$(\mathcal{K}3)$ $\mathcal{K}(u, v, \sigma) = \mathcal{K}(v, u, 1 - \sigma)$,

$(\mathcal{K}4)$ $\varrho(\mathcal{K}(u, w, \sigma), \mathcal{K}(v, z, \sigma)) \leq (1 - \sigma)\varrho(u, v) + \sigma\varrho(w, z)$,

for all $u, v, w, z \in Q$ and $\sigma, \tau \in [0, 1]$.

If a space satisfies only $(\mathcal{K}1)$, it corresponds to the convex metric space in [29]. The notion of hyperbolic space in [17] is more confining than the hyperbolic type in [11] and more general than the concept of hyperbolic space in [23].

A subset S of a KHS Q is *convex* if $\mathcal{K}(u, v, \sigma) = (1 - \sigma)u + \sigma v \in S$ for all $u, v \in S$ and $\sigma \in [0, 1]$. The following equalities are procurable even for the more general setting of convex metric space (see [29]): for all $u, v \in Q$ and $\sigma \in [0, 1]$,

$$\varrho(v, \mathcal{K}(u, v, \sigma)) = (1 - \sigma)\varrho(u, v) \text{ and } \varrho(u, \mathcal{K}(u, v, \sigma)) = \sigma\varrho(u, v).$$

So,

$$\mathcal{K}(u, v, 0) = u \quad \text{and} \quad \mathcal{K}(u, v, 1) = v. \tag{3.2}$$

A KHS $(Q, \varrho, \mathcal{K})$ is *uniformly convex* (see [25]) if for any $l > 0$ and $\varepsilon \in (0, 2]$, there is a constant $\delta \in (0, 1]$ such that, for all $r, u, v \in Q$,

$$\varrho\left(\mathcal{K}\left(u, v, \frac{1}{2}\right), r\right) \leq (1 - \delta)l,$$

with $\varrho(u, r) \leq l$, $\varrho(v, r) \leq l$, and $\varrho(u, v) \geq \varepsilon l$.

A map $\mu : (0, \infty) \times (0, 2] \to (0, 1]$ is called a *modulus of uniform convexity (MUC)* if $\delta = \mu(r, \varepsilon)$ for given $r > 0$ and $\varepsilon \in (0, 2]$. The map μ is *monotone* if it decreases with r for fix ε.

Leustean demonstrated in [19] that CAT(0) spaces are UCHS with the MUC $\mu(r, \epsilon) = \frac{\epsilon^2}{8}$ quadratic in ϵ. As an outcome, UCHS is a natural generalization of uniformly convex Banach space and CAT(0) space.

Let $\{u_j\}$ be a bounded sequence in a non-empty subset S of a metric space (Q, ϱ). Consider a continuous functional $r(., \{u_j\}) : Q \to [0, \infty)$ defined by

$$r(u, \{u_j\}) = \limsup_{j \to \infty} \varrho(u, u_j), \quad u \in Q.$$

Then, the infimum of $r(., \{u_j\})$ over S is said to be the *asymptotic radius* of $\{u_j\}$ with regard to S and is denoted by $r(S, \{u_j\})$.

A point $w \in S$ is said to be an *asymptotic center* of the sequence $\{u_j\}$ with regard to S if

$$r\left(w, \{u_j\}\right) = \inf\{r(u, \{u_j\}) : u \in S\},$$

and the set of all asymptotic centers of $\{u_j\}$ with regard to S is denoted by $A(S, \{u_j\})$. This set may be empty, a singleton, or contain an unlimited number of points.

When the asymptotic radius and center are considered in relation to Q, they are denoted by $r(Q, \{u_j\}) = r(\{u_j\})$ and $A(Q, \{u_j\}) = A(\{u_j\})$, respectively.

Lemma 3.1 *(see [20]) Let $(Q, \varrho, \mathcal{K})$ be a complete UCHS with the monotone MUC μ and S be a closed, convex, non-empty subset of Q. Then every bounded sequence $\{u_j\}$ in Q has a unique asymptotic center with regard to S.*

We know that a sequence $\{u_j\}$ in Q is said to be Δ-*convergent* to $u \in Q$ if u is the unique asymptotic center of $\{u_{j_k}\}$ for every subsequence $\{u_{j_k}\}$ of $\{u_j\}$.

Lemma 3.2 *(see [18]) Let $(Q, \varrho, \mathcal{K})$ be a UCHS with the monotone MUC μ and $u \in Q$ and $\{\sigma_j\}$ be a sequence in $[e, f]$ for some $e, f \in (0, 1)$. If $\{u_j\}$ and $\{v_j\}$ are sequences in Q such that, for some $c \geq 0$,*

$$\limsup_{j\to\infty} \varrho\left(u_j, u\right) \leq c, \limsup_{j\to\infty} \varrho\left(v_j, u\right) \leq c, \lim_{j\to\infty} \varrho\left(\mathcal{K}\left(u_j, v_j, \sigma_j\right), u\right) = c$$

then

$$\lim_{j\to\infty} \varrho\left(u_j, v_j\right) = 0.$$

Lemma 3.3 *(see [21]) Let $\{u_j\}, \{\xi_j\}$ and $\{\pi_j\}$ be non-negative real numbers sequences such that*

$$u_{j+1} \leq (1 + \pi_j)u_j + \xi_j, \ j \geq 1.$$

If $\sum \pi_j < \infty$ and $\sum \xi_j < \infty$, then $\lim_{j\to\infty} u_j$ exists.

3.3 The Weak h-w^2-Stability and Data Dependence Outcomes

First, by use of $(\mathcal{K}3)$ and (3.2) in (3.1), we expand the repetition method (3.1) into the KHS, as follows:

$$\begin{cases} u_1 \in S, \\ w_j = h(\mathcal{K}(u_j, hu_j, \tau_j)), \\ v_j = h(\mathcal{K}(w_j, hw_j, \sigma_j)), \\ u_{j+1} = hv_j, \quad j \geq 1, \end{cases} \tag{3.3}$$

where S is a non-empty convex subset of a KHS Q, h is a self map on S and $\{\sigma_j\}, \{\tau_j\}$ are two real sequences in [0, 1].

Theorem 2.1 in [8] is extended to KHS in the theorem below.

Theorem 3.1 *Let S be a non-empty closed convex subset of a KHS Q, $h : S \to S$ be a contraction map with $F(h) \neq \emptyset$ and $\{u_j\}$ be the repetitive method (3.3) with real sequences $\{\sigma_j\}$ and $\{\tau_j\}$ in [0, 1] satisfying $\sum \tau_j = \infty$. Then $\{u_j\}$ converges strongly to a FP of h.*

Proof It is obvious that the FP of h is unique when the contraction map h has a FP. Let $t^\star$ be the unique FP. By use of $(\mathscr{K}1)$, (3.3) and the contraction map h, we have

$$\varrho(u_{j+1}, t^\star) = \varrho(hv_j, t^\star) \leq a\varrho(v_j, t^\star), \tag{3.4}$$

$$\begin{aligned}\varrho(v_j, t^\star) &= \varrho(h(\mathscr{K}(w_j, hw_j, \sigma_j)), t^\star) \\ &\leq a\varrho(\mathscr{K}(w_j, hw_j, \sigma_j), t^\star) \\ &\leq a\left[(1-\sigma_j)\varrho(w_j, t^\star) + \sigma_j\varrho(hw_j, t^\star)\right] \\ &\leq a\left[(1-\sigma_j)\varrho(w_j, t^\star) + \sigma_j a\varrho(w_j, t^\star)\right] \\ &= a(1-\sigma_j(1-a))\varrho(w_j, t^\star) \leq a\varrho(w_j, t^\star)\end{aligned} \tag{3.5}$$

and

$$\begin{aligned}\varrho(w_j, t^\star) &= \varrho(h(\mathscr{K}(u_j, hu_j, \tau_j)), t^\star) \\ &\leq a\varrho(\mathscr{K}(u_j, hu_j, \tau_j), t^\star) \\ &\leq a\left[(1-\tau_j)\varrho(u_j, t^\star) + \tau_j\varrho(hu_j, t^\star)\right] \\ &\leq a\left[(1-\tau_j)\varrho(u_j, t^\star) + \tau_j a\varrho(u_j, t^\star)\right] \\ &= a(1-\tau_j(1-a))\varrho(u_j, t^\star).\end{aligned} \tag{3.6}$$

Combining (3.4), (3.5) and (3.6), we obtain

$$\begin{aligned}\varrho(u_{j+1}, t^\star) &\leq a^3(1-\tau_j(1-a))\varrho(u_j, t^\star) \\ &\leq a^3(1-\tau_j(1-a))a^3(1-\tau_{j-1}(1-a))\varrho(u_{j-1}, t^\star) \\ &\vdots \\ &\leq \left(a^3\right)^j \prod_{k=1}^{j}(1-\tau_k(1-a))\varrho(u_1, t^\star).\end{aligned} \tag{3.7}$$

Based the classical analysis, it is known $1 - u \leq e^{-u}$, $\forall u \in [0, 1]$. Taking this and (3.7) into consideration, we get

$$\varrho(u_{j+1}, t^\star) \leq \left(a^3\right)^j e^{-(1-a)\sum_{k=1}^{j}\tau_k}\varrho(u_1, t^\star).$$

Since $\sum \tau_j = \infty$ and $a \in [0, 1)$, then we have

$$\lim_{j \to \infty} \varrho(u_{j+1}, t^\star) = 0.$$

So, we obtain $u_j \to t^\star \in F(h)$. □

Remark 3.1 If the situation $\sum \tau_j = \infty$ replace with $\sum \sigma_j = \infty$ in Theorem 3.1, then we can re-write (3.7) as

$$\varrho(u_{j+1}, t^\star) \leq \left(a^3\right)^j \prod_{k=1}^{j} (1 - \sigma_k(1 - a))\varrho(u_1, t^\star).$$

Therefore, we get the desired outcome.

We call that the sequence $\{u_j\}$ is *h-stable* or *stable with respect to h* if the sequence $\{u_j\}$ converges strongly to a FP $t^\star$ of h, then an approximate sequence $\{v_j\}$ converges strongly to $t^\star$. This concept was defined by Urabe [34]. Harder and Hicks [12] provide a formal stability definition for a general repetitive procedure.

Definition 3.3 (see [7]) Let (Q, ϱ) be a metric space, h be a self map on Q and $\{v_j\}$ be an arbitrary sequence in Q. Then a repetitive sequence $\{u_j\} \subset Q$ produced by

$$\begin{cases} u_1 \in Q, \\ u_{j+1} = f(h, u_j), \quad j \geq 1, \end{cases} \tag{3.8}$$

which is convergent strongly to a FP $t^\star$ of h, it is said to be h-stable if

$$\lim_{j \to \infty} \varrho\left(v_{j+1}, f(h, v_j)\right) = 0 \Longleftrightarrow \lim_{j \to \infty} v_j = t^\star.$$

Definition 3.4 (see [10]) Two sequences $\{u_j\}$ and $\{v_j\}$ are equivalent sequences if

$$\lim_{j \to \infty} \varrho(u_j, v_j) = 0.$$

Using equivalent sequences in place of arbitrary sequences in Definition 3.3, Timiş [30] defined the concept of weak w^2-stability as follows.

Definition 3.5 Let (Q, ϱ) be a metric space, h be a self map on Q and $\{u_j\} \subset Q$ be the repetitive sequence (3.8) converging strongly to $t^\star \in F(h)$. If for any equivalent sequence $\{v_j\} \subset Q$ of $\{u_j\}$,

$$\lim_{j \to \infty} \varrho\left(v_{j+1}, f(h, v_j)\right) = 0 \Longrightarrow \lim_{j \to \infty} v_j = t^\star,$$

then the repetitive sequence $\{u_j\}$ is said to be weak w^2-stable with regard to h or weak h-w^2-stable.

Next, we give a theorem related to the weak h-w^2-stability of the altered repetition procedure (3.3).

Theorem 3.2 *Assume that all of the conditions of Theorem 3.1 are satisfied. Then the repetition procedure (3.3) is weak h-w^2-stable.*

Proof Suppose that $\{u_j\}$ be the repetitive process given by (3.3) and $\{k_j\} \subset S$ be an equivalent sequence of $\{u_j\}$. Set

$$\varepsilon_j = \varrho(k_{j+1}, hl_j)$$

where $l_j = h(\mathcal{K}(m_j, hm_j, \sigma_j))$ with $m_j = h(\mathcal{K}(k_j, hk_j, \tau_j))$. Assume that $\lim_{j\to\infty} \varepsilon_j = 0$. By using ($\mathcal{K}4$) and (3.3), we have

$$\begin{aligned} \varrho(k_{j+1}, t^\star) &\leq \varrho(k_{j+1}, u_{j+1}) + \varrho(u_{j+1}, t^\star) \\ &\leq \varrho(k_{j+1}, hl_j) + \varrho(hl_j, hv_j) + \varrho(u_{j+1}, t^\star) \\ &\leq \varepsilon_j + a\varrho(v_j, l_j) + \varrho(u_{j+1}, t^\star), \end{aligned} \tag{3.9}$$

$$\begin{aligned} \varrho(v_j, l_j) &= \varrho(h(\mathcal{K}(w_j, hw_j, \sigma_j)), h(\mathcal{K}(m_j, hm_j, \sigma_j))) \\ &\leq a\varrho(\mathcal{K}(w_j, hw_j, \sigma_j), \mathcal{K}(m_j, hm_j, \sigma_j)) \\ &\leq a\left[(1-\sigma_j)\varrho(w_j, m_j) + \sigma_j\varrho(hw_j, hm_j)\right] \\ &\leq a\left[(1-\sigma_j)\varrho(w_j, m_j) + \sigma_j a\varrho(w_j, m_j)\right] \\ &= a(1-\sigma_j(1-a))\varrho(w_j, m_j) \\ &\leq a\varrho(w_j, m_j) \end{aligned} \tag{3.10}$$

and

$$\begin{aligned} \varrho(w_j, m_j) &= \varrho(h(\mathcal{K}(u_j, hu_j, \tau_j)), h(\mathcal{K}(k_j, hk_j, \tau_j))) \\ &\leq a\varrho(\mathcal{K}(u_j, hu_j, \tau_j), \mathcal{K}(k_j, hk_j, \tau_j)) \\ &\leq a\left[(1-\tau_j)\varrho(u_j, k_j) + \tau_j\varrho(hu_j, hk_j)\right] \\ &\leq a\left[(1-\tau_j)\varrho(u_j, k_j) + \tau_j a\varrho(u_j, k_j)\right] \\ &= a(1-\tau_j(1-a))\varrho(u_j, k_j). \end{aligned} \tag{3.11}$$

Combining (3.9), (3.10) and (3.11), we have

$$\varrho(k_{j+1}, t^\star) \leq \varepsilon_j + a^3(1-\tau_j(1-a))\varrho(u_j, k_j) + \varrho(u_{j+1}, t^\star). \tag{3.12}$$

From Theorem 3.1, it follows that $\lim_{j\to\infty} \varrho(u_{j+1}, t^\star) = 0$. Since $\{u_j\}$ and $\{k_j\}$ are equivalent sequences, $\lim_{j\to\infty} \varrho(u_j, k_j) = 0$. Now, if we take the limit of both sides of (3.12) and keep $\lim_{j\to\infty} \varepsilon_j = 0$ in mind, then it brings up $\lim_{j\to\infty} \varrho(k_{j+1}, t^\star) = 0$. Thus $\{u_j\}$ is weak h-w^2-stable. □

Many analytical methods may fail to detect a map's FP. As a result, rather than computing the FP of the map, we approximate it with the help of another whose FPs are easily determined. This strategy is known as 'data dependence', and it has recently attracted a lot of attention due to its hopeful and fascinating applications.

Now we give some information, which is necessary for the data dependence outcome of the altered repetition procedure (3.3).

Definition 3.6 (see [7]) Let $h, \widetilde{h} : Q \to Q$ be two operators. $\widetilde{h}$ is said to be an approximate operator of h if $\varrho(hu, \widetilde{h}u) \leq \varepsilon$ for all $u \in Q$ and for a fixed $\varepsilon > 0$.

Lemma 3.4 *(see [28]) Let $\{x_j\}$ be a non-negative sequence for which it is assumed that there is a $j_0 \in \mathbb{N}$ such that the following inequality is satisfied, for all $j \geq j_0$,*

$$x_{j+1} \leq (1 - r_j)x_j + r_j t_j$$

where $r_j \in (0, 1)$, $\forall j \in \mathbb{N}$, $\sum r_j = \infty$ and $t_j \geq 0$, $\forall j \in \mathbb{N}$. Then,

$$0 \leq \limsup_{j\to\infty} x_j \leq \limsup_{j\to\infty} t_j.$$

Now, we prove the data dependence outcome for the altered repetition procedure (3.3).

Theorem 3.3 *Suppose that Q, S and h be the same as in Theorem 3.1 and $\widetilde{h} : S \to S$ be an approximate operator of h for given ε. Let $\{u_j\}$ and $\{\widetilde{u}_j\}$ be repetitive sequences defined by (3.3) as*

$$\begin{cases} \widetilde{u}_1 \in S, \\ \widetilde{w}_j = \widetilde{h}(\mathscr{K}(\widetilde{u}_j, \widetilde{h}\widetilde{u}_j, \tau_j)), \\ \widetilde{v}_j = \widetilde{h}(\mathscr{K}(\widetilde{w}_j, \widetilde{h}\widetilde{w}_j, \sigma_j)), \\ \widetilde{u}_{j+1} = \widetilde{h}\widetilde{v}_j, \quad j \geq 1, \end{cases} \tag{3.13}$$

respectively, where $\{\sigma_j\}$ and $\{\tau_j\}$ are real sequences in $[0, 1]$ satisfying $\sum \tau_j = \infty$. If $t^\star = ht^\star$ and $\widetilde{t^\star} = \widetilde{h}\widetilde{t^\star}$, then we have

$$\varrho(t^\star, \widetilde{t^\star}) \leq \frac{(a^3 + 2a^2 + a + 1)\varepsilon}{1 - a^3}$$

where $a \in [0, 1)$.

Proof It follows from (3.3) and (3.13), we get

$$\begin{aligned} \varrho(u_{j+1}, \widetilde{u}_{j+1}) &= \varrho(hv_j, \widetilde{h}\widetilde{v}_j) \\ &\leq \varrho(hv_j, h\widetilde{v}_j) + \varrho(h\widetilde{v}_j, \widetilde{h}\widetilde{v}_j) \\ &\leq a\varrho(v_j, \widetilde{v}_j) + \varepsilon, \end{aligned} \tag{3.14}$$

$$
\begin{aligned}
\varrho(v_j, \widetilde{v}_j) &= \varrho(h(\mathcal{K}(w_j, hw_j, \sigma_j)), \widetilde{h}(\mathcal{K}(\widetilde{w}_j, \widetilde{h}\widetilde{w}_j, \sigma_j))) \\
&\leq \varrho(h(\mathcal{K}(w_j, hw_j, \sigma_j)), h(\mathcal{K}(\widetilde{w}_j, \widetilde{h}\widetilde{w}_j, \sigma_j))) \\
&+ \varrho(h(\mathcal{K}(\widetilde{w}_j, \widetilde{h}\widetilde{w}_j, \sigma_j)), \widetilde{h}(\mathcal{K}(\widetilde{w}_j, \widetilde{h}\widetilde{w}_j, \sigma_j))) \\
&\leq a\varrho(\mathcal{K}(w_j, hw_j, \sigma_j), \mathcal{K}(\widetilde{w}_j, \widetilde{h}\widetilde{w}_j, \sigma_j)) + \varepsilon \\
&\leq a\left[(1-\sigma_j)\varrho(w_j, \widetilde{w}_j) + \sigma_j\varrho(hw_j, \widetilde{h}\widetilde{w}_j)\right] + \varepsilon \\
&\leq a(1-\sigma_j)\varrho(w_j, \widetilde{w}_j) + a\sigma_j\left[\varrho(hw_j, h\widetilde{w}_j) + \varrho(h\widetilde{w}_j, \widetilde{h}\widetilde{w}_j)\right] + \epsilon \\
&\leq a(1-\sigma_j)\varrho(w_j, \widetilde{w}_j) + a\sigma_j[a\varrho(w_j, \widetilde{w}_j) + \epsilon] + \varepsilon \\
&= a(1-\sigma_j(1-a))\varrho(w_j, \widetilde{w}_j) + a\sigma_j\varepsilon + \varepsilon
\end{aligned} \tag{3.15}
$$

and

$$
\begin{aligned}
\varrho(w_j, \widetilde{w}_j) &= \varrho(h(\mathcal{K}(u_j, hu_j, \tau_j)), \widetilde{h}(\mathcal{K}(\widetilde{u}_j, \widetilde{h}\widetilde{u}_j, \tau_j))) \\
&\leq \varrho(h(\mathcal{K}(u_j, hu_j, \tau_j)), h(\mathcal{K}(\widetilde{u}_j, \widetilde{h}\widetilde{u}_j, \tau_j))) \\
&+ \varrho(h(\mathcal{K}(\widetilde{u}_j, \widetilde{h}\widetilde{u}_j, \tau_j)), \widetilde{h}(\mathcal{K}(\widetilde{u}_j, \widetilde{h}\widetilde{u}_j, \tau_j))) \\
&\leq a\varrho(\mathcal{K}(u_j, hu_j, \tau_j), \mathcal{K}(\widetilde{u}_j, \widetilde{h}\widetilde{u}_j, \tau_j)) + \varepsilon \\
&\leq a\left[(1-\tau_j)\varrho(u_j, \widetilde{u}_j) + \tau_j\varrho(hu_j, \widetilde{h}\widetilde{u}_j)\right] + \varepsilon \\
&\leq a(1-\tau_j)\varrho(u_j, \widetilde{u}_j) + a\tau_j\left[\varrho(hu_j, h\widetilde{u}_j) + \varrho(h\widetilde{u}_j, \widetilde{h}\widetilde{u}_j)\right] + \epsilon \\
&\leq a(1-\tau_j)\varrho(u_j, \widetilde{u}_j) + a\tau_j[a\varrho(u_j, \widetilde{u}_j) + \epsilon] + \varepsilon \\
&= a(1-\tau_j(1-a))\varrho(u_j, \widetilde{u}_j) + a\tau_j\varepsilon + \varepsilon.
\end{aligned} \tag{3.16}
$$

Combining (3.14), (3.15) and (3.16), we get

$$
\begin{aligned}
&\varrho(u_{j+1}, \widetilde{u}_{j+1}) \\
&\leq a^3(1-\sigma_j(1-a))(1-\tau_j(1-a))\varrho(u_j, \widetilde{u}_j) + a^3(1-\sigma_j(1-a))\tau_j\varepsilon \\
&\quad + a^2(1-\sigma_j(1-a))\varepsilon + a^2\sigma_j\varepsilon + a\varepsilon + \varepsilon.
\end{aligned} \tag{3.17}
$$

If $a^3 \in (0, 1)$, then there is a real number $k \in (0, 1)$ such that

$$
a^3 = 1 - k. \tag{3.18}
$$

Regarding (3.18) and using the facts of $\sigma_j \leq 1$, $\tau_j \leq 1$, $1 - \sigma_j(1-a) \leq 1$ and $1 - \tau_j(1-a) \leq 1$ for all $j \geq 1$, we can re-write (3.17) as

$$
\varrho(u_{j+1}, \widetilde{u}_{j+1}) \leq (1-k)\varrho(u_j, \widetilde{u}_j) + k\frac{a^3\varepsilon + 2a^2\varepsilon + a\varepsilon + \varepsilon}{k}.
$$

By Lemma 3.4, we have

$$\varrho(t^\star, \widetilde{t^\star}) \leq \frac{(a^3 + 2a^2 + a + 1)\varepsilon}{1 - a^3}.$$

If $a^3 = 0$, from (3.17), we get $\varrho(t^\star, \widetilde{t^\star}) \leq \varepsilon$. So the proof ends. □

Remark 3.2 In the proof above, we can also re-write (3.17) as

$$\varrho(u_{j+1}, \widetilde{u}_{j+1}) \leq (1 - k)\varrho(u_j, \widetilde{u}_j) + k\frac{a^3\tau_j\varepsilon + a^2\varepsilon + a^2\sigma_j\varepsilon + a\varepsilon + \varepsilon}{1 - a^3}.$$

If the condition $\lim_{j\to\infty} \sigma_j = \lim_{j\to\infty} \tau_j = 0$ is added for the sequences $\{\sigma_j\}$ and $\{\tau_j\}$ in the hypotheses of Theorem 3.3, then we get that

$$\varrho(t^\star, \widetilde{t^\star}) \leq \frac{\varepsilon}{1 - a}.$$

3.4 Some New Convergence Outcomes of a Generalized σ-NM

We first give the following.

Lemma 3.5 *Let S be a non-empty convex subset of a KHS Q and $h : S \to S$ be a generalized σ-NM with $F(h) \neq \emptyset$. Let $\{u_j\}$ be the repetitive sequence (3.3). Then $\lim_{j\to\infty} \varrho(u_j, t^\star)$ exists for each $t^\star \in F(h)$.*

Proof By Proposition 3.1 (i), we get

$$\varrho(u_{j+1}, t^\star) = \varrho(hv_j, t^\star) \leq \varrho(v_j, t^\star), \tag{3.19}$$

$$\begin{aligned} \varrho(v_j, t^\star) &= \varrho(h(\mathcal{K}(w_j, hw_j, \sigma_j)), t^\star) \\ &\leq \varrho(\mathcal{K}(w_j, hw_j, \sigma_j), t^\star) \\ &\leq (1 - \sigma_j)\varrho(w_j, t^\star) + \sigma_j\varrho(hw_j, t^\star) \\ &\leq (1 - \sigma_j)\varrho(w_j, t^\star) + \sigma_j\varrho(w_j, t^\star) \\ &= \varrho(w_j, t^\star) \end{aligned} \tag{3.20}$$

and

$$\begin{aligned}\varrho(w_j, t^\star) &= \varrho(h(\mathscr{K}(u_j, hu_j, \tau_j)), t^\star)\\ &\leq \varrho(\mathscr{K}(u_j, hu_j, \tau_j), t^\star)\\ &\leq (1-\tau_j)\varrho(u_j, t^\star) + \tau_j \varrho(hu_j, t^\star)\\ &\leq (1-\tau_j)\varrho(u_j, t^\star) + \tau_j \varrho(u_j, t^\star)\\ &= \varrho(u_j, t^\star). \end{aligned} \tag{3.21}$$

Using (3.19), (3.20) and (3.21), we obtain

$$\varrho(u_{j+1}, t^\star) \leq \varrho(u_j, t^\star).$$

From this, $\{\varrho(u_j, t^\star)\}$ is a non-increasing and bounded below sequence. Thus $\lim_{j\to\infty} \varrho(u_j, t^\star)$ exists for all $t^\star \in F(h)$. The proof is completed. □

The theorem below generalizes Theorem 3.2 in [8].

Theorem 3.4 *Let Q be a complete UCHS with the monotone MUC μ, S be a non-empty closed convex subset of Q and $h : S \to S$ be a generalized σ-NM. Let $\{u_j\}$ be the repetitive method (3.3) with real sequences $\{\sigma_j\}$ and $\{\tau_j\}$ in $[e, f]$ for some $e, f \in (0, 1)$. Then, $F(h) \neq \emptyset$ if and only if $\{u_j\}$ is bounded and $\lim_{j\to\infty} \varrho(u_j, hu_j) = 0$.*

Proof Suppose $F(h) \neq \emptyset$ and choose $t^\star \in F(h)$. Then, by Lemma 3.5, $\lim_{j\to\infty} \varrho(u_j, t^\star)$ exists and $\{u_j\}$ is bounded. Let

$$\lim_{j\to\infty} \varrho(u_j, t^\star) = c \quad \text{for some } c \geq 0. \tag{3.22}$$

Noting

$$\varrho(hu_j, t^\star) \leq \varrho(u_j, t^\star),$$

by (3.22) we have

$$\limsup_{j\to\infty} \varrho(hu_j, t^\star) \leq c. \tag{3.23}$$

If we take the limit supremum of both sides of (3.21), we get

$$\limsup_{j\to\infty} \varrho(w_j, t^\star) \leq c. \tag{3.24}$$

By using (3.19) and (3.20), we have

$$\varrho(u_{j+1}, t^\star) \leq \varrho(w_j, t^\star)$$

which holds that

$$c \leq \liminf_{j\to\infty} \varrho(w_j, t^\star). \tag{3.25}$$

From the estimates of (3.24) and (3.25), we have that

$$\lim_{j\to\infty} \varrho(w_j, t^\star) = c. \tag{3.26}$$

Thus, from (3.21), (3.22) and (3.26), we obtain

$$\lim_{j\to\infty} \varrho(\mathscr{K}(u_j, hu_j, \tau_j), t^\star) = c. \tag{3.27}$$

By use of (3.22), (3.23), (3.27) and Lemma 3.2, we get $\lim_{j\to\infty} \varrho(u_j, hu_j) = 0$.

Conversely, we suppose that $\{u_j\}$ is bounded and $\lim_{j\to\infty} \varrho(u_j, hu_j) = 0$. Let $t^\star \in A(S, \{u_j\})$. By using Proposition 3.1 (ii), we have

$$\begin{aligned} r(ht^\star, \{u_j\}) &= \limsup_{j\to\infty} \varrho(u_j, ht^\star) \\ &\leq \left(\frac{3+\sigma}{1-\sigma}\right) \limsup_{j\to\infty} \varrho(u_j, hu_j) + \limsup_{j\to\infty} \varrho(u_j, t^\star) \\ &= \limsup_{j\to\infty} \varrho(u_j, t^\star) \\ &= r(t^\star, \{u_j\}). \end{aligned}$$

Hence, we conclude that $ht^\star \in A(S, \{u_j\})$. Since $\{u_j\}$ is a bounded sequence, by Lemma 3.1, $A(S, \{u_j\})$ occurs a unique element. Hence, we obtain $ht^\star = t^\star$, that is, $F(h) \neq \emptyset$. □

We now prove the Δ-convergence theorem of the altered repetitive procedure (3.3) for a generalized σ-NM in a KHS.

Theorem 3.5 *Assume that Q, S and $\{u_j\}$ be the same as in Theorem 3.4 and $h : S \to S$ be a generalized σ-NM with $F(h) \neq \emptyset$. Then $\{u_j\}$ is Δ-convergent to a FP of h.*

Proof By Lemma 3.1, the sequence $\{u_j\}$ has a unique asymptotic center $A(S, \{u_j\}) = \{u\}$. Let $\{u_{j_k}\}$ be any subsequence of $\{u_j\}$ such that $A(S, \{u_{j_k}\}) = \{x\}$. Then, by Theorem 3.4, we have

$$\lim_{k\to\infty} \varrho(u_{j_k}, hu_{j_k}) = 0. \tag{3.28}$$

It is similarly seen from the proof of Theorem 3.4 that x is a FP of h. Next, we prove that the FP x is the unique asymptotic center for each subsequence $\{u_{j_k}\}$ of $\{u_j\}$. Assume on the contrary, that is, $u \neq x$. Since $\lim_{j\to\infty} \varrho(u_j, x)$ exists, therefore by the uniqueness of the asymptotic center, we have

$$\begin{aligned} \limsup_{k\to\infty} \varrho(u_{j_k}, x) &< \limsup_{k\to\infty} \varrho(u_{j_k}, u) \\ &\leq \limsup_{j\to\infty} \varrho(u_j, u) \\ &< \limsup_{j\to\infty} \varrho(u_j, x) \\ &= \limsup_{k\to\infty} \varrho(u_{j_k}, x). \end{aligned}$$

This is a contradiction. Hence $u = x$. Since $\{u_{j_k}\}$ is an arbitrary subsequence of $\{u_j\}$, therefore $A(\{u_{j_k}\}) = \{x\}$ for all subsequences $\{u_{j_k}\}$ of $\{u_j\}$. This proves that the sequence $\{u_j\}$ is Δ-convergent to a FP of h. □

The strong convergence theorems are proved as follows.

Theorem 3.6 *If S is a compact subset of Q, then $\{u_j\}$ converges strongly to a FP of h under the assumptions of Theorem 3.5.*

Proof Since $F(h) \neq \emptyset$, then from Theorem 3.4, we get $\lim_{j\to\infty} \varrho(u_j, hu_j) = 0$. Since S is compact, so there is a subsequence $\{u_{j_k}\}$ of $\{u_j\}$ such that $\{u_{j_k}\}$ converges strongly to $t^\star$ for some $t^\star \in S$. Using Proposition 3.1 (ii), we get

$$\lim_{k\to\infty} \varrho(u_{j_k}, ht^\star) \leq \left(\frac{3+\sigma}{1-\sigma}\right) \lim_{k\to\infty} \varrho(u_{j_k}, hu_{j_k}) + \lim_{k\to\infty} \varrho(u_{j_k}, t^\star) = 0.$$

Then, we obtain $ht^\star = t^\star$, that is, $t^\star \in F(h)$. It follows from Lemma 3.5 that $\lim_{j\to\infty} \varrho(u_j, t^\star)$ exists for every $t^\star \in F(h)$ and hence $\{u_j\}$ is convergent strongly to $t^\star$. □

Theorem 3.7 *Suppose that Q, S, h and $\{u_j\}$ be the same as in Theorem 3.5. Then $\{u_j\}$ converges strongly to a FP of h if and only if*

$$\liminf_{j\to\infty} \varrho(u_j, F(h)) = 0$$

where $\varrho(u, F(h)) = \inf\{\varrho(u, t^\star) : t^\star \in F(h)\}$.

Proof If $\{u_j\}$ converges strongly to $t^\star \in F(h)$, then $\lim_{j\to\infty} \varrho(u_j, t^\star) = 0$. Since $0 \leq \varrho(u_j, F(h)) \leq \varrho(u_j, t^\star)$, we have $\liminf_{j\to\infty} \varrho(u_j, F(h)) = 0$.

Conversely, assume that $\liminf_{j\to\infty} \varrho(u_j, F(h)) = 0$. It is seen from Lemma 3.5 that $\lim_{j\to\infty} \varrho(u_j, F(h))$ exists and hence $\lim_{j\to\infty} \varrho(u_j, F(h)) = 0$. Therefore, there is a subsequence $\{u_{j_k}\}$ of $\{u_j\}$ and a sequence $\{t_k^\star\}$ in $F(h)$ such that

$$\varrho(u_{j_k}, t_k^\star) < \frac{1}{2^k} \quad \text{for all } k \geq 1.$$

By Lemma 3.5, we have

$$\varrho(u_{j_{k+1}}, t_{k+1}^\star) \leq \varrho(u_{j_k}, t_k^\star) < \frac{1}{2^k},$$

which implies that

$$\begin{aligned} \varrho(t_{k+1}^\star, t_k^\star) &\leq \varrho(t_{k+1}^\star, u_{j_{k+1}}) + \varrho(u_{j_{k+1}}, t_k^\star) \\ &< \frac{1}{2^{k+1}} + \frac{1}{2^k} < \frac{1}{2^{k-1}} \to 0 \text{ as } k \to \infty. \end{aligned}$$

Hence, we deduce that $\{t_k^\star\}$ is a Cauchy sequence in $F(h)$ and so it converges strongly to some $t^\star$. By Proposition 3.1(iii), $F(h)$ is closed and so $t^\star \in F(h)$. By Lemma 3.5, $\lim_{j\to\infty} \varrho(u_j, t^\star)$ exists and hence $t^\star$ is the strong limit of $\{u_j\}$. □

Remark 3.3 The condition $\liminf_{j\to\infty} \varrho(u_j, F(h)) = 0$ in Theorem 3.7 can be replaced with $\limsup_{j\to\infty} \varrho(u_j, F(h)) = 0$.

The definition of condition (I) was given by Senter and Dotson [24]. A map $h : S \to S$ is said to satisfy *condition* (I) if there is a non-decreasing self function f on $[0, \infty)$ with $f(0) = 0$ and $f(r) > 0$ for all $r \in (0, \infty)$ such that

$$\varrho(u, hu) \geq f(\varrho(u, F(h))), \quad \forall u \in S. \tag{3.29}$$

We will now use condition (I) to show the strong convergence theorem.

Theorem 3.8 *If h holds condition (I), then $\{u_j\}$ converges strongly to a FP of h under the assumptions of Theorem 3.5.*

Proof By Theorem 3.4 and (3.29), we have

$$\lim_{j\to\infty} f(\varrho(u_j, F(h))) \leq \lim_{j\to\infty} \varrho(u_j, hu_j) = 0.$$

Therefore, we get that $\lim_{j\to\infty} f(\varrho(u_j, F(h))) = 0$. Since f is a non-decreasing function satisfying $f(0) = 0$ and $f(r) > 0$ for all $r \in (0, \infty)$, we obtain

$$\lim_{j\to\infty} \varrho(u_j, F(h)) = 0.$$

The remainder of the proof appears in the way of Theorem 3.7. □

3.5 Some New Convergence Outcomes for a Finite Class of TANM

Now, we first modify the repetition procedure (3.1) for a finite class of maps into KHS:

$$\begin{cases} u_1 \in S, \\ w_j = h_i^j(\mathscr{K}(u_j, h_i^j u_j, \tau_j)), \\ v_j = h_i^j(\mathscr{K}(w_j, h_i^j w_j, \sigma_j)), \\ u_{j+1} = h_i^j v_j, \quad j \geq 1, \end{cases} \tag{3.30}$$

where $h_i = h_{i(mod\, N)}$ (here the function $mod\, N$ takes values in $\{1, 2, ..., N\}$) and for each $i = 1, 2, ..., N$, $h_i : S \to S$ is a uniformly L_i-Lipschitzian and $(\{\theta_j^{(i)}\}, \{\mu_j^{(i)}\}, \zeta^{(i)})$-TANM.

Remark 3.4 In fact, letting

$$L = \max\{L_i; i = 1, 2, ..., N\}, \theta_j = \max\{\theta_j^{(i)}; i = 1, 2, ..., N\},$$
$$\mu_j = \max\{\mu_j^{(i)}; i = 1, 2, ..., N\}, \zeta = \max\{\zeta^{(i)}; i = 1, 2, ..., N\},$$

then $\{h_i\}_{i=1}^N$ is a finite class of uniformly L-Lipschitzian and $(\{\theta_j\}, \{\mu_j\}, \zeta)$-TANM.

From now on for a finite class $\{h_i\}_{i=1}^N$, we set $F = \bigcap_{i=1}^{N} F(h_i) \neq \emptyset$.

We give some convergence theorems of the altered repetition sequence $\{u_j\}$ defined by (3.30) for a finite class of TANM in a KHS.

Theorem 3.9 *Let Q be a complete UCHS with the monotone MUC μ and S be a non-empty closed convex subset of Q. Let $\{h_i\}_{i=1}^N$ be a finite class of uniformly L-Lipschitzian and $(\{\theta_j\}, \{\mu_j\}, \zeta)$-TAN self maps on S. If*

(i) $\sum \theta_j < \infty$ and $\sum \mu_j < \infty$;

(ii) there is constants $e, f \in (0, 1)$ such that $\{\sigma_j\}, \{\tau_j\} \subset [e, f]$;

(iii) there is a constant $M > 0$ such that $\zeta(r) \leq Mr, \forall r \geq 0$;

then

(a) the sequence $\{u_j\}$ defined by (3.30), Δ-convergent to a point in F.

(b) the sequence $\{u_j\}$ converges strongly to $t^\star \in F$ if and only if $\liminf_{j\to\infty} \varrho(u_j, F) = 0$.

Proof (a) Let $t^\star \in F$. Since $\{h_i\}_{i=1}^N$ is a finite class of TAN maps, by the condition (iii), then we obtain

$$\begin{aligned}
\varrho(w_j, t^\star) &= \varrho(h_i^j(\mathscr{K}(u_j, h_i^j u_j, \tau_j)), t^\star) \\
&\leq \varrho(\mathscr{K}(u_j, h_i^j u_j, \tau_j), t^\star) + \theta_j \zeta(\varrho(\mathscr{K}(u_j, h_i^j u_j, \tau_j), t^\star)) + \mu_j \\
&\leq (1 + \theta_j M)\varrho(\mathscr{K}(u_j, h_i^j u_j, \tau_j), t^\star) + \mu_j \\
&\leq (1 + \theta_j M)\left[(1 - \tau_j)\varrho(u_j, t^\star) + \tau_n \varrho(h_i^j u_j, t^\star)\right] + \mu_j \\
&\leq (1 + \theta_j M)\left[(1 - \tau_j)\varrho(u_j, t^\star) + \tau_n\{\varrho(u_j, t^\star) + \theta_j \zeta(\varrho(u_j, t^\star)) + \mu_j\}\right] + \mu_j \\
&\leq (1 + \theta_j M)\left[(1 + \tau_j \theta_j M)\varrho(u_j, t^\star) + \tau_j \mu_j\right] + \mu_j \\
&\leq (1 + \theta_j M)^2 \varrho(u_j, t^\star) + (2 + \theta_j M)\mu_j.
\end{aligned} \tag{3.31}$$

Similarly, we get

$$\begin{aligned}
\varrho(v_j, t^\star) &= \varrho(h_i^j(\mathscr{K}(w_j, h_i^j w_j, \sigma_j)), t^\star) \\
&\leq (1 + \theta_j M)\varrho(\mathscr{K}(w_j, h_i^j w_j, \sigma_j), t^\star) + \mu_j \\
&\leq (1 + \theta_j M)^2 \varrho(w_j, t^\star) + (2 + \theta_j M)\mu_j.
\end{aligned} \tag{3.32}$$

Substituting (3.31) into (3.32), we have

$$\varrho(v_j, t^\star) \leq (1+\theta_j M)^4 \varrho(u_j, t^\star) + (2+\theta_j M)(1+(1+\theta_j M)^2)\mu_j. \tag{3.33}$$

Also, we obtain

$$\begin{aligned}\varrho(u_{j+1}, t^\star) &= \varrho(h_i^j v_j, t^\star) \\ &\leq \varrho(v_j, t^\star) + \theta_j \zeta(\varrho(v_j, t^\star)) + \mu_j \\ &\leq (1+\theta_j M)\varrho(v_j, t^\star) + \mu_j. \end{aligned} \tag{3.34}$$

Combining (3.33) and (3.34), we have

$$\varrho(u_{j+1}, t^\star) \leq (1+\pi_j)\varrho(u_j, t^\star) + \xi_j, \; j \geq 1 \text{ and } t^\star \in F(h), \tag{3.35}$$

where $\pi_j = 5\left(\theta_j M\right) + 10\left(\theta_j M\right)^2 + 10\left(\theta_j M\right)^3 + 5\left(\theta_j M\right)^4 + \left(\theta_j M\right)^5$ and $\xi_j = 1 + (1+\theta_j M)(2+\theta_j M)(1+(1+\theta_j M)^2)$. By virtue of the condition (i),

$$\sum \pi_j < \infty \text{ and } \sum \xi_j < \infty.$$

By using Lemma 3.3,

$$\lim_{j\to\infty} \varrho(u_j, t^\star) \text{ exists for each } t^\star \in F. \tag{3.36}$$

We may suppose that

$$\lim_{j\to\infty} \varrho(u_j, t^\star) = c \geq 0. \tag{3.37}$$

If we take limit supremum of both sides in the inequality (3.31), we obtain

$$\limsup_{j\to\infty} \varrho(w_j, t^\star) \leq c. \tag{3.38}$$

Since

$$\begin{aligned}\varrho(h_i^j w_j, t^\star) &\leq \varrho(w_j, t^\star) + \theta_j \zeta(\varrho(w_j, t^\star)) + \mu_j \\ &\leq (1+\theta_j M)\varrho(w_j, t^\star) + \mu_j, \; j \geq 1,\end{aligned}$$

we have

$$\limsup_{j\to\infty} \varrho(h_i^j w_j, t^\star) \leq c. \tag{3.39}$$

In the same way, we get

$$\limsup_{j\to\infty} \varrho(h_i^j u_j, t^\star) \leq c. \tag{3.40}$$

Now, we can write

$$\begin{aligned}\varrho(u_{j+1}, t^\star) &\leq (1+\theta_j M)\varrho(v_j, t^\star) + \mu_j \\ &\leq (1+\theta_j M)\left[(1+\theta_j M)^2\varrho(w_j, t^\star) + (2+\theta_j M)\mu_j\right] + \mu_j \\ &= (1+\theta_j M)^3\varrho(w_j, t^\star) + \left[1 + (1+\theta_j M)(2+\theta_j M)\right]\mu_j.\end{aligned}$$

Taking limit infimum of both sides of this inequality, we have

$$\liminf_{j\to\infty} \varrho(w_j, t^\star) \geq c.$$

Combined with (3.38), it satisfies that

$$\lim_{j\to\infty} \varrho(w_j, t^\star) = c. \tag{3.41}$$

On the other hand, since

$$\begin{aligned}\lim_{j\to\infty} \varrho(w_j, t^\star) &\leq \lim_{j\to\infty} \varrho(h_i^j(\mathscr{K}(u_j, h_i^j u_j, \tau_j)), t^\star) \\ &\leq \lim_{j\to\infty}\left[(1+\theta_j M)\varrho(\mathscr{K}(u_j, h_i^j u_j, \tau_j), t^\star) + \mu_j\right] \\ &= \lim_{j\to\infty} \varrho(\mathscr{K}(u_j, h_i^j u_j, \tau_j), t^\star) \\ &\leq \lim_{j\to\infty}\left[(1-\tau_j)\varrho(u_j, t^\star) + \tau_j\varrho(h_i^j u_j, t^\star)\right] \\ &\leq \lim_{j\to\infty}\left[(1+\tau_j\theta_j M)\varrho(u_j, t^\star) + \tau_j\mu_j\right] \\ &= \lim_{j\to\infty} \varrho(u_j, t^\star),\end{aligned}$$

we have

$$\lim_{j\to\infty} \varrho(\mathscr{K}(u_j, h_i^j u_j, \tau_j), t^\star) = c. \tag{3.42}$$

By using Lemma 3.2 with (3.37), (3.40), (3.42), we get

$$\lim_{j\to\infty} \varrho(u_j, h_i^j u_j) = 0. \tag{3.43}$$

From (3.33) and (3.34), we obtain that

$$\limsup_{j\to\infty} \varrho(v_j, t^\star) \leq c \text{ and } \liminf_{j\to\infty} \varrho(v_j, t^\star) \geq c,$$

respectively. Hence, we have

$$\lim_{j\to\infty} \varrho(v_j, t^\star) = c.$$

On the other hand, since

$$\begin{aligned}\lim_{j\to\infty} \varrho(v_j, t^\star) &\le \lim_{n\to\infty} \varrho(h_i^j(\mathscr{K}(w_j, h_i^j w_j, \sigma_j)), t^\star)\\ &\le \lim_{j\to\infty} \Big[(1+\theta_j M)\varrho(\mathscr{K}(w_j, h_i^j w_j, \sigma_j), t^\star) + \mu_j\Big]\\ &= \lim_{j\to\infty} \varrho(\mathscr{K}(w_j, h_i^j w_j, \sigma_j), t^\star)\\ &\le \lim_{j\to\infty} \Big[(1-\sigma_j)\varrho(w_j, t^\star) + \sigma_j \varrho(h_i^j w_j, t^\star)\Big]\\ &\le \lim_{j\to\infty} \big[(1+\sigma_j\theta_j M)\varrho(w_j, t^\star) + \sigma_j\mu_j\big]\\ &= \lim_{j\to\infty} \varrho(w_j, t^\star),\end{aligned}$$

we have

$$\lim_{j\to\infty} \varrho(\mathscr{K}(w_j, h_i^j w_j, \sigma_j), t^\star) = c. \tag{3.44}$$

Again, by using Lemma 3.2 with (3.39), (3.41), (3.44), we get

$$\lim_{j\to\infty} \varrho(w_j, h_i^j w_j) = 0. \tag{3.45}$$

By (3.43) and (3.45), we have

$$\begin{aligned}&\varrho(h_i^j u_j, h_i^j w_j)\\ \le&\varrho(u_j, w_j) + \theta_j\zeta(\varrho(u_j, w_j)) + \mu_j\\ \le&(1+\theta_j M)\varrho(u_j, h_i^j(\mathscr{K}(u_j, h_i^j u_j, \tau_j))) + \mu_j\\ \le&(1+\theta_j M)\Big[\varrho(u_j, h_i^j u_j) + \varrho(h_i^j u_j, h_i^j(\mathscr{K}(u_j, h_i^j u_j, \tau_j)))\Big] + \mu_j\\ \le&(1+\theta_j M)\varrho(u_j, h_i^j u_j) + (1+\theta_j M)[\varrho(u_j, \mathscr{K}(u_j, h_i^j u_j, \tau_j))\\ &+\theta_j\zeta(\varrho(u_j, \mathscr{K}(u_j, h_i^j u_j, \tau_j))) + \mu_j] + \mu_j\\ \le&(1+\theta_j M)\varrho(u_j, h_i^j u_j) + (1+\theta_j M)^2\varrho(u_j, \mathscr{K}(u_j, h_i^j u_j, \tau_j)) + (2+\theta_j M)\mu_j\\ \le&(1+\theta_j M)\varrho(u_j, h_i^j u_j) + (1+\theta_j M)^2\tau_j\varrho(u_j, h_i^j u_j) + (2+\theta_j M)\mu_j\\ \to&0 \quad \text{as} \quad j\to\infty\end{aligned} \tag{3.46}$$

and

$$\begin{aligned}&\varrho(h_i^j w_j, h_i^j v_j)\\ \le&\varrho(w_j, v_j) + \theta_j\zeta(\varrho(w_j, v_j)) + \mu_j\\ \le&(1+\theta_j M)\varrho(w_j, h_i^j(\mathscr{K}(w_j, h_i^j w_j, \sigma_j))) + \mu_j\\ \le&(1+\theta_j M)\Big[\varrho(w_j, h_i^j w_j) + \varrho(h_i^j w_j, h_i^j(\mathscr{K}(w_j, h_i^j w_j, \sigma_j)))\Big] + \mu_j\end{aligned}$$

$$
\begin{aligned}
&\le (1+\theta_j M)\varrho(w_j, h_i^j w_j) + (1+\theta_j M)[\varrho(w_j, \mathscr{K}(w_j, h_i^j w_j, \sigma_j)) \\
&\quad + \theta_j \zeta(\varrho(u_j, \mathscr{K}(w_j, h_i^j w_j, \sigma_j))) + \mu_j] + \mu_j \\
&\le (1+\theta_j M)\varrho(w_j, h_i^j w_j) + (1+\theta_j M)^2 \varrho(w_j, \mathscr{K}(w_j, h_i^j w_j, \sigma_j)) + (2+\theta_j M)\mu_j \\
&\le (1+\theta_j M)\varrho(w_j, h_i^j w_j) + (1+\theta_j M)^2 \sigma_j \varrho(w_j, h_i^j w_j) + (2+\theta_j M)\mu_j \\
&\to 0 \quad \text{as} \quad j \to \infty, \qquad (3.47)
\end{aligned}
$$

respectively. From (3.43), (3.46) and (3.47), we get

$$
\begin{aligned}
\varrho(u_j, u_{j+1}) &= \varrho(u_j, h_i^j v_j) \\
&\le \varrho(u_j, h_i^j u_j) + \varrho(h_i^j u_j, h_i^j w_j) + \varrho(h_i^j w_j, h_i^j v_j) \\
&\to 0 \text{ as } j \to \infty. \qquad (3.48)
\end{aligned}
$$

Since $\{h_i\}_{i=1}^N$ is a finite class of uniformly L-Lipschitzian, therefore we obtain

$$
\begin{aligned}
\varrho(u_j, hu_j) &\le \varrho(u_j, u_{j+1}) + \varrho(u_{j+1}, h_i^{j+1} u_{j+1}) + \varrho(h_i^{j+1} u_{j+1}, h_i^{j+1} u_j) \\
&\quad + \varrho(h_i^{j+1} u_j, h_i u_j) \\
&\le (1+L)\varrho(u_j, u_{j+1}) + \varrho(u_{j+1}, h_i^{j+1} u_{j+1}) + L\varrho(h_i^j u_j, u_j).
\end{aligned}
$$

Hence, (3.43) and (3.48) imply that

$$
\lim_{j\to\infty} \varrho(u_j, h_i u_j) = 0 \quad \text{for each } i = 1, 2, ..., N. \qquad (3.49)
$$

The remainder of the proof concludes from the pattern of Theorem 3.4 in [16].

(b) We just need to prove sufficiency because the necessity is evident. It follows from (3.36) that $\lim_{j\to\infty} \varrho(u_j, F)$ exists. Moreover, $\liminf_{j\to\infty} \varrho(u_j, F) = 0$ implies that $\lim_{j\to\infty} \varrho(u_j, F) = 0$. Because the remainder of the proof is similar to Theorem 4 in [35], it is not included. □

Remark 3.5 The condition $\liminf_{j\to\infty} \varrho(u_j, F) = 0$ in Theorem 3.9 can be replaced with $\limsup_{j\to\infty} \varrho(u_j, F) = 0$.

It is known that a self map h on a subset S of a metric space (Q, ϱ) is *semi-compact* if every bounded sequence $\{u_j\} \subset S$ such that $\varrho(u_j, hu_j) \to 0$ as $j \to \infty$ has a strongly convergent subsequence.

Using this concept, we give the strong convergence theorem below.

Theorem 3.10 *If one of the maps in the class* $\{h_i\}_{i=1}^N$ *is semi-compact, then* $\{u_j\}$ *converges strongly to a point in* F *under the assumptions of Theorem 3.9.*

Proof Suppose that the map h_k in $\{h_i\}_{i=1}^N$ is semi-compact. By (3.49) and semi-compactness of h_k, there is a subsequence $\{u_{j_k}\} \subset \{u_j\}$ such that $\{u_{j_k}\}$ converges strongly to some point $t^\star \in S$. Moreover, by the uniform continuity of $\{h_i\}_{i=1}^N$, we have

$$\varrho(t^\star, h_i t^\star) = \lim_{k\to\infty} \varrho(u_{j_k}, h_i u_{j_k}) = 0 \text{ for each } i = 1, 2, ..., N.$$

This requires that $t^\star \in F$. It follows from (3.36) that $\lim_{j\to\infty} \varrho(u_j, t^\star)$ exists and hence $\lim_{j\to\infty} \varrho(u_j, t^\star) = 0$. As an outcome, $\{u_j\}$ converges strongly to a point $t^\star$ in F. □

Next, we give a new condition which is defined by Khan et al. [18, p. 9] for the strong convergence of the repetition procedure (3.30) as follows.

Let f be a non-decreasing self map on $[0, \infty)$ with $f(0) = 0$ and $f(r) > 0$ for all $r \in (0, \infty)$. Then a class $\{h_i\}_{i=1}^N$ of self maps on S with $F = \bigcap_{i=1}^N F(h_i) \neq \emptyset$ is said to satisfy *condition* (A) if

$$\varrho(u, h_k u) \geq f(\varrho(u, F)) \quad \text{for all} \quad u \in S,$$

holds for at least one $h_k \in \{h_i\}_{i=1}^N$ or

$$\max\{\varrho(u, h_i u) : i = 1, 2, ...N\} \geq f(\varrho(u, F)) \quad \text{for all} \quad u \in S, \tag{3.50}$$

holds.

We now present a strong convergence outcome under condition (A).

Theorem 3.11 *If the class* $\{h_i\}_{i=1}^N$ *satisfies the condition* (A), *then* $\{u_j\}$ *converges strongly to a point in* F *under the assumptions of Theorem 3.9.*

Proof Since the class $\{h_i\}_{i=1}^N$ satisfiesthe condition (A), then we can assume that the inequality (3.50) holds. Thus, from (3.49) and (3.50), we obtain $\lim_{j\to\infty} f(\varrho(u_j, F)) = 0$. Since f is a non-decreasing map with $f(0) = 0$ and $f(r) > 0, \forall r > 0$, we have $\lim_{j\to\infty} \varrho(u_j, F) = 0$. The conclusion now follows from Theorem 3.9. □

Remark 3.6 Theorems 3.9, 3.10, 3.11 generalize the outcomes of Izhar-ud-din et al. [15] in two aspects: (1) from a TANM to a finite class of these maps, (2) from CAT(0) space to UCHS.

3.6 An Application to Nonlinear Integral Equations

This section is devoted to find the approximate solutions of non-linear integral equations by using our outcomes. In this manner, this section depicts the utility of our theoretical outcomes.

Consider the non-linear integral equation below

$$u(t) = m(t) + \lambda \int_e^f G(t, s)g(s, u(s))ds, \tag{3.51}$$

where $\lambda \in [0, 1]$, $G : [e, f] \times [e, f] \to [0, \infty)$, $g : [e, f] \times \mathbb{R} \to \mathbb{R}$ and $m : [e, f] \to \mathbb{R}$ are continuous functions.

Theorem 3.12 *Let $Q = C([e, f], \mathbb{R})$ be the space of continuous functions equipped with supremum metric defined as $\varrho(u, v) = \sup_{s\in[e,f]} |u(s) - v(s)|$ for $u, v \in Q$ and map given by $\mathscr{K}(u, v, \sigma) = (1 - \sigma)u + \sigma v$ for $u, v \in Q$ and $\sigma \in [0, 1]$. Define ordered relation $\leq$ as, for all $u, v \in Q$, $u \leq v$ if and only if $u(s) \leq v(s)$ for all $s \in [e, f]$. Let S be a compact subset of Q and $h : S \to S$ be a map given by*

$$hu(t) = m(t) + \lambda \int_e^f G(t, s)g(s, u(s))ds$$

where $t, s \in [e, f]$. Furthermore, suppose that the following conditions hold:
(i) There is a continuous map $k : S \times S \to [0, \infty)$ such that

$$|g(s, u(s)) - g(s, v(s))| \leq k(u, v)\, |u(s) - v(s)|$$

for all $s \in [e, f]$ and $u, v \in S$.
(ii) There is a constant $w \in [0, 1]$ such that

$$\int_e^f G(t, s)k(u, v)ds \leq w.$$

If the integral equation (3.51) has a solution, then the repetitive sequence produced by (3.3) converges strongly to the solution.

Proof Without loss of generality, we assume that $u \leq v$, so that

$$\sup\{|v(s) - u(s)| : s \in [e, f]\} \geq \sup\{|hu(s) - u(s)| : s \in [e, f]\},$$

which implies that

$$\frac{1}{2}\varrho(hu, u) \leq \varrho(hu, u) \leq \varrho(v, u).$$

We notice that

$$\begin{aligned}
&|hu(s) - hv(s)| \\
=&\left|m(t) + \lambda \int_e^f G(t, s)g(s, u(s))ds - m(t) - \lambda \int_e^f G(t, s)g(s, v(s))ds\right| \\
=&\lambda\left|\int_e^f G(t, s)[g(s, u(s)) - g(s, v(s))]ds\right|
\end{aligned}$$

$$
\begin{aligned}
&\leq \lambda \int_e^f G(t,s)\left|g(s,u(s)) - g(s,v(s))\right| ds \\
&\leq \lambda \int_e^f G(t,s)k(u,v)\left|u(s) - v(s)\right| ds \\
&\leq \sup_{s\in[e,f]} \left|u(s) - v(s)\right| \lambda \int_e^f G(t,s)k(u,v)ds \\
&\leq \lambda w \varrho(u,v) \leq \varrho(u,v).
\end{aligned}
$$

Hence, we get that, for $\frac{1}{2}\varrho(u,hu) \leq \varrho(u,v)$,

$$\varrho(hu,hv) \leq \varrho(u,v).$$

This shows that h is a generalized σ-NM with $\sigma = 0$. As an outcome, all of Theorem 3.6's assumptions have been met, and the repetitive method produced by (3.3) converges strongly to the solution of the integral equation (3.51). □

Example 3.1 If we take $m(t) = t$, $G(t,s) = ts + s^2$, $g(s,u(s)) = u(s)$ and $\lambda = \frac{1}{3}$, then (3.51) will be a linear Fredholm integral equation in the following form:

$$u(t) = t + \frac{1}{3}\int_0^1 (ts + s^2)u(s)ds, \quad t,s \in [0,1]. \tag{3.52}$$

Now, for arbitrary $u, v \in C([0,1],\mathbb{R})$ with $u(s) \leq v(s)$, we have

$$\left|g(s,u(s)) - g(s,v(s))\right| \leq \left|u(s) - v(s)\right|.$$

Then, there is the constant map defined as $k(u,v) = 1$. Also, for $t \in [0,1]$

$$
\begin{aligned}
\int_0^1 G(t,s)k(u,v)ds &= \int_0^1 (ts + s^2)ds \\
&= \left(t\frac{s^2}{2} + \frac{s^3}{3}\right)\Bigg|_0^1 \\
&= \frac{t}{2} + \frac{1}{3} \leq \frac{1}{2} + \frac{1}{3} = \frac{5}{6}.
\end{aligned}
$$

So, here $w = \frac{5}{6} \in [0,1]$. As an outcome, all of Theorem 3.12's assumptions are met. The exact solution of the integral equation (3.52) is $u(t) = \frac{576}{503}t + \frac{54}{503}$ where $t \in [0,1]$. In the view of Theorem 3.12, the repetitive method $\{u_j(t)\}$ produced by (3.3) converges strongly to the FP of map given by

$$hu(t) = t + \frac{1}{3}\int_0^1 (ts + s^2)u(s)ds$$

and hence to the exact solution of the given integral equation.

3.7 Conclusion

In the above sections, we have altered a repetition procedure into the KHS, and have used it to prove weak h-w^2-stability, data dependence outcomes for contraction maps and some convergence outcomes for generalized σ-NM. Further, we have extended this repetition procedure for a finite class of TANM in KHS and have presented some convergence theorems of the altered repetition method. Also, we have provided the reader with an application to illustrate the utility of our outcomes. These findings generalize the corresponding theorems of Bhutia and Tiwary [8] for uniformly convex Banach spaces and Izhar-ud-din et al. [15] for CAT(0) spaces, and many more in this line.

In [4], the concept of generalized (σ, τ)-NM is defined and some FP outcomes for this group of maps in Banach spaces are proved. As an outcome, in the future, the FP problem for generalized (σ, τ)-NM in KHS can be studied using similar approaches.

References

1. Aggarwal, S., Izhar-ud-din.: Convergence and stability of Fibonacci-Mann iteration for a monotone non-Lipschitzian mapping. Demonstr. Math. **52**(1), 388–396 (2019)
2. Aggarwal, S., Izhar-ud-din, Nieto, J.J.: A fixed-point theorem for monotone nearly asymptotically nonexpansive mappings. J. Fixed Point Theo. Appl. **21**(91), 11 (2019)
3. Aggarwal, S., Izhar-ud-din, Nieto, J.J.: Existence and convergence of fixed point for a G-non-Lipschitzian mapping. Miskolc Math. Notes **23**(1), 29–40 (2022)
4. Akutsah, F., Narain, O.K.: On generalized (α, β)-nonexpansive mappings in Banach spaces with applications. Nonlinear Funct. Anal. Appl. **26**(4), 663–684 (2021)
5. Alber, Y.I., Chidume, C.E., Zegeye, H.: Approximating fixed points of total asymptotically nonexpansive mappings. Fixed Point Theo. Appl. **2006**(10673), 20 (2006)
6. Aoyama, K., Kohsaka, F.: Fixed point theorem for α-nonexpansive mappings in Banach spaces. Nonlinear Anal. **74**(13), 4387–4391 (2011)
7. Berinde, V.: Iterative Approximation of Fixed Points. Springer, Berlin (2007)
8. Bhutia, J.D., Tiwary, K.: New iteration process for approximating fixed points in Banach spaces. J. Linear Topol. Algebra **8**(4), 237–250 (2019)
9. Bridson, M., Haefliger, A.: Metric Spaces of Non-Positive Curvature. Springer, Berlin (1999)
10. Cardinali, T., Rubbioni, P.: A generalization of the Caristi fixed point theorem in metric spaces. Fixed Point Theory **11**(1), 3–10 (2010)
11. Goebel, K., Kirk, W.A.: Iteration processes for nonexpansive mappings. In: Singh, S.P., Thomeier, S., Watson, B (eds.) Topological Methods in Nonlinear Functional Analysis. Contemp. Math., vol. 21, pp. 115–123. Amer. Math. Soc., Providence (1983)
12. Harder, A.M., Hicks, T.L.: Stability results for fixed point iteration procedures. Math. Japonica **33**(5), 693–706 (1988)
13. Hussain, N., Ullah, K., Arshad, M.: Fixed point approximation for Suzuki generalized nonexpansive mappings via new iteration process. J. Nonlinear Convex Anal. **19**(8), 1383–1393 (2018)
14. Izhar-ud-din, Aggarwal, S., Abdou, A.A.N.: Approximation of endpoints for α-Reich-Suzuki nonexpansive mappings in hyperbolic metric spaces. Math **9**(14), 1692, 8 (2021)
15. Izhar-ud-din, Khatoon, S., Mlaiki, N., Abdeljawad, T.: A modified iteration for total asymptotically nonexpansive mappings in Hadamard spaces. AIMS Math. **6**(5), 4758–4770 (2021)

16. Khan, M.A.A., Fukhar-ud-din, H., Kalsoom, A.: Existence and higher arity iteration for total asymptotically nonexpansive mappings in uniformly convex hyperbolic spaces. Fixed Point Theory Appl. **2016**(3), 18 (2016)
17. Kohlenbach, U.: Some logical metatheorems with applications in functional analysis. Trans. Am. Math. Soc. **357**(1), 89–128 (2004)
18. Khan, A.R., Fukhar-ud-din, H., Khan, M.A.A.: An implicit algorithm for two finite families of nonexpansive maps in hyperbolic spaces. Fixed Point Theo. Appl. **2012**(54), 12 (2012)
19. Leustean, L.: A quadratic rate of asymptotic regularity for CAT(0) spaces. J. Math. Anal. Appl. **325**(1), 386–399 (2007)
20. Leustean, L.: Nonexpansive iterations in uniformly convex W-hyperbolic spaces. In: Leizarowitz, A., Mordukhovich, B.S., Shafrir, I., Zaslavski, A (eds.) Nonlinear Analysis and Optimization I: Nonlinear Analysis, Contemp. Math., vol. 513, pp. 193–209. Amer. Math. Soc. (2010)
21. Liu, Q.: Iterative sequences for asymptotically quasi-nonexpansive mappings with error member. J. Math. Anal. **259**, 18–24 (2001)
22. Pant, D., Shukla, R.: Approximating fixed points of generalized α-nonexpansive mappings in Banach spaces. Numer. Funct. Anal. Optim. **38**(2), 248–266 (2017)
23. Reich, S., Shafrir, I.: Nonexpansive iterations in hyperbolic spaces. Nonlinear Anal. **15**, 537–558 (1990)
24. Senter, H.F., Dotson, W.G.: Approximating fixed points of nonexpansive mappings. Proc. Am. Math. Soc. **44**, 375–380 (1974)
25. Shimizu, T., Takahashi, W.: Fixed points of multivalued mappings in certain convex metric spaces. Topol. Methods Nonlinear Anal. **8**, 197–203 (1996)
26. Suzuki, T.: Fixed point theorems and convergence theorems for some generalized nonexpansive mappings. J. Math. Anal. Appl. **340**, 1088–1095 (2008)
27. Şahin, A., Başarır, M.: Some new results for the J-iterative scheme in Kohlenbach hyperbolic space. Konuralp J. Math. **10**(2), 210–219 (2022)
28. Şoltuz, ŞM., Grosan, T.: Data dependence for Ishikawa iteration when dealing with contractive like operators. Fixed Point Theo. Appl. **2008**(242916), 7 (2008)
29. Takahashi, W.: A convexity in metric spaces and nonexpansive mappings. Kodai Math. Semin. Rep. **22**, 142–149 (1970)
30. Timiş, I.: On the weak stability of Picard iteration for some contractive type mappings. Annal. Uni. Craiova Math. Comput. Sci. Ser. **37**(2), 106–114 (2010)
31. Ullah, K., Arshad, M.: New iteration process and numerical reckoning fixed point in Banach spaces. U.P.B. Sci. Bull. (Ser. A) **79**(4), 113–122 (2017)
32. Ullah, K., Arshad, M.: New three-step iteration process and fixed point approximation in Banach spaces. J. Linear Topol. Algebra **7**(2), 87–100 (2018)
33. Ullah, K., Arshad, M.: Numerical reckoning fixed points for Suzuki's generalized nonexpansive mappings via new iteration process. Filomat **32**(1), 187–196 (2018)
34. Urabe, M.: Convergence of numerical iteration in solution of equations. J. Sci. Hiroshima Univ. A **19**, 479–489 (1956)
35. Wan, L.L.: Demiclosed principle and convergence theorems for total asymptotically nonexpansive nonself mappings in hyperbolic spaces. Fixed Point Theo. Appl. **2015**(4), 10 (2015)

Chapter 4
On Common Fixed Point Results for Integral Type Contractive Conditions in Ω-Metric Spaces and Application to Integral Equations

G. S. Saluja, Reena Jain, and Hemant Kumar Nashine

Abstract Sedghi and Dung (Fixed point theorems on Ω-metric spaces, 66(1) (2014), 113–124) demonstrated that $d(\vartheta, \varkappa) = \Omega(\vartheta, \vartheta, \varkappa)$ may be used to derive the results of d-metric spaces from Ω-metric spaces. In this study, no ideas that correspond to the results of metric spaces are explored. In the framework of complete Ω-metric spaces, we construct standard fixed point theorems for integral type contractions involving rational terms and analyse their ramifications. In addition, we provide an example to support the outcome. Our findings extend and generalise a number of previously released findings. Finally, an application of the Fredhlom integral equation is discussed.

Keywords Common fixed point · Integral type contractions involving rational terms · Ω-metric space · Fredhlom integral equation

4.1 Introduction

Denote $\mathbb{R} :=$ the set of real numbers, $\mathbb{R}_+ := [0, +\infty)$.

The Banach contraction principle [1] is a significant milestone in the development of fixed point theory. Its relevance derives from its wide applicability to a multitude

G. S. Saluja
H.N. 3/1005, Geeta Nagar, Raipur 492001, Chhattisgarh, India
e-mail: saluja1963@gmail.com

R. Jain (✉) · H. K. Nashine
Mathematics Division, School of Advanced Sciences and Languages, VIT Bhopal University, Bhopal-Indore Highway, Kothrikalan, Sehore 466114, Madhya Pradesh, India
e-mail: reena.jain@vitbhopal.ac.in; reenakhatod@gmail.com

H. K. Nashine
e-mail: hemantkumar.nashine@vitbhopal.ac.in

H. K. Nashine
Department of Mathematics and Applied Mathematics, University of Johannesburg, Kingsway Campus, Auckland Park 2006, South Africa

B. Hazarika et al. (eds.), *Advances in Functional Analysis and Fixed-Point Theory*, Industrial and Applied Mathematics,
https://doi.org/10.1007/978-981-99-9207-2_4

of mathematical fields, such as the theory of existence of solutions for nonlinear differential, integral, functional equations, variational inequalities, optimization theory, and approximation theory. Numerous generalisations exist for this Banach contraction concept. These generalisations are achieved by employing various contractive conditions or by placing an extra constraint on ambient spaces. In contrast, a variety of generalisations of metric spaces have been accomplished, one of which is a Ω-metric space.

In 2012, Sedghi et al. [23] has been introduced the concept of an Ω-metric as a generalisation of a metric. The notion of an Ω-metric is as follows (see, [23]).

Definition 4.1 ([23]) Let Ξ be a nonempty set and let $\Omega : \Xi^3 \to \mathbb{R}_+$ be a function satisfying the following conditions for all $\vartheta, \varkappa, \varpi, \upsilon \in \Omega$:

$(S1)$ $\Omega(\vartheta, \varkappa, \varpi) = 0$ if and only if $\vartheta = \varkappa = \varpi$;

$(S2)$ $\Omega(\vartheta, \varkappa, \varpi) \leq \Omega(\vartheta, \vartheta, \upsilon) + \Omega(\varkappa, \varkappa, \upsilon) + \Omega(\varpi, \varpi, \upsilon)$.

Then the function Ω is called an Ω-metric on Ξ and the pair (Ξ, Ω) is called an Ω-metric space (in short SMS).

Example 4.1 ([23]) Let $\Xi = \mathbb{R}^n$ and $\|.\|$ a norm on Ω, then $\Omega(\vartheta, \varkappa, \varpi) = \|\varkappa + \varpi - 2\vartheta\| + \|\varkappa - \varpi\|$ is an Ω-metric on Ω.

Example 4.2 ([23]) Let $\Xi = \mathbb{R}^n$ and $\|.\|$ a norm on Ω, then $\Omega(\vartheta, \varkappa, \varpi) = \|\vartheta - \varpi\| + \|\varkappa - \varpi\|$ is an Ω-metric on Ω.

Example 4.3 ([24]) Let $\Xi = \mathbb{R}$ be the real line. Then $\Omega(\vartheta, \varkappa, \varpi) = |\vartheta - \varpi| + |\varkappa - \varpi|$ for all $\vartheta, \varkappa, \varpi \in \mathbb{R}$ is an Ω-metric on Ω. This Ω-metric on Ξ is called the usual Ω-metric on Ξ.

Several fixed point and common fixed point theorems for self-mappings satisfying various contractive conditions on a Ω-metric space have been given in the literature. (see [3–5, 7–10, 18–24]). Among these works, the Banach's contraction principle on a complete Ω-metric space, which was generalised by Sedghi et al. [23]. The statement of the result is as follows.

Theorem 4.1 ([23]) *Let (Ξ, Ω) be a complete Ω-metric space and let $\Im : \Xi \to \Xi$ be a self-mapping on Ξ such that*

$$\Omega(\Im\vartheta, \Im\varkappa, \Im\varpi) \leq L\, \Omega(\vartheta, \varkappa, \varpi), \tag{4.1}$$

for all $\vartheta, \varkappa, \varpi \in \Xi$, where $L \in (0, 1)$ is a constant. Then $Fix(\Im)$ is singleton in Ω.

Various contractive conditions have been studied using fixed point theory recently. Integral type contraction is, in fact, one of them.

The following theorem was developed by Branciari [2] in 2002 after he examined the existence of fixed points for mappings defined on complete metric spaces that satisfied a general contractive condition of integral type.

Denote $\Phi = \{\chi : \mathbb{R}_+ \to \mathbb{R}_+$ is a Lebesgue-integrable mapping which is summable on each compact subset of $\mathbb{R}_+$, nonnegative and such that for each $\varepsilon > 0$, $\int_0^{\varepsilon} \chi(\varsigma)d\varsigma > 0\}$.

Theorem 4.2 ([2]) *Let (Ξ, d) be a complete metric space, $h \in (0,1)$, $\chi \in \Phi$ and let $\Im: \Xi \to \Xi$ such that*

$$\int_0^{d(\Im\vartheta,\Im\varkappa)} \chi(\varsigma)d\varsigma \le h \int_0^{d(\vartheta,\varkappa)} \chi(\varsigma)d\varsigma, \tag{4.2}$$

for all $\vartheta, \varkappa \in \Xi$. Then $\Im$ has a unique fixed point in $\rho \in \Xi$ such that $\lim_{n\to+\infty} \Im^n \omega = \rho$, for each $\omega \in \Xi$.

Following this finding, numerous studies have been conducted on generalising integral type contractive conditions for various contractive mappings that satisfy a number of well-known properties (see [14, 15]). Similar work has been done by Rhoades [17], who extended Theorem 4.2 [2]'s conclusion by substituting the condition (4.3) with the following condition:

$$\int_0^{d(\Im\vartheta,\Im\varkappa)} \chi(\varsigma)d\varsigma \le k \int_0^{\max\left\{d(\vartheta,\varkappa),d(\vartheta,\Im\vartheta),d(\varkappa,\Im\varkappa),\frac{d(\vartheta,\Im\varkappa)+d(\varkappa,\Im\vartheta)}{2}\right\}} \chi(\varsigma)d\varsigma, \tag{4.3}$$

for each $k \in [0, 1)$ and $\vartheta, \varkappa \in \Xi$.

Recently, Rahman et al. [13] have established a common fixed point result of Altman integral type for four self-mappings in the context of Ω-metric spaces and gave an example in support of the result. Very recently, Ozgur and Tas [11] have studied new contractive conditions of integral type on Ω-metric spaces and established some fixed point theorems for various contractive conditions of integral type and gave examples in support of the established results. Also they obtained an application to the Fredholm integral equation. Rashwan and Hammad [16] recently established some common fixed point theorems for weakly compatible mappings satisfying a general contractive condition of integral type involving rational terms in complete metric spaces and provided examples.

Motivated by Ozgur and Tas [11] and Rashwan and Hammad [16], we demonstrate common fixed point theorems for contractive conditions of integral type using rational expressions in the setting of complete Ω-metric spaces in a manner that cannot be converted to metric space findings. To justify the effort, supporting examples are provided. The conclusions of this research extend and generalise several previously reported findings. Finally we discuss an application to the Fredhlom integral equation.

4.2 Preliminaries

We now recall some basic properties and auxiliary results about Ω-metric spaces.

Definition 4.2 ([23]) Let (Ξ, Ω) be an Ω-metric space.

(s_1) A sequence $\{\zeta_n\}$ in Ω converges to $\zeta \in \Xi$ if and only if $\Omega(\zeta_n, \zeta_n, \zeta) \to 0$ as $n \to +\infty$, that is, for each $\varepsilon > 0$, there exists $n_0 \in \mathbb{N}$ such that for all $n \geq n_0$ we have $\Omega(\zeta_n, \zeta_n, \zeta) < \varepsilon$. We denote this by $\lim_{n\to+\infty} \zeta_n = \zeta$ or $\zeta_n \to \zeta$ as $n \to +\infty$.

(s_2) A sequence $\{\zeta_n\}$ in Ω is called a Cauchy sequence if $\Omega(\zeta_n, \zeta_n, \zeta_m) \to 0$ as $n, m \to +\infty$, that is, for each $\varepsilon > 0$, there exists $n_0 \in \mathbb{N}$ such that for all $n, m \geq n_0$ we have $\Omega(\zeta_n, \zeta_n, \zeta_m) < \varepsilon$.

(s_3) The Ω-metric space (Ξ, Ω) is called complete if every Cauchy sequence in Ω is convergent in Ω.

Lemma 4.1 ([23], Lemma 2.5) *Let (Ξ, Ω) be an Ω-metric space. Then, we have $\Omega(\nu, \nu, \zeta) = \Omega(\zeta, \zeta, \nu)$ for all $\nu, \zeta \in \Xi$.*

The preceding Lemma 4.1 can be thought of as a symmetry condition for a Ω-metric space.

Lemma 4.2 ([23], Lemma 2.12) *Let (Ξ, Ω) be an Ω-metric space. If $\zeta_n \to \zeta$ and $\xi_n \to \xi$ as $n \to +\infty$ then $\Omega(\zeta_n, \zeta_n, \xi_n) \to \Omega(\zeta, \zeta, \xi)$ as $n \to +\infty$.*

The relationship between a metric and a Ω-metric is demonstrated in the following lemma.

Lemma 4.3 ([4]) *Let (Ξ, d) be a metric space. Then the following properties are satisfied:*

(1) $\Omega_d(\vartheta, \varkappa, \varpi) = d(\vartheta, \varpi) + d(\varkappa, \varpi)$ for all $\vartheta, \varkappa, \varpi \in \Xi$ is an Ω-metric on Ω.

(2) $\zeta_n \to \zeta$ in (Ξ, d) if and only if $\zeta_n \to \zeta$ in (Ξ, Ω_d).

(3) $\{\zeta_n\}$ is Cauchy in (Ξ, d) if and only if $\{\zeta_n\}$ is Cauchy in (Ξ, Ω_d).

(4) (Ξ, d) is complete if and only if (Ξ, Ω_d) is complete.

The metric d generates the Ω-metric Ω_d in Lemma 4.3 (1). A non-metric Ω-metric is found in [4, 10].

4.3 Main Results

In this section, we shall prove some unique common fixed point theorems for integral type contractive conditions in the setting of complete Ω-metric spaces.

We denote $Fix(\Im_1, \Im_2)$:= set of common fixed points of $\Im_1, \Im_2$, and $Fix(\Im)$:= set of fixed points of $\Im$.

Theorem 4.3 *Let (Ξ, Ω) be a complete Ω-metric space, $\chi \in \Phi$ and let $\Im_1, \Im_2 \colon \Xi \to \Xi$ such that*

$$\begin{aligned}\int_0^{\Omega(\Im_1\vartheta,\Im_1\varkappa,\Im_2\varpi)}\chi(\varsigma)d\varsigma \le\, & h_1\int_0^{\Omega(\vartheta,\varkappa,\varpi)}\chi(\varsigma)d\varsigma + h_2\int_0^{\Omega(\vartheta\,\varkappa,\Im_1\vartheta)}\chi(\varsigma)d\varsigma\\ & + h_3\int_0^{\Omega(\varpi,\varkappa,\Im_2\varpi)}\chi(\varsigma)d\varsigma + h_4\int_0^{\Omega(\vartheta,\varkappa,\Im_2\varpi)}\chi(\varsigma)d\varsigma\\ & + h_5\int_0^{\Omega(\varpi,\varkappa,\Im_1\vartheta)}\chi(\varsigma)d\varsigma + h_6\int_0^{\max\{\Omega(\vartheta,\varkappa,\varpi),\Omega(\varpi,\varkappa,\Im_1\vartheta)\}}\chi(\varsigma)d\varsigma,\end{aligned} \tag{4.4}$$

for all $\vartheta, \varkappa, \varpi \in \Xi$, *where* $h_1, h_2, h_3, h_4, h_5, h_6$ *are nonnegative reals such that* $h_1 + h_2 + 2h_3 + 3h_4 + h_5 + h_6 < 1$. *Then* $Fix(\Im_1, \Im_2)$ *is singleton in* Ξ.

Proof Let $\vartheta_0 \in \Xi$ and the sequence $\{\vartheta_n\}$ be defined as $\vartheta_{2n+1} = \Im_1\vartheta_{2n}$ and $\vartheta_{2n+2} = \Im_2\vartheta_{2n+1}$ for $n = 1, 2, \ldots$. Then from inequality (4.4), and using the condition $(S1)$, $(S2)$ and Lemma 4.1, we have

$$\begin{aligned}\int_0^{\Omega(\vartheta_{2n},\vartheta_{2n},\vartheta_{2n+1})}\chi(\varsigma)d\varsigma =& \int_0^{\Omega(\Im_1\vartheta_{2n-1},\Im_1\vartheta_{2n-1},\Im_2\vartheta_{2n})}\chi(\varsigma)d\varsigma\\ \le & h_1\int_0^{\Omega(\vartheta_{2n-1},\vartheta_{2n-1},\vartheta_{2n})}\chi(\varsigma)d\varsigma\\ & + h_2\int_0^{\Omega(\vartheta_{2n-1},\vartheta_{2n-1},\Im_1\vartheta_{2n-1})}\chi(\varsigma)d\varsigma\\ & + h_3\int_0^{\Omega(\vartheta_{2n},\vartheta_{2n-1},\Im_2\vartheta_{2n})}\chi(\varsigma)d\varsigma\\ & + h_4\int_0^{\Omega(\vartheta_{2n-1},\vartheta_{2n-1},\Im_2\vartheta_{2n})}\chi(\varsigma)d\varsigma\\ & + h_5\int_0^{\Omega(\vartheta_{2n},\vartheta_{2n-1},\Im_1\vartheta_{2n-1})}\chi(\varsigma)d\varsigma\\ & + h_6\int_0^{\max\{\Omega(\vartheta_{2n-1},\vartheta_{2n-1},\vartheta_{2n}),\Omega(\vartheta_{2n},\vartheta_{2n-1},\Im_1\vartheta_{2n-1})\}}\chi(\varsigma)d\varsigma\\ =& h_1\int_0^{\Omega(\vartheta_{2n-1},\vartheta_{2n-1},\vartheta_{2n})}\chi(\varsigma)d\varsigma\\ & + h_2\int_0^{\Omega(\vartheta_{2n-1},\vartheta_{2n-1},\vartheta_{2n})}\chi(\varsigma)d\varsigma\\ & + h_3\int_0^{\Omega(\vartheta_{2n},\vartheta_{2n-1},\vartheta_{2n+1})}\chi(\varsigma)d\varsigma\\ & + h_4\int_0^{\Omega(\vartheta_{2n-1},\vartheta_{2n-1},\vartheta_{2n+1})}\chi(\varsigma)d\varsigma\\ & + h_5\int_0^{\Omega(\vartheta_{2n},\vartheta_{2n-1},\vartheta_{2n})}\chi(\varsigma)d\varsigma\\ & + h_6\int_0^{\max\{\Omega(\vartheta_{2n-1},\vartheta_{2n-1},\vartheta_{2n}),\Omega(\vartheta_{2n},\vartheta_{2n-1},\vartheta_{2n})\}}\chi(\varsigma)d\varsigma\end{aligned}$$

$$
\begin{aligned}
&\le (h_1 + h_2 + h_3 + h_4 + h_5 + h_6) \int_0^{\Omega(\vartheta_{2n-1}, \vartheta_{2n-1}, \vartheta_{2n})} \chi(\varsigma) d\varsigma \\
&\quad + (h_3 + 2h_4) \int_0^{\Omega(\vartheta_{2n}, \vartheta_{2n}, \vartheta_{2n+1})} \chi(\varsigma) d\varsigma, \qquad (4.5)
\end{aligned}
$$

which implies

$$
\begin{aligned}
&\int_0^{\Omega(\vartheta_{2n}, \vartheta_{2n}, \vartheta_{2n+1})} \chi(\varsigma) d\varsigma \\
&\le \Big(\frac{h_1 + h_2 + h_3 + h_4 + h_5 + h_6}{1 - h_3 - 2h_4}\Big) \int_0^{\Omega(\vartheta_{2n-1}, \vartheta_{2n-1}, \vartheta_{2n})} \chi(\varsigma) d\varsigma. \qquad (4.6)
\end{aligned}
$$

If we take $\lambda = \frac{h_1+h_2+h_3+h_4+h_5+h_6}{1-h_3-2h_4}$, then $\lambda < 1$ as $h_1 + h_2 + 2h_3 + 3h_4 + h_5 + h_6 < 1$. Using the inequality (4.6) again, we obtain

$$
\int_0^{\Omega(\vartheta_{2n}, \vartheta_{2n}, \vartheta_{2n+1})} \chi(\varsigma) d\varsigma \le \lambda^n \int_0^{\Omega(\vartheta_0, \vartheta_0, \vartheta_1)} \chi(\varsigma) d\varsigma. \qquad (4.7)
$$

Thus in general, for $n = 0, 1, 2, \ldots$, we have

$$
\int_0^{\Omega(\vartheta_n, \vartheta_n, \vartheta_{n+1})} \chi(\varsigma) d\varsigma \le \lambda^n \int_0^{\Omega(\vartheta_0, \vartheta_0, \vartheta_1)} \chi(\varsigma) d\varsigma. \qquad (4.8)
$$

Taking the limit as $n \to +\infty$, in inequality (4.8), we get

$$
\lim_{n \to +\infty} \int_0^{\Omega(\vartheta_n, \vartheta_n, \vartheta_{n+1})} \chi(\varsigma) d\varsigma = 0, \qquad (4.9)
$$

since $0 < \lambda < 1$. The condition (4.9) implies

$$
\lim_{n \to +\infty} \Omega(\vartheta_n, \vartheta_n, \vartheta_{n+1}) = 0. \qquad (4.10)
$$

Now, we have to show that the sequence $\{\vartheta_n\}$ is a Cauchy sequence. Assume that $\{\vartheta_n\}$ is not a Cauchy sequence. Then there exists an $\varepsilon > 0$ for which we can find subsequences $\{\vartheta_{2m(\ell)}\}$ and $\{\vartheta_{2n(\ell)}\}$ of $\{\vartheta_{2n}\}$ and increasing sequences of integers $\{2m(\ell)\}$ and $\{2n(\ell)\}$ such that $n(\ell)$ is smallest index for which,

$$
n(\ell) > m(\ell) > k, \qquad (4.11)
$$

$$
\Omega(\vartheta_{2m(\ell)}, \vartheta_{2m(\ell)}, \vartheta_{2n(\ell)}) \ge \varepsilon. \qquad (4.12)
$$

Further corresponding to $m(\ell)$, we can choose $n(\ell)$ in such a way that it is the smallest integer with $n(\ell) > m(\ell)$ and satisfying (4.11). Then

$$\Omega(\vartheta_{2m(\ell)}, \vartheta_{2m(\ell)}, \vartheta_{2n(\ell)-1}) < \varepsilon. \tag{4.13}$$

Now, using (4.12), $(S2)$ and Lemma 4.1, we have

$$\begin{aligned}
\varepsilon &\leq \Omega(\vartheta_{2m(\ell)}, \vartheta_{2m(\ell)}, \vartheta_{2n(\ell)}) \\
&= \Omega(\vartheta_{2n(\ell)}, \vartheta_{2n(\ell)}, \vartheta_{2m(\ell)}) \\
&\leq 2\Omega(\vartheta_{2n(\ell)}, \vartheta_{2n(\ell)}, \vartheta_{2n(\ell)-1}) \\
&\quad + \Omega(\vartheta_{2m(\ell)}, \vartheta_{2m(\ell)}, \vartheta_{2n(\ell)-1}) \\
&\leq \varepsilon + 2\Omega(\vartheta_{2n(\ell)}, \vartheta_{2n(\ell)}, \vartheta_{2n(\ell)-1}). \text{ (by (4.13))}
\end{aligned} \tag{4.14}$$

Letting $k \to +\infty$ in Eq. (4.14) and using (4.10), we get

$$\lim_{k \to +\infty} \Omega(\vartheta_{2m(\ell)}, \vartheta_{2m(\ell)}, \vartheta_{2n(\ell)}) = \varepsilon. \tag{4.15}$$

Again, with the help of $(S2)$ and Lemma 4.1, we have

$$\begin{aligned}
\Omega(\vartheta_{2m(\ell)}, \vartheta_{2m(\ell)}, \vartheta_{2n(\ell)}) &\leq 2\Omega(\vartheta_{2m(\ell)}, \vartheta_{2m(\ell)}, \vartheta_{2m(\ell)-1}) \\
&\quad + \Omega(\vartheta_{2n(\ell)}, \vartheta_{2n(\ell)}, \vartheta_{2m(\ell)-1}) \\
&= 2\Omega(\vartheta_{2m(\ell)}, \vartheta_{2m(\ell)}, \vartheta_{2m(\ell)-1}) \\
&\quad + \Omega(\vartheta_{2m(\ell)-1}, \vartheta_{2m(\ell)-1}, \vartheta_{2n(\ell)}).
\end{aligned} \tag{4.16}$$

Also, with the help of $(S2)$, Lemma 4.1, we have

$$\begin{aligned}
\Omega(\vartheta_{2m(\ell)-1}, \vartheta_{2m(\ell)-1}, \vartheta_{2n(\ell)-1}) &\leq 2\Omega(\vartheta_{2m(\ell)-1}, \vartheta_{2m(\ell)-1}, \vartheta_{2m(\ell)}) \\
&\quad + \Omega(\vartheta_{2n(\ell)-1}, \vartheta_{2n(\ell)-1}, \vartheta_{2m(\ell)}) \\
&= 2\Omega(\vartheta_{2m(\ell)}, \vartheta_{2m(\ell)}, \vartheta_{2m(\ell)-1}) \\
&\quad + \Omega(\vartheta_{2m(\ell)}, \vartheta_{2m(\ell)}, \vartheta_{2n(\ell)-1}),
\end{aligned} \tag{4.17}$$

and

$$\lim_{k \to +\infty} \int_0^{\Omega(\vartheta_{2m(\ell)-1}, \vartheta_{2m(\ell)-1}, \vartheta_{2n(\ell)-1})} \chi(\varsigma) d\varsigma \leq \int_0^{\varepsilon} \chi(\varsigma) d\varsigma. \tag{4.18}$$

Using the inequality (4.4), (4.6), (4.12) and (4.18), we obtain

$$\begin{aligned}
\int_0^{\varepsilon} \chi(\varsigma) d\varsigma &\leq \int_0^{\Omega(\vartheta_{2m(\ell)}, \vartheta_{2m(\ell)}, \vartheta_{2n(\ell)})} \chi(\varsigma) d\varsigma \\
&= \int_0^{\Omega(\Im_1 \vartheta_{2m(\ell)-1}, \Im_1 \vartheta_{2m(\ell)-1}, \Im_2 \vartheta_{2n(\ell)-1})} \chi(\varsigma) d\varsigma
\end{aligned}$$

$$\leq \lambda \int_0^{\Omega(\vartheta_{2m(\ell)-1},\vartheta_{2m(\ell)-1},\vartheta_{2n(\ell)-1})} \chi(\varsigma)d\varsigma$$
$$\leq \lambda \int_0^{\varepsilon} \chi(\varsigma)d\varsigma, \tag{4.19}$$

which is a contradiction with $0 < \lambda < 1$. Thus, the sequence $\{\vartheta_n\}$ is a Cauchy sequence in Ω. Hence by completeness hypothesis of Ω, it is convergent to the point, say, $u \in \Xi$, that is, $\lim_{n\to+\infty} \vartheta_n = u$.

Now, from the given inequality (4.4), we find

$$\begin{aligned}
\int_0^{\Omega(\Im_1\rho,\Im_1\rho,\vartheta_{2n+1})} \chi(\varsigma)d\varsigma &= \int_0^{\Omega(\Im_1\rho,\Im_1\rho,\Im_2\vartheta_{2n})} \chi(\varsigma)d\varsigma \\
&\leq h_1 \int_0^{\Omega(\rho,\rho,\vartheta_{2n})} \chi(\varsigma)d\varsigma \\
&\quad + h_2 \int_0^{\Omega(\rho,\rho,\Im_1\rho)} \chi(\varsigma)d\varsigma \\
&\quad + h_3 \int_0^{\Omega(\vartheta_{2n},\rho,\Im_2\vartheta_{2n})} \chi(\varsigma)d\varsigma \\
&\quad + h_4 \int_0^{\Omega(\rho,\rho,\Im_2\vartheta_{2n})} \chi(\varsigma)d\varsigma \\
&\quad + h_5 \int_0^{\Omega(\vartheta_{2n},\rho,\Im_1\rho)} \chi(\varsigma)d\varsigma \\
&\quad + h_6 \int_0^{\max\{\Omega(\rho,\rho,\vartheta_{2n}),\Omega(\vartheta_{2n},\rho,\Im_1\rho)\}} \chi(\varsigma)d\varsigma \\
&= h_1 \int_0^{\Omega(\rho,\rho,\vartheta_{2n})} \chi(\varsigma)d\varsigma \\
&\quad + h_2 \int_0^{\Omega(\rho,\rho,\Im_1\rho)} \chi(\varsigma)d\varsigma \\
&\quad + h_3 \int_0^{\Omega(\vartheta_{2n},\rho,\vartheta_{2n+1})} \chi(\varsigma)d\varsigma \\
&\quad + h_4 \int_0^{\Omega(\rho,\rho,\vartheta_{2n+1})} \chi(\varsigma)d\varsigma \\
&\quad + h_5 \int_0^{\Omega(\vartheta_{2n},\rho,\Im_1\rho)} \chi(\varsigma)d\varsigma \\
&\quad + h_6 \int_0^{\max\{\Omega(\rho,\rho,\vartheta_{2n}),\Omega(\vartheta_{2n},\rho,\Im_1\rho)\}} \chi(\varsigma)d\varsigma.
\end{aligned} \tag{4.20}$$

Taking the limit as $n \to +\infty$ in (4.20) and using the condition $(S1)$ and Lemma 4.1, we get

$$\int_0^{\Omega(\Im_1\rho,\Im_1\rho,\rho)} \chi(\varsigma)d\varsigma \leq (h_2 + h_5 + h_6) \int_0^{\Omega(\Im_1\rho,\Im_1\rho,\rho)} \chi(\varsigma)d\varsigma, \tag{4.21}$$

which implies $\Omega(\Im_1\rho, \Im_1\rho, \rho) = 0$, that is, $\Im_1\rho = \rho$ since $h_2 + h_5 + h_6 < 1$. This shows that $\rho \in Fix(\Im_1)$. Similarly, we can show that $\rho \in Fix(\Im_2)$. Thus $\rho \in Fix(\Im_1, \Im_2)$.

To prove $\rho \in Fix(\Im_1, \Im_2)$ is singleton, let $\rho_1 \neq \rho$ such that $\Im_1\rho_1 = \rho_1 = \Im_2\rho_1$ with $\rho_1 \neq \rho$. Using the inequality (4.23) and Lemma 4.1, we obtain

$$\begin{aligned}
\int_0^{\Omega(\rho,\rho,\rho_1)} \chi(\varsigma)d\varsigma &= \int_0^{\Omega(\Im_1\rho,\Im_1\rho,\Im_2\rho_1)} \chi(\varsigma)d\varsigma \\
&\leq h_1 \int_0^{\Omega(\rho,\rho,\rho_1)} \chi(\varsigma)d\varsigma + h_2 \int_0^{\Omega(\rho,\rho,\Im_1\rho)} \chi(\varsigma)d\varsigma \\
&\quad + h_3 \int_0^{\Omega(\rho_1,\rho,\Im_2\rho_1)} \chi(\varsigma)d\varsigma + h_4 \int_0^{\Omega(\rho,\rho,\Im_2\rho_1)} \chi(\varsigma)d\varsigma \\
&\quad + h_5 \int_0^{\Omega(\rho_1,\rho,\Im_1\rho)} \chi(\varsigma)d\varsigma + h_6 \int_0^{\max\{\Omega(\rho,\rho,\rho_1),\Omega(\rho_1,\rho,\Im_1\rho\}} \chi(\varsigma)d\varsigma \\
&= h_1 \int_0^{\Omega(\rho,\rho,\rho_1)} \chi(\varsigma)d\varsigma + h_2 \int_0^{\Omega(\rho,\rho,\rho)} \chi(\varsigma)d\varsigma \\
&\quad + h_3 \int_0^{\Omega(\rho_1,\rho,\rho_1)} \chi(\varsigma)d\varsigma + h_4 \int_0^{\Omega(\rho,\rho,\rho_1)} \chi(\varsigma)d\varsigma \\
&\quad + h_5 \int_0^{\Omega(\rho_1,\rho,\rho)} \chi(\varsigma)d\varsigma + h_6 \int_0^{\max\{\Omega(\rho,\rho,\rho_1),\Omega(\rho_1,\rho,u\}} \chi(\varsigma)d\varsigma \\
&= (h_1 + h_3 + h_4 + h_5 + h_6) \int_0^{\Omega(\rho,\rho,\rho_1)} \chi(\varsigma)d\varsigma,
\end{aligned} \tag{4.22}$$

which implies $\Omega(\rho, \rho, \rho_1) = 0$, that is, $\rho = \rho_1$ since $h_1 + h_3 + h_4 + h_5 + h_6 < 1$. Thus, $Fix(\Im_1, \Im_2)$ is singleton in Ω. □

If we take $\Im_1 = \Im_2 = \Im$ in Theorem 4.3, then the result is as follows:

Corollary 4.1 *Let (Ξ, Ω) be a complete Ω-metric space, $\chi \in \Phi$ and let $\Im : \Xi \to \Xi$ be a self-mapping such that*

$$\begin{aligned}
\int_0^{\Omega(\Im\vartheta,\Im\varkappa,\Im\varpi)} \chi(\varsigma)d\varsigma \leq& h_1 \int_0^{\Omega(\vartheta,\varkappa,\varpi)} \chi(\varsigma)d\varsigma + h_2 \int_0^{\Omega(\vartheta,\varkappa,\Im\vartheta)} \chi(\varsigma)d\varsigma \\
&+ h_3 \int_0^{\Omega(\varpi,\varkappa,\Im\varpi)} \chi(\varsigma)d\varsigma + h_4 \int_0^{\Omega(\vartheta,\varkappa,\Im\varpi)} \chi(\varsigma)d\varsigma \\
&+ h_5 \int_0^{\Omega(\varpi,\varkappa,\Im\vartheta)} \chi(\varsigma)d\varsigma + h_6 \int_0^{\max\{\Omega(\vartheta,\varkappa,\varpi),\Omega(\varpi,\varkappa,\Im\vartheta)\}} \chi(\varsigma)d\varsigma,
\end{aligned}$$

for all $\vartheta, \varkappa, \varpi \in \Xi$, *where* $h_1, h_2, h_3, h_4, h_5, h_6$ *are nonnegative reals such that* $h_1 + h_2 + 2h_3 + 3h_4 + h_5 + h_6 < 1$. *Then* $\rho \in Fix(\Im)$ *for* $\rho \in \Xi$.

If we take $h_1 = h$ and $h_2 = h_3 = h_4 = h_5 = h_6 = 0$ in Corollary 4.1, then the result is as follows:

Corollary 4.2 ([11]) *Let* (Ξ, Ω) *be a complete* Ω*-metric space,* $\chi \in \Phi$ *and let* $\Im\colon \Xi \to \Xi$ *be a self-mapping such that*

$$\int_0^{\Omega(\Im\vartheta,\Im\varkappa,\Im\varpi)} \chi(\varsigma)d\varsigma \le h \int_0^{\Omega(\vartheta,\varkappa,\varpi)} \chi(\varsigma)d\varsigma,$$

for all $\vartheta, \varkappa, \varpi \in \Xi$, *where* $h \in (0, 1)$ *is a constant. Then* $\rho \in Fix(\Im)$ *is unique in* Ξ *and* $\Im^n\omega = \rho$ *for each* $\omega \in \Xi$.

Remark 4.1 (1) Corollary 4.2 is a generalisation of Branciari [2] fixed point result from complete metric space to the set up of complete Ω-metric space.
(2) In Corollary 4.2, if we set the function $\chi : \mathbb{R}_+ \to \mathbb{R}_+$ as $\chi(\varsigma) = 1$ for all $t \in \mathbb{R}_+$, then we obtain Theorem 3.1 of [23] for $n = 1$.
(3) Corollary 4.2 is also a generalisation of Sedghi's et al. result [23] to the case of integral type contraction condition.
(4) Theorem 4.3 is a generalisation of Theorem 2.4 of Özgür and Tas [11]. Indeed, if we take $\Im_1 = \Im_2 = \Im$, $h_1 = h$ and $h_2 = h_3 = h_4 = h_5 = h_6 = 0$ in Theorem 4.3, then we get Theorem 2.4 of [11].
(5) Corollary 4.1 is also a generalisation of Theorem 2.4 of Özgür and Tas [11]. Indeed, if we take $h_1 = h$ and $h_2 = h_3 = h_4 = h_5 = h_6 = 0$ in Corollary 4.1, then we get Theorem 2.4 of [11].
(6) Due to the fact that a Ω-metric space is a generalisation of a metric space. The Corollary 4.1 is a generalisation of the classical fixed point theorem of Banach [1].
(7) If we set the Ω-metric as $\Omega\colon \Xi^3 \to \mathbb{C}$ and take the function $\chi : \mathbb{R}_+ \to \mathbb{R}_+$ as $\chi(\varsigma) = 1$ for all $t \in \mathbb{R}_+$ in Corollary 4.2, then we get Theorem 3.1 in [12] and Corollary 2.5 in [6] for $n = 1$.

Example 4.4 Let $\Xi = [0, 1]$ and the function $\Omega\colon \Xi^3 \to \mathbb{R}_+$ be defined as

$$\Omega(\vartheta, \varkappa, \varpi) = |\varkappa - \varpi| + |\varkappa + \varpi - 2\vartheta|,$$

for all $\vartheta, \varkappa, \varpi \in \Xi$. The function Ω is readily apparent to be a complete Ω-metric space. Now we demonstrate that this Ω-metric cannot be derived from any d-metric. In contrast, we assume there exists a metric d such that

$$\Omega(\vartheta, \varkappa, \varpi) = d(\vartheta, \varpi) + d(\varkappa, \varpi), \tag{4.23}$$

for all $\vartheta, \varkappa, \varpi \in \Xi$.

Let us define the self-mapping $\Im_1, \Im_2 \colon \Xi \to \Xi$ as $\Im_1(\vartheta) = \frac{\vartheta}{3}$, $\Im_2(\vartheta) = \frac{2\vartheta}{3}$ for all $\vartheta \in \Xi$ and the function $\chi : \mathbb{R}_+ \to \mathbb{R}_+$ as $\chi(\varsigma) = 3\varsigma^2$ for all $\varsigma \in \mathbb{R}_+$. Then we get

$$\int_0^{\varepsilon} \chi(\varsigma)d\varsigma = \int_0^{\varepsilon} 3\varsigma^2 d\varsigma = \varepsilon^3 > 0,$$

for each $\varepsilon > 0$.

Now

$$\Omega(\vartheta, \varkappa, \Im_1\vartheta) = |\varkappa - \frac{\vartheta}{3}| + |\varkappa - \frac{5\vartheta}{3}|, \Omega(\varpi, \varkappa, \Im_2\varpi) = |\varkappa - \frac{2\varpi}{3}| + |\varkappa - \frac{4\varpi}{3}|$$

$$\Omega(\vartheta, \varkappa, \Im_2\varpi) = |\varkappa - \frac{2\varpi}{3}| + |\varkappa + \frac{2\varpi}{3} - 2\vartheta|, \Omega(\varpi, \varkappa, \Im_1\vartheta) = |\varkappa - \frac{\vartheta}{3}| + |\varkappa + \frac{\vartheta}{3} - 2\varpi|$$

$$\max\{\Omega(\vartheta, \varkappa, \varpi), \Omega(\varpi, \varkappa, \Im_1\vartheta)\} = \{|\varkappa - \varpi| + |\varkappa + \varpi - 2\vartheta|, |\varkappa - \frac{\vartheta}{3}| + |\varkappa + \frac{\vartheta}{3} - 2\varpi|\}$$

$$\Omega(\Im_1\vartheta, \Im_1\varkappa, \Im_2\varpi) = |\frac{\varkappa}{3} - \frac{2\varpi}{3}| + |\varkappa + \frac{2\varpi}{3} - \frac{2\vartheta}{3}|.$$

Case (I): Suppose $\vartheta = \varkappa \neq \varpi$. Then $\Omega(\vartheta, \varkappa, \varpi) = 2|\vartheta - \varpi|$, $\Omega(\vartheta, \varkappa, \Im_1\vartheta) = \frac{4\vartheta}{3}$, $\Omega(\varpi, \varkappa, \Im_2\varpi) = |\vartheta - \frac{2\varpi}{3}| + |\vartheta - \frac{4\varpi}{3}|$, $\Omega(\vartheta, \varkappa, \Im_2\varpi) = 2|\vartheta - \frac{2\varpi}{3}|$, $\Omega(\varpi, \varkappa, \Im_1\vartheta) = \frac{2\vartheta}{3} + |\frac{4\vartheta}{3} - 2\varpi|$, $\max\{\Omega(\vartheta, \varkappa, \varpi), \Omega(\varpi, \vartheta, \Im_1\vartheta)\} = \{2|\vartheta - \varpi| + \frac{2\vartheta}{3}| + |\frac{4\vartheta}{3} - 2\varpi|\}$, $\Omega(\Im_1\vartheta, \Im_1\varkappa, \Im_2\varpi) = \frac{1}{3}[|\vartheta - 2\varpi| + |\vartheta + 2\varpi|]$.

- If $\vartheta = 1, \varpi = 0$. Then
  $\Omega(\vartheta, \varkappa, \varpi) = 2$, $\Omega(\vartheta, \varkappa, \Im_1\vartheta) = \frac{4}{3}$, $\Omega(\vartheta, \varkappa, \Im_2\varpi) = 2$, $\Omega(\varpi, \varkappa, \Im_2\varpi) = 2$, $\Omega(\varpi, \varkappa, \Im_1\vartheta) = 2$, $\max\{\Omega(\vartheta, \varkappa, \varpi), \Omega(\vartheta, \varkappa, \Im_1\vartheta)\} = 2$, $\Omega(\vartheta, \varkappa, \Im_2\varpi) = 2$, $\Omega(\Im_1\vartheta, \Im_1\varkappa, \Im_2\varpi) = \frac{2}{3}$.
  Therefore $\Im_1, \Im_2$ satisfy (4.23).
- If $\vartheta = 0, \varpi = 1$. Then similarly $\Im_1, \Im_2$ satisfy (4.23).

Case (II): Suppose $\varkappa = \varpi \neq \vartheta$. Then $\Omega(\vartheta, \varkappa, \varpi)=2|\vartheta-\varkappa|$, $\Omega(\vartheta, \varkappa, \Im_1\vartheta) = |\varkappa - \frac{\vartheta}{3}| + |\varkappa - \frac{5\vartheta}{3}|$, $\Omega(\varpi, \varkappa, \Im_2\varpi) = \frac{2\varkappa}{3}$, $\Omega(\vartheta, \varkappa, \Im_2\varpi) = \frac{\varkappa}{3} + |2\vartheta - \frac{4\varkappa}{3}|$, $\Omega(\varpi, \varkappa, \Im_1\vartheta)=2|\varkappa-\frac{\vartheta}{3}|$, $\max\{\Omega(\vartheta, \varkappa, \varpi), \Omega(\varpi, \varkappa, \Im_1\vartheta)\} = \{2|\varkappa - \vartheta|, 2|\frac{\vartheta}{3} - \varkappa|\}$, $\Omega(\Im_1\vartheta, \Im_1\varkappa, \Im_2\varpi) = \frac{1}{3}[\varkappa + |\varkappa - \frac{2\vartheta}{3}|]$.

- If $\varkappa = 1, \varpi = 0$. Then
  $\Omega(\vartheta, \varkappa, \varpi) = 2$, $\Omega(\vartheta, \varkappa, \Im_1\vartheta) = 2$, $\Omega(\varpi, \varkappa, \Im_2\varpi) = \frac{2}{3}$, $\Omega(\vartheta, \varkappa, \Im_2\varpi) = \frac{7}{3}$, $\Omega(\varpi, \varkappa, \Im_1\vartheta) = 2$, $\max\{\Omega(\vartheta, \varkappa, \varpi), \Omega(\varpi, \varkappa, \Im_1\vartheta)\} = 2$, $\Omega(\Im_1\vartheta, \Im_1\varkappa, \Im_2\varpi) = \frac{2}{3}$.
  Therefore $\Im_1, \Im_2$ satisfy (4.23).
- If $\vartheta = 1, \varkappa = 0$. Then similarly $\Im_1, \Im_2$ satisfy (4.23).

Case (II) : Suppose $\vartheta \neq \varkappa \neq \varpi$. In particular, $\vartheta = 1$, $\varkappa = 1/2$, $\varpi = 0$. Then $\Omega(\vartheta, \varkappa, \varpi) = 1$, $\Omega(\vartheta, \varkappa, \Im_1\vartheta) = \frac{5}{3}$, $\Omega(\varpi, \vartheta, \Im_2\varpi) = 1$, $\Omega(\vartheta, \varkappa, \Im_2\varpi) = 2$, $\Omega(\varpi\varpi, \varkappa, \Im_1\vartheta) = 1$, $\max\{\Omega(\vartheta, \varkappa, \varpi), \Omega(\varpi, \varkappa, \Im_1\vartheta)\} = 2$, $\Omega(\Im_1\vartheta, \Im_1\varkappa, \Im_2\varpi) = \frac{2}{3}$.

Therefore $\Im_1, \Im_2$ satisfy (4.23). Similarly, we can discuss other cases and show that (4.23) is satisfy. Hence all the assumptions in Theorem 4.3 are satisfied for suitable value of h_i such that $\sum_{i=1}^{5} h_i < 1$. Indeed $Fix(\Im_1, \Im_2) = \{0\}$ in Ω.

Theorem 4.4 *Let (Ξ, Ω) be a complete Ω-metric space, $\chi \in \Phi$ and let $\Im_1, \Im_2 \colon \Xi \to \Xi$ be two self-mappings such that*

$$\begin{aligned}\int_0^{\Omega(\Im_1\vartheta,\Im_1\varkappa,\Im_2\varpi)} \chi(\varsigma)d\varsigma \leq & h_1 \int_0^{\Omega(\vartheta,\varkappa,\varpi)} \chi(\varsigma)d\varsigma + h_2 \int_0^{\Omega(\vartheta,\varkappa,\Im_1\vartheta)} \chi(\varsigma)d\varsigma \\ & + h_3 \int_0^{\Omega(\varpi,\varkappa,\Im_2\varpi)} \chi(\varsigma)d\varsigma \\ & + h_4 \int_0^{\frac{\Omega(\varpi,\varkappa,\Im_2\varpi)[1+\Omega(\vartheta,\varkappa,\Im_1\vartheta)]}{[1+\Omega(\vartheta,\varkappa,\varpi)]}} \chi(\varsigma)d\varsigma \\ & + h_5 \int_0^{\max\{\Omega(\vartheta,\varkappa,\Im_1\vartheta),\Omega(\varpi,\varkappa,\Im_2\varpi\}} \chi(\varsigma)d\varsigma, \quad (4.24)\end{aligned}$$

for all $\vartheta, \varkappa, \varpi \in \Xi$, where h_1, h_2, h_3, h_4, h_5 are nonnegative reals such that $h_1 + h_2 + 2h_3 + 2h_4 + 2h_5 < 1$. Then $\rho \in Fix(\Im_1, \Im_2)$ for $\rho \in \Xi$.

Proof Let $\vartheta_0 \in \Xi$ and the sequence $\{\vartheta_n\}$ be defined as $\vartheta_{2n+1} = \Im_1\vartheta_{2n}$ and $\vartheta_{2n+2} = \Im_2\vartheta_{2n+1}$ for $n = 1, 2, \ldots$. Then from inequality (4.24), and using the condition $(S1)$, $(S2)$ and Lemma 4.1, we have

$$\begin{aligned}\int_0^{\Omega(\vartheta_{2n},\vartheta_{2n},\vartheta_{2n+1})} \chi(\varsigma)d\varsigma = & \int_0^{\Omega(\Im_1\vartheta_{2n-1},\Im_1\vartheta_{2n-1},\Im_2\vartheta_{2n})} \chi(\varsigma)d\varsigma \\ \leq & h_1 \int_0^{\Omega(\vartheta_{2n-1},\vartheta_{2n-1},\vartheta_{2n})} \chi(\varsigma)d\varsigma \\ & + h_2 \int_0^{\Omega(\vartheta_{2n-1},\vartheta_{2n-1},\Im_1\vartheta_{2n-1})} \chi(\varsigma)d\varsigma \\ & + h_3 \int_0^{\Omega(\vartheta_{2n},\vartheta_{2n-1},\Im_2\vartheta_{2n})} \chi(\varsigma)d\varsigma \\ & + h_4 \int_0^{\frac{\Omega(\vartheta_{2n},\vartheta_{2n-1},\Im_2\vartheta_{2n})[1+\Omega(\vartheta_{2n-1},\vartheta_{2n-1},\Im_1\vartheta_{2n-1})]}{[1+\Omega(\vartheta_{2n-1},\vartheta_{2n-1},\vartheta_{2n})]}} \chi(\varsigma)d\varsigma \\ & + h_5 \int_0^{\max\{\Omega(\vartheta_{2n-1},\vartheta_{2n-1},\Im_1\vartheta_{2n-1}),\Omega(\vartheta_{2n},\vartheta_{2n-1},\Im_2\vartheta_{2n})\}} \chi(\varsigma)d\varsigma\end{aligned}$$

$$
\begin{aligned}
&= h_1 \int_0^{\Omega(\vartheta_{2n-1},\vartheta_{2n-1},\vartheta_{2n})} \chi(\varsigma)d\varsigma \\
&\quad + h_2 \int_0^{\Omega(\vartheta_{2n-1},\vartheta_{2n-1},\vartheta_{2n})} \chi(\varsigma)d\varsigma \\
&\quad + h_3 \int_0^{\Omega(\vartheta_{2n},\vartheta_{2n-1},\vartheta_{2n+1})} \chi(\varsigma)d\varsigma \\
&\quad + h_4 \int_0^{\frac{\Omega(\vartheta_{2n},\vartheta_{2n-1},\vartheta_{2n+1})[1+\Omega(\vartheta_{2n-1},\vartheta_{2n-1},\vartheta_{2n})]}{[1+\Omega(\vartheta_{2n-1},\vartheta_{2n-1},\vartheta_{2n})]}} \chi(\varsigma)d\varsigma \\
&\quad + h_5 \int_0^{\max\{\Omega(\vartheta_{2n-1},\vartheta_{2n-1},\vartheta_{2n}),\Omega(\vartheta_{2n},\vartheta_{2n-1},\vartheta_{2n+1})\}} \chi(\varsigma)d\varsigma \\
&\leq (h_1 + h_2 + h_3 + h_4 + h_5) \int_0^{\Omega(\vartheta_{2n-1},\vartheta_{2n-1},\vartheta_{2n})} \chi(\varsigma)d\varsigma \\
&\quad + (h_3 + h_4 + h_5) \int_0^{\Omega(\vartheta_{2n},\vartheta_{2n},\vartheta_{2n+1})} \chi(\varsigma)d\varsigma, \qquad (4.25)
\end{aligned}
$$

which implies

$$
\int_0^{\Omega(\vartheta_{2n},\vartheta_{2n},\vartheta_{2n+1})} \chi(\varsigma)d\varsigma \leq \left(\frac{h_1 + h_2 + h_3 + h + 4 + h_5}{1 - h_3 - h_4 - h_5}\right) \int_0^{\Omega(\vartheta_{2n-1},\vartheta_{2n-1},\vartheta_{2n})} \chi(\varsigma)d\varsigma. \tag{4.26}
$$

If we take $\Lambda = \left(\frac{h_1+h_2+h_3+h_4+h_5}{1-h_3-h_4-h_5}\right)$, then we find $\Lambda < 1$ since $h_1 + h_2 + 2h_3 + 2h_4 + 2h_5 < 1$. Using the inequality (4.24) again, we obtain

$$
\int_0^{\Omega(\vartheta_{2n},\vartheta_{2n},\vartheta_{2n+1})} \chi(\varsigma)d\varsigma \leq \Lambda^n \int_0^{\Omega(\vartheta_0,\vartheta_0,\vartheta_1)} \chi(\varsigma)d\varsigma. \tag{4.27}
$$

Thus in general, for $n = 0, 1, 2, \ldots,$ we have

$$
\int_0^{\Omega(\vartheta_n,\vartheta_n,\vartheta_{n+1})} \chi(\varsigma)d\varsigma \leq \Lambda^n \int_0^{\Omega(\vartheta_0,\vartheta_0,\vartheta_1)} \chi(\varsigma)d\varsigma. \tag{4.28}
$$

Taking the limit as $n \to +\infty$, in inequality (4.28), we get

$$
\lim_{n\to+\infty} \int_0^{\Omega(\vartheta_n,\vartheta_n,\vartheta_{n+1})} \chi(\varsigma)d\varsigma = 0, \tag{4.29}
$$

since $0 < b(\varsigma) < 1$. The condition (4.29) implies

$$
\lim_{n\to+\infty} \Omega(\vartheta_n, \vartheta_n, \vartheta_{n+1}) = 0. \tag{4.30}
$$

By analogous arguments used in Theorem 4.3, we demonstrate that the sequence $\{\vartheta_n\}$ in Ω is a Cauchy sequence. Hence by completeness of the space Ω, it is convergent to the point, say, $\rho \in \Xi$, that is, $\lim_{n\to+\infty} \vartheta_n = u$.

Now, from the given inequality (4.24), we find

$$\begin{aligned}
\int_0^{\Omega(\Im_1\rho,\Im_1\rho,\vartheta_{2n+1})} \chi(\varsigma)d\varsigma &= \int_0^{\Omega(\Im_1\rho,\Im_1\rho,\Im_2\vartheta_{2n})} \chi(\varsigma)d\varsigma \\
&\leq h_1 \int_0^{\Omega(\rho,\rho,\vartheta_{2n})} \chi(\varsigma)d\varsigma + h_2 \int_0^{\Omega(\rho,\rho,\Im_1\rho)} \chi(\varsigma)d\varsigma \\
&\quad + h_3 \int_0^{\Omega(\vartheta_{2n},\vartheta_{2n},\Im_2\vartheta_{2n})} \chi(\varsigma)d\varsigma \\
&\quad + h_4 \int_0^{\frac{\Omega(\vartheta_{2n},\vartheta_{2n},G\vartheta_{2n})[1+\Omega(\rho,\rho,\Im_1\rho)]}{[1+\Omega(\rho,\rho,\vartheta_{2n})]}} \chi(\varsigma)d\varsigma \\
&\quad + h_5 \int_0^{\max\{\Omega(\rho,\rho,\Im_1\rho),\Omega(\vartheta_{2n},\vartheta_{2n},\Im_2\vartheta_{2n})\}} \chi(\varsigma)d\varsigma \\
&= h_1 \int_0^{\Omega(\rho,\rho,\vartheta_{2n})} \chi(\varsigma)d\varsigma + h_2 \int_0^{\Omega(\rho,\rho,\Im_1\rho)} \chi(\varsigma)d\varsigma \\
&\quad + h_3 \int_0^{\Omega(\vartheta_{2n},\vartheta_{2n},\vartheta_{2n+1})} \chi(\varsigma)d\varsigma \\
&\quad + h_4 \int_0^{\frac{\Omega(\vartheta_{2n},\vartheta_{2n},\vartheta_{2n+1})[1+\Omega(\rho,\rho,\Im_1\rho)]}{[1+\Omega(\rho,\rho,\vartheta_{2n})]}} \chi(\varsigma)d\varsigma \\
&\quad + h_5 \int_0^{\max\{\Omega(\rho,\rho,\Im_1\rho),\Omega(\vartheta_{2n},\vartheta_{2n},\vartheta_{2n+1})\}} \chi(\varsigma)d\varsigma.
\end{aligned} \tag{4.31}$$

Taking the limit as $n \to +\infty$ in (4.31) and using the condition $(S1)$ and Lemma 4.1, we get

$$\int_0^{\Omega(\Im_1\rho,\Im_1\rho,\rho)} \chi(\varsigma)d\varsigma \leq (h_2 + h_5) \int_0^{\Omega(\Im_1\rho,\Im_1\rho,\rho)} \chi(\varsigma)d\varsigma, \tag{4.32}$$

which implies $\Omega(\Im_1\rho, \Im_1\rho, \rho) = 0$, that is, $\Im_1\rho = \rho$ since $h_2 + h_5 < 1$. This shows that $\rho \in Fix(\Im_1)$. Similarly, we can show that $\rho \in Fix(\Im_2)$. Thus $\rho \in Fix(\Im_1, \Im_2)$

To prove singleness of $Fix(\Im_1, \Im_2)$, on the contrary, we take $\rho_1 \in Fix(\Im_1, \Im_2)$ such that $\Im_1\rho_1 = \rho_1 = \Im_2\rho_1$ with $\rho_1 \neq \rho$. Using the inequality (4.24), the condition $(S1)$ and Lemma 4.1, we obtain

$$
\begin{aligned}
\int_0^{\Omega(\rho,\rho,\rho_1)} \chi(\varsigma)d\varsigma = \int_0^{\Omega(\Im_1\rho,\Im_1\rho,\Im_2\rho_1)} \chi(\varsigma)d\varsigma
&\leq h_1 \int_0^{\Omega(\rho,\rho,\rho_1)} \chi(\varsigma)d\varsigma + h_2 \int_0^{\Omega(\rho,\rho,\Im_1\rho)} \chi(\varsigma)d\varsigma \\
&\quad + h_3 \int_0^{\Omega(\rho_1,\rho_1,\Im_2\rho_1)} \chi(\varsigma)d\varsigma \\
&\quad + h_4 \int_0^{\frac{\Omega(\rho_1,\rho_1,\Im_2\rho_1)[1+\Omega(\rho,\rho,\Im_1\rho)]}{[1+\Omega(\rho,\rho,\rho_1)]}} \chi(\varsigma)d\varsigma \\
&\quad + h_5 \int_0^{\max\{\Omega(\rho,\rho,\Im_1\rho),\Omega(\rho_1,\rho_1,\Im_2\rho_1)\}} \chi(\varsigma)d\varsigma \\
&= h_1 \int_0^{\Omega(\rho,\rho,\rho_1)} \chi(\varsigma)d\varsigma + h_2 \int_0^{\Omega(\rho,\rho,\rho)} \chi(\varsigma)d\varsigma \\
&\quad + h_3 \int_0^{\Omega(\rho_1,\rho_1,\rho_1)} \chi(\varsigma)d\varsigma \\
&\quad + h_4 \int_0^{\frac{\Omega(\rho_1,\rho_1,\rho_1)[1+\Omega(\rho,\rho,\rho)]}{[1+\Omega(\rho,\rho,\rho_1)]}} \chi(\varsigma)d\varsigma \\
&\quad + h_5 \int_0^{\max\{\Omega(\rho,\rho,\rho),\Omega(\rho_1,\rho_1,\rho_1)\}} \chi(\varsigma)d\varsigma \\
&= h_1 \int_0^{\Omega(\rho,\rho,\rho_1)} \chi(\varsigma)d\varsigma, \qquad (4.33)
\end{aligned}
$$

which implies $\Omega(\rho, \rho, \rho_1) = 0$, that is, $\rho = \rho_1$ since $h_1 < 1$. Thus, $Fix(\Im_1, \Im_2)$ is singleton. □

If we take $\Im_1 = \Im_2 = \Im$ in Theorem 4.4, then the result is as follows:

Corollary 4.3 *Let (Ξ, Ω) be a complete Ω-metric space, $\chi \in \Phi$ and let $\Im : \Xi \to \Xi$ be a self-mapping such that*

$$
\begin{aligned}
\int_0^{\Omega(\Im\vartheta,\Im\varkappa,\Im\varpi)} \chi(\varsigma)d\varsigma
&\leq h_1 \int_0^{\Omega(\vartheta,\varkappa,\varpi)} \chi(\varsigma)d\varsigma + h_2 \int_0^{\Omega(\vartheta,\varkappa,\Im\vartheta)} \chi(\varsigma)d\varsigma \\
&\quad + h_3 \int_0^{\Omega(\varpi,\varkappa,\Im\varpi)} \chi(\varsigma)d\varsigma \\
&\quad + h_4 \int_0^{\frac{\Omega(\varpi,\varkappa,\Im\varpi)[1+\Omega(\vartheta,\varkappa,\Im\vartheta)]}{[1+\Omega(\vartheta,\varkappa,\varpi)]}} \chi(\varsigma)d\varsigma \\
&\quad + h_5 \int_0^{\max\{\Omega(\vartheta,\varkappa,\Im\vartheta),\Omega(\varpi,\varkappa,\Im\varpi\}} \chi(\varsigma)d\varsigma,
\end{aligned}
$$

for all $\vartheta, \varkappa, \varpi \in \Xi$, where h_1, h_2, h_3, h_4, h_5 are nonnegative reals such that $h_1 + h_2 + h_3 + h_4 + 2h_5 < 1$. Then $\rho \in \Xi$ such that $\rho \in Fix(\Im)$.

Remark 4.2 (1) Theorem 4.4 is a generalisation of Theorem 2.4 of Özgür and Tas [11]. Indeed, if we take $\Im_1 = \Im_2 = \Im$, $h_1 = h$ and $h_2 = h_3 = h_4 = h_5 = 0$ in Theorem 4.4, then we get Theorem 2.4 of [11].
(2) Corollary 4.3 is also a generalisation of Theorem 2.4 of Özgür and Tas [11]. Indeed, if we take $h_1 = h$ and $h_2 = h_3 = h_4 = h_5 = 0$ in Corollary 4.3, then we get Theorem 2.4 of [11].
(3) Since an Ω-metric space is a generalisation of a metric space. Corollary 4.3 is a generalisation of the classical Banach fixed point theorem [1].

4.4 An Application to Fredhlom Integral Equation

In this section, we give an application of the contraction condition of Corollary 4.2 to the Fredhlom integral equation

$$\wp(\rho) = h(\rho) + \mu \int_a^b K(\rho, \varsigma)\wp(\varsigma)d\varsigma, \tag{4.34}$$

where $\wp\colon [a, b] \to \mathbb{R}$ with $-\infty < a < b < +\infty$, $K(\rho, \varsigma)$ is called kernel of the integral equation (4.34) is continuous on the squared region $[a, b] \times [a, b]$ with $|K(\rho, \varsigma)| \le \mathscr{M}$ ($\mathscr{M} > 1$) and $h(\rho)$ is continuous on $[a, b]$.

Let $C[a, b] = \{f | f\colon [a, b] \to \mathbb{R}$ is a continuous function$\}$. Now,we define the function $\Omega\colon C[a, b] \times C[a, b] \times C[a, b] \to \mathbb{R}_+$ by

$$\Omega(\vartheta, \varkappa, \varpi) = \sup_{\varsigma\in[a,b]} |\varkappa(\varsigma) - \varpi(\varsigma)| + \sup_{\varsigma\in[a,b]} |\varkappa(\varsigma) + \varpi(\varsigma) - 2\vartheta(\varsigma)|, \tag{4.35}$$

for all $\vartheta, \varkappa, \varpi \in C[a, b]$. Then the function Ω is an Ω-metric. Now, we show that this Ω-metric can not be generated by and metric d. We assume that this Ω-metric is generated by any metric d, that is, there exists a metric d such that

$$\Omega(\vartheta, \varkappa, \varpi) = d(\vartheta, \varpi) + d(\varkappa, \varpi), \tag{4.36}$$

for all $\vartheta, \varkappa, \varpi \in C[a, b]$. Then we get

$$\Omega(\vartheta, \vartheta, \varpi) = 2d(\vartheta, \varpi) = 2 \sup_{\varsigma\in[a,b]} |\vartheta(\varsigma) - \varpi(\varsigma)|,$$

and

$$d(\vartheta, \varpi) = \sup_{\varsigma\in[a,b]} |\vartheta(\varsigma) - \varpi(\varsigma)|. \tag{4.37}$$

Similarly, we obtain

$$\Omega(\varkappa, \varkappa, \varpi) = 2d(\varkappa, \varpi) = 2 \sup_{\varsigma \in [a,b]} |\varkappa(\varsigma) - \varpi(\varsigma)|,$$

and

$$d(\varkappa, \varpi) = \sup_{\varsigma \in [a,b]} |\varkappa(\varsigma) - \varpi(\varsigma)|. \tag{4.38}$$

From Eqs. (4.36), (4.37) and (4.38), we get

$$\sup_{\varsigma \in [a,b]} |\varkappa(\varsigma) - \varpi(\varsigma)| + \sup_{\varsigma \in [a,b]} |\varkappa(\varsigma) + \varpi(\varsigma) - 2\vartheta(\varsigma)| = \sup_{\varsigma \in [a,b]} |\vartheta(\varsigma) - \varpi(\varsigma)| + \sup_{\varsigma \in [a,b]} |\varkappa(\varsigma) - \varpi(\varsigma)|,$$

which is a contradiction. Hence this Ω-metric is not generated by any metric d. Thus, $(C[a, b], \Omega))$ is a complete Ω-metric space.

Theorem 4.5 (Proposition 3.1, [11]) *Let $(C[a, b], \Omega))$ be a complete Ω-metric space with the Ω-metric defined in (4.35) and μ be a real number such that*

$$|\mu| < \frac{1}{\mathcal{M}(b-a)}.$$

Then the Fredhlom integral equation (4.34) has a unique solution $y\colon [a, b] \to \mathbb{R}$.

Proof Define the function $\Im\colon C[a, b] \to C[a, b]$ as

$$\Im(\wp(\rho)) = h(\rho) + \mu \int_a^b K(\rho, \varsigma)\wp(\varsigma)d\varsigma.$$

Now we show that $\Im$ satisfies the contractive condition of Corollary 4.2. We get

$$\begin{aligned}
\Omega(\Im\varkappa_1, \Im\varkappa_1, \Im\varkappa_2) &= 2 \sup_{\upsilon \in [a,b]} |\Im\varkappa_1(\rho) - \Im\varkappa_2(\rho)| \\
&= 2 \sup_{\upsilon \in [a,b]} \left|\mu \int_a^b K(\rho, t)[\varkappa_1(\rho) - \varkappa_2(\rho)]d\varsigma\right| \\
&\le 2|\mu|\mathcal{M} \sup_{\upsilon \in [a,b]} \left|\int_a^b [\varkappa_1(\rho) - \varkappa_2(\rho)]d\varsigma\right| \\
&\le 2|\mu|\mathcal{M} \sup_{\upsilon \in [a,b]} |\varkappa_1(\rho) - \varkappa_2(\rho)| \left|\int_a^b d\varsigma\right| \\
&\le |\mu|\mathcal{M}(b-a)\Omega(\varkappa_1, \varkappa_1, \varkappa_2) \\
&< \Omega(\varkappa_1, \varkappa_1, \varkappa_2),
\end{aligned}$$

which implies

$$\int_0^{\Omega(\Im\varkappa_1,\Im\varkappa_1,\Im\varkappa_2)} \chi(\varsigma)d\varsigma < \int_0^{\Omega(\varkappa_1,\varkappa_1,\varkappa_2)} \chi(\varsigma)d\varsigma.$$

Consequently, the contractive condition of Corollary 4.2 is satisfied and the Fredhlom integral equation (4.34) has a unique solution $\wp$. □

4.5 Conclusion

In this work, we demonstrate certain unique common fixed point theorems for contractive conditions of integral type in the context of entire Ω-metric spaces, as well as their corollaries. In addition, an illustration is supplied to confirm the outcome. Also gained is an application of the Fredhlom integral equation. The conclusions of this research generalise and expand a number of previously published discoveries.

References

1. Banach, S.: Sur les operation dans les ensembles abstraits et leur application aux equation integrals. Fund. Math. **3**, 133–181 (1922)
2. Branciari, A.: A fixed point theorem for mappings satisfying a general contractive condition of integral type. Int. J. Math. Math. Sci. **29**(9), 531–536 (2002)
3. Gupta, A.: Cyclic contraction on S-metric space. Int. J. Anal. Appl. **3**(2), 119–130 (2013)
4. Hieu, N.T., Ly, N.T., Dung, N.V.: A generalization of Ciric quasi-contractions for maps on S-metric spaces. Thai J. Math. **13**(2), 369–380 (2015)
5. Kim, J.K., Sedghi, S., Gholidahneh, A., Rezaee, M.M.: Fixed point theorems in S-metric spaces. East Asian Math. J. **32**(5), 677–684 (2016)
6. Mlaiki, N.M.: Common fixed points in complex S-metric space. Adv. Fixed Point Theory **4**(4), 509–524 (2014)
7. Nashine, H.K., Saluja, G.S., Ibrahim, R.W.: Some fixed point theorems for $(\psi - \phi)$-almost weak contractions in S-metric spaces solving conformable differential equations. J. Inequalities Appl. **2020**, 139 (2020)
8. Özgür, N.Y., Tas, N.: Some generalizations of fixed point theorems on S-metric spaces, Essays in Mathematics and Its Applications in Honor of Vladimir Arnold. Springer, New York (2016)
9. Özgür, N.Y., Tas, N.: Some fixed point theorems on S-metric spaces. Mat. Vesnik **69**(1), 39–52 (2017)
10. Özgür, N.Y., Tas, N.: Some new contractive mappings on S-metric spaces and their relationships with the mapping. Math. Sci. **11**(7), 7–16 (2017)
11. Özgür, N.Y., Tas, N.: New contractive conditions of integral type on S-metric spaces. Math. Sci. **11**, 231–240 (2017)
12. Özgür, N.Y., Tas, N.: Common fixed point results on complex valued S-metric spaces. (Submitted)
13. Rahman, M.U., Sarwar, M., Rahman, M.U.: Fixed point results of Altman integral type mappings in S-metric spaces. Int. J. Anal. Appl. **10**(1), 58–63 (2016)
14. Rashwan, R.A., Hammad, H.A.: Common fixed point theorems for weak contraction conditions of integral type. In: Proceedings of the 1st International Conference on New Horizons in Basic and Applied Sciences, Hurghada-Egypt, Anglisticum Journal, Tetovo, Macedonia, vol. 1, pp. 133–148 (2013)

15. Rashwan, R.A., Hammad, H.A.: A common fixed point theorems for a pair of self maps satisfying a general contractive condition of integral type. Int. J. Res. Engin. Appl. Sci. **4**, 1–12 (2014)
16. Rashwan, R.A., Hammad, H.A.: Common fixed point theorems for weakly compatible mappings satisfying a general contractive conditions of integral type. Palestine J. Math. **8**(2), 114–126 (2019)
17. Rhoades, B.E.: Some theorems on weakly contractive maps. Nonlinear Anal. (TMA) **74**, 2683–2693 (2001)
18. Saluja, G.S.: Common fixed point for generalized $(\psi - \phi)$-weak contractions in S-metric spaces. The Aligarh Bull. Math. **38**(1–2), 41–62 (2019)
19. Saluja, G.S.: Some fixed point theorems under implicit relation on S-metric spaces. Bull. Int. Math. Virtual Instt. **11**(2), 327–340 (2021)
20. Saluja, G.S., Kim, J.K., Lim, W.H.: Common fixed point theorems under generalized $(\psi - \varphi)$-weak contractions in S-metric spaces with applications. Nonlinear Funct. Anal. Appl. **26**(1), 13–33 (2021)
21. Saluja, G.S.: Some fixed point results on S-metric spaces with well-posedness and limit shadowing property related to a fixed point problem. J. Adv. Math. Stud. **14**(3), 431–446 (2021)
22. Saluja, G.S.: Fixed point theorems for cyclic contractions in S-metric spaces involving C-class function. Math. Moravica **26**(1), 57–76 (2022)
23. Sedghi, S., Shobe, N., Aliouche, A.: A generalization of fixed point theorems in S-metric spaces. Mat. Vesnik **64**(3), 258–266 (2012)
24. Sedghi, S., Dung, N.V.: Fixed point theorems on S-metric spaces. Mat. Vesnik **66**(1), 113–124 (2014)

Chapter 5
On (f, λ)—Harmonic Summability

Mikail Et, Hacer Şengül Kandemir, and Nazlım Deniz Aral

Abstract The concepts of $(H, 1)_\lambda$—summability, λ—strong harmonic summability (denoted by $[SH_\lambda]$), λ—statistical logarithmic convergence (denoted by $[LH_\lambda]$) and λ—statistically logarithmic boundedness of sequences of real numbers are introduced and tried to investigate some relations between the λ—strongly harmonically summability and λ—statistical logarithmic convergence in this work. We also establish some connections between $[SH_\mu]$ and $[LH_\lambda]$. It is shown that if a sequence is bounded then it is λ—statistically logarithmic boundedness. In addition, we define (f, λ)—harmonic summability and (f, λ)—statistical logarithmic convergence of real number sequences and provide some inclusion connections between these spaces.

Keywords Statistical convergence · Harmonic convergence · Modulus function

5.1 Introduction

Statistical convergence has become an active research area since 1951 when it was introduced by Steinhaus [29] and Fast [12]. In 1959, Schoenberg [27] established several fundamental characteristics of statistical convergence and also investigated as a method of summability. Later, Altin et al. [3], Altinok et al. [4], Cakalli [6, 7], Çolak [9], Et et al. [8], Connor [10], Di Maio and Kočinac [11], Fridy [13], Gungor et al. [14, 15], Isik et al. [2, 16, 17], Salat [25], Savas and Et [26], Sengul et al. [5, 30–34] continued to explore statistical convergence from the perspective of the sequence space and connected it with summability theory. With the help

M. Et (✉)
Department of Mathematics, Fırat University, 23119 Elazığ, Turkey
e-mail: mikailet68@gmail.com

H. Şengül Kandemir
Faculty of Education, Harran University, Osmanbey Campus, 63190 Şanlıurfa, Turkey

N. Deniz Aral
Department of Mathematics, Bitlis Eren University, 13000 Bitlis, Turkey

B. Hazarika et al. (eds.), *Advances in Functional Analysis and Fixed-Point Theory*,
Industrial and Applied Mathematics,
https://doi.org/10.1007/978-981-99-9207-2_5

of the sequence $\lambda = (\lambda_n)$, Mursaleen [21] first described the idea of λ-statistical convergence and Malkowsky et al. [20] first described various λ—sequence spaces that were defined by a modulus. Moricz [22] introduced and investigated statistical harmonic summability. Nuray defined lacunary strong harmonic summability in [23].

If the set $T(\varepsilon) = \{t \leq r : |x_t - A| \geq \varepsilon\}$ has natural density zero for every $\varepsilon > 0$, i.e.,

$$\lim_{r\to\infty} \frac{1}{r} |T(\varepsilon)| = 0,$$

then the number sequence $x = (x_t)$ is said to be statistically convergent to A.

Let $\lambda = (\lambda_r)$ be a non-decreasing sequence of positive number tending to ∞ such that $\lambda_{r+1} \leq \lambda_r + 1$ and $\lambda_1 = 1$. By Λ, we refer to the category of any non-decreasing positive real number sequences that tend to ∞ and are such that $\lambda_{r+1} \leq \lambda_r + 1$ and $\lambda_1 = 1$.

$$t_r(x) = \frac{1}{\lambda_r} \sum_{s \in I_r} x_s$$

defines the generalized de la Vallee-Poussin mean, where $I_r = [r - \lambda_r + 1, r]$.

If

$$t_r(x) \to A \text{ as } r \to \infty,$$

a sequence $x = (x_s)$ is called (V, λ) −summable to a number A (see [18]). If $\lambda_r = r$ for each $r \in \mathbb{N}$, then (V, λ)—summability reduces to $(C, 1)$—summability. The sets of sequences $x = (x_s)$ which are strongly Cesaro summable to A;

$$[C, 1] = \left\{ x = (x_s) \in w : \exists\, A \in \mathbb{C},\ \lim_{r\to\infty} \frac{1}{r} \sum_{s=1}^{r} |x_s - A| = 0 \right\},$$

i.e., $x_s \to A\,[C, 1]$ and strongly (V, λ)—summable to A;

$$[V, \lambda] = \left\{ x = (x_s) \in w : \exists\, A \in \mathbb{C},\ \lim_{r\to\infty} \frac{1}{\lambda_r} \sum_{s \in I_r} |x_s - A| = 0 \right\}$$

i.e., $x_s \to A\,[V, \lambda]$.

$$\tau_r = \frac{1}{\ell_r} \sum_{t=1}^{r} \frac{x_t}{t},$$

defines the harmonic mean of the sequence $x = (x_t)$ where $\ell_r = \sum_{t=1}^{r} \frac{1}{t} \approx \log r$ for $r = 1, 2, \ldots$.

If

$$\lim_{r\longrightarrow\infty} \frac{1}{\ell_r} \sum_{t=1}^{r} \frac{x_t}{t} = A,$$

a sequence $x = (x_t)$ is called $(H, 1)$—summable to A.

H stands for the collection of all $(H, 1)$—summable real sequences. The set of all real sequences that $(H, 1)$—summable to 0 shall be designated as H^0. It is well known that harmonic summability follows naturally from ordinary convergence and that the inverse conclusion holds only under additional circumstances.

If

$$\lim_{r\longrightarrow\infty} \frac{1}{\ell_r} \sum_{t=1}^{r} \frac{|x_t - A|}{t} = 0,$$

then a sequence $x = (x_t)$ is called strongly harmonically summable to A. The $[H]$ stands for the collection of all strongly harmonically summable real sequences. $[H^0]$ stands for the collection of all strongly harmonically summable real sequences to 0. The study of harmonic summability can be found in [22, 23, 28].

The term *log* refers to the natural logarithm throughout the study. In addition, we write $x_t \sim y_t$ if $\lim_{i\rightarrow\infty} x_t/y_t = 1$ given two sequences of positive values $x = (x_t)$ and $y = (y_t)$ (apart from perhaps a finite number of terms).

The definition of the λ—harmonic means of the sequence (x_t) is

$$S_r = \frac{1}{E_r} \sum_{t\in I_r} \frac{x_t}{t}$$

where $E_r = \sum\limits_{t\in I_r} \frac{1}{t}$, $r = 1, 2, \ldots$ and $I_r = [r - \lambda_r + 1, r]$.

5.2 On λ—Harmonic Summability

We give the ideas of $(H, 1)_\lambda$—summability, λ—strong harmonic summability, λ—statistical logarithmic convergence of sequences and study relationships between these spaces in this part.

Definition 5.1 A sequence $x = (x_t)$ of real numbers is called $(H, 1)_\lambda$—summable to the number A, if there is a real number A such that

$$\lim_{r\rightarrow\infty} \frac{1}{E_r} \sum_{t\in I_r} \frac{x_t}{t} = A.$$

In this instance, we use $H_\lambda - \lim x_t = A$. The set of every sequence that is $(H, 1)_\lambda$—summable sequences shall be referred to as H_λ. When $\lambda = (r)$ is a specific instance, we will use H instead of H_λ.

Definition 5.2 A sequence $x = (x_t)$ of real numbers is called λ—strongly harmonically summable to the number A, if there is a real number A such that

$$\lim_{r\to\infty} \frac{1}{E_r} \sum_{t\in I_r} \frac{|x_t - A|}{t} = 0.$$

We write $[SH_\lambda] - \lim x_t = A$ in this instance.

The set of every sequence that is λ—strongly harmonically summable sequences shall be referred to as $[SH_\lambda]$. In the special case $\lambda = (r)$, we should use $[SH]$ rather than $[SH_\lambda]$. If $A = 0$, then $[SH_\lambda^0]$ should be used in place of instead of $[SH_\lambda]$.

Theorem 5.1 *$[SH_\lambda]$ and H_λ are linear spaces.*

Proof Omitted. □

Theorem 5.2 *The sequence space $[SH_\lambda]$ is a Banach space normed by*

$$\|x\| = \sup_r \frac{1}{E_r} \sum_{t\in I_r} \frac{|x_t|}{t}. \tag{5.1}$$

Proof The existence of $[SH_\lambda]$ as a normed space, normed by (5.1), is routinely verified. (x^k) be a Cauchy sequence so $x^k = (x_i^s)_i = (x_1^s, x_2^s, ...) \in [SH_\lambda]$ for each $s \in \mathbb{N}$. Then, we obtain

$$\left\|x^k - x^i\right\| \to 0 \text{ as s}, i \to \infty.$$

Give us $\varepsilon > 0$, then there exists a positive integer $r_0(\varepsilon)$ such that $\left\|x^k - x^i\right\| \leq \varepsilon$ for all $k, i > r_0$. So we have

$$\frac{1}{E_r} \sum_{t\in I_r} \frac{\left|x_t^k - x_t^i\right|}{t} \leq \varepsilon, \text{ for all } k, i > r_0.$$

From this, we get $\left|x_t^k - x_t^i\right| \to 0$ as $k, i \to \infty$, and for all $t \in \mathbb{N}$. Following from the sequence $\left(x_t^k\right)_t = \left(x_t^1, x_t^2, ...\right)$ is a Cauchy in $\mathbb{R}$ for each $t \in \mathbb{N}$. Additionally, it converges, because $\mathbb{R}$ is complete. Consider that $\lim_k x_t^k = x_t$ for each $t \in \mathbb{N}$. Then for every $\varepsilon > 0$, there exists a number $r_0 = r_0(\varepsilon)$ such that

$$\left|x^k - x^i\right| < \varepsilon$$

for all $k, i \geq r_0$ because of (x^s) is a Cauchy sequence. From this, we have

$$\frac{1}{E_r}\sum_{t\in I_r}\frac{\left|x_t^k - x_t^i\right|}{t} < \varepsilon$$

for all $r \in \mathbb{N}$ and all $k, i \geq r_0$. Taking limit as $i \to \infty$ in the last inequalities, we have

$$\lim_i \frac{1}{E_r}\sum_{t\in I_r}\frac{\left|x_t^k - x_t^i\right|}{t} = \frac{1}{E_r}\sum_{t\in I_r}\frac{\left|x_t^k - x_t\right|}{t} < \varepsilon$$

for $k \geq r_0$ and all $r \in \mathbb{N}$. This implies that

$$\left\|x^k - x\right\| = \sup_r \frac{1}{E_r}\sum_{t\in I_r}\frac{\left|x_t^k - x_t\right|}{t} \to 0,$$

and hence $x^k \to x$ as $k \to \infty$. As

$$\frac{1}{E_r}\sum_{t\in I_r}\frac{|x_t - a|}{t} \leq \frac{1}{E_r}\sum_{t\in I_r}\frac{\left|x_t^{r_0} - a\right|}{t} + \frac{1}{E_r}\sum_{t\in I_r}\frac{\left|x_t^{r_0} - x_t\right|}{t},$$

we get $x \in [SH_\lambda]$. □

Theorem 5.3 $\liminf_r \dfrac{\ell_r}{\ell_{r-\lambda_r}} > 1$ *if and only if* $[SH] \subseteq [SH_\lambda]$.

Proof Consider that $\liminf_r \dfrac{\ell_r}{\ell_{r-\lambda_r}} > 1$. So, a $Y > 0$ exists with the property that $\dfrac{\ell_r}{\ell_{r-\lambda_r}} \geq 1 + Y$ for $r \geq 1$ is large enough which suggests that

$$\frac{\ell_r}{E_r} \leq \frac{1+Y}{Y} \text{ and } \frac{\ell_{r-\lambda_r}}{E_r} \leq \frac{1}{Y}$$

where $(E_r) = \ell_r - \ell_{r-\lambda_r}$.

If $x \in [SH^0]$, we have for sufficiently large r and every $\varepsilon > 0$

$$\begin{aligned} S_r &= \frac{1}{E_r}\sum_{t\in I_r}\frac{|x_t|}{t} = \frac{1}{E_r}\left(\sum_{t=1}^{r}\frac{|x_t|}{t} - \sum_{t=1}^{r-\lambda_r}\frac{|x_t|}{t}\right) \\ &= \frac{\ell_r}{E_r}\left(\frac{1}{\ell_r}\sum_{t=1}^{r}\frac{|x_t|}{t}\right) - \frac{\ell_{r-\lambda_r}}{E_r}\left(\frac{1}{\ell_{r-\lambda_r}}\sum_{t=1}^{r-\lambda_r}\frac{|x_t|}{t}\right), \end{aligned}$$

this proves the sufficiency.

Now assume that $\liminf_r \dfrac{\ell_r}{\ell_{r-\lambda_r}} = 1$. Since λ is a sequence, we can select a subsequence (r_j) satisfying

$$\frac{\ell_{r_j}}{\ell_{r_j-\lambda_{r_j}}} < \frac{1}{j} \text{ and } \frac{\ell_{r_j-\lambda_{r_j}}}{\ell_{r_{j-1}-\lambda_{r_{j-1}}}} > j$$

where $r_j \geq r_{j-1} + 2$.

Define $x = (x_t)$ by

$$x_t = \begin{cases} 1, & t \in I_{r_j}, \text{ for some } j = 1, 2, ... \\ 0, & \text{otherwise} \end{cases}.$$

If so, for any real A,

$$\frac{1}{E_{r_j}} \sum_{t \in I_{r_j}} \frac{|x_t - A|}{t} = \frac{E_{r_j}}{E_{r_j}} |1 - A| = |1 - A|,$$

and for $r \neq r_j$

$$\frac{1}{E_r} \sum_{t \in I_r} \frac{|x_t - A|}{t} = \frac{E_r}{E_r} |A| = |A|.$$

Therefore, $x \notin [SH_\lambda]$. However, $x \in [SH^0]$ since r is a large enough integer, we have a unique j for which $\ell_{r_j-\lambda_{r_j}} < \ell_i \leq \ell_{r_{j+1}-\lambda_{r_{j+1}}}$ and write

$$\frac{1}{\ell_i} \sum_{t=1}^{i} \frac{|x_t|}{t} \leq \frac{\ell_{r_{j-1}-\lambda_{r_{j-1}}} + E_{r_j}}{\ell_{r_j-\lambda_{r_j}}} < \frac{\ell_{r_{j-1}-\lambda_{r_{j-1}}}}{\ell_{r_j-\lambda_{r_j}}} + \frac{\ell_{r_j}}{\ell_{r_j-\lambda_{r_j}}} < \frac{2}{j}.$$

As $i \to \infty$ it follows also $j \to \infty$. So, $x \in [SH^0]$. □

Let $\nu = (\nu_r), \eta = (\eta_r) \in \Lambda$ be two sequences such that $I_r \subseteq I'_r$ for all $r \in \mathbb{N}$ in the next theorems.

Theorem 5.4 *We have*

(i) If

$$\lim_{r\to\infty} \inf \frac{E_r}{E'_r} > 0 \tag{5.2}$$

holds then $[SH_\eta] \subseteq [SH_\nu]$,

(ii) If

$$\lim_{r\to\infty} \frac{E'_r}{E_r} = 1 \tag{5.3}$$

holds and $x \in \ell_\infty$ *then* $[SH_\nu] \subseteq [SH_\eta]$, *where* $I_r = [r - \nu_r + 1, r]$, $I'_r = [r - \eta_r + 1, r]$.

Proof (*i*) Let $x \in [SH_\eta]$ and suppose that (5.2) holds. We get

$$\frac{1}{E'_r}\sum_{t\in I'_r}\frac{|x_t - A|}{t} \geq \frac{E_r}{E_r}\frac{1}{E'_r}\sum_{t\in I_r}\frac{|x_r - A|}{t}$$

for all $r \in \mathbb{N}$. Therefore, $[SH_\eta] \subseteq [SH_\nu]$.

(ii) Let $x \in [SH_\nu]$ and assume that (5.3) holds. So, there exists some $Z > 0$ such that $\frac{|x_t - A|}{t} \leq |x_t - A| \leq Z$ for all t because of $x \in \ell_\infty$. Since $I_r \subseteq I'_r$ and $E_r \leq E'_r$ for all $r \in \mathbb{N}$, we obtain

$$\begin{aligned}
\frac{1}{E'_r}\sum_{t\in I'_r}\frac{|x_t - A|}{t} &= \frac{1}{E'_r}\sum_{t\in I'_r - I_r}\frac{|x_t - A|}{t} + \frac{1}{E'_r}\sum_{t\in I_r}\frac{|x_t - A|}{t} \\
&\leq \frac{E'_r - E_r}{E'_r} Z + \frac{1}{E'_r}\sum_{t\in I_r}\frac{|x_t - A|}{t} \\
&\leq \frac{E'_r - E_r}{E_r} Z + \frac{1}{E_r}\sum_{t\in I_r}\frac{|x_t - A|}{t} \\
&\leq \left(\frac{E'_r}{E_r} - 1\right) Z + \frac{1}{E_r}\sum_{t\in I_r}\frac{|x_t - A|}{t}
\end{aligned}$$

for every $r \in \mathbb{N}$. Therefore, $\ell_\infty \cap [SH_\nu] \subseteq [SH_\eta]$. □

Definition 5.3 If there is a real number A such that for every $\varepsilon > 0$,

$$\lim_{r\to\infty}\frac{1}{E_r}\left|\left\{t \in I_r : \frac{1}{t}|x_t - A| \geq \varepsilon\right\}\right| = 0.$$

a sequence $x = (x_t)$ of real numbers is called λ—statistically logarithmic convergent to the number A.

We write $[LH_\lambda] - \lim x_t = A$ in this instance. The set of every sequence that is λ—statistically logarithmic convergent shall be referred to as $[LH_\lambda]$. In the special case $\lambda = (r)$, we should use $[LH]$ rather than $[LH_\lambda]$.

We present the following three theorems without proof.

Theorem 5.5 *$[LH_\lambda]$ is linear space.*

Theorem 5.6 *Let $\lambda \in \Lambda$, then $[SH_\lambda] \subseteq [LH_\lambda]$.*

Theorem 5.7 *If* (5.2) *holds, then $[LH_\eta] \subseteq [LH_\nu]$.*

Theorem 5.8 *Let* (5.2) *holds, then $[SH_\eta] \subseteq [LH_\nu]$,*

Proof Assume that $x \in [SH_\eta]$. We obtain

$$\frac{1}{E'_r}\sum_{t\in I'_r}\frac{|x_t-A|}{t}=\frac{1}{E'_r}\sum_{\substack{i\in I'_r\\ \frac{|x_t-A|}{t}\geq\varepsilon}}\frac{|x_t-A|}{t}+\frac{1}{E'_r}\sum_{\substack{i\in I'_r\\ \frac{|x_t-A|}{t}<\varepsilon}}\frac{|x_t-A|}{t}$$

$$\geq\frac{1}{E'_r}\sum_{\substack{t\in I_r\\ \frac{|x_t-A|}{t}\geq\varepsilon}}\frac{|x_t-A|}{t}+\frac{1}{E'_r}\sum_{\substack{t\in I_r\\ \frac{|x_t-A|}{t}<\varepsilon}}\frac{|x_t-A|}{t}$$

$$\geq\frac{E_r}{E'_r}\frac{1}{E_r}\sum_{\substack{t\in I_r\\ \frac{|x_t-A|}{t}\geq\varepsilon}}\frac{|x_t-A|}{t}$$

$$\geq\frac{1}{E_r}\left|\left\{t\in I_r:\frac{|x_t-A|}{t}\geq\varepsilon\right\}\right|.\varepsilon.$$

Consequently, $[LH_\nu]\subseteq[SH_\eta]$. □

Definition 5.4 If there exists such a nonnegative number B that

$$\lim_{r\to\infty}\frac{1}{E_r}\left|\left\{t\in I_r:\frac{1}{t}|x_t|>B\right\}\right|=0$$

the sequence $x=(x_t)$ of real numbers is called λ—statistically logarithmic bounded.

The set of every sequence that is λ—statistically logarithmic bounded sequences shall be referred to as SH^b_λ.

Theorem 5.9 *λ—statistically logarithmic convergence implies λ—statistically logarithmic bounded.*

Proof Let $x=(x_t)\in[LH_\lambda]$. Thefore, there exists an $A\in\mathbb{R}$ such that

$$\lim_{r\to\infty}\frac{1}{E_r}\left|\left\{t\in I_r:\frac{1}{t}|x_t-A|\geq\varepsilon\right\}\right|=0,$$

for all $\varepsilon>0$. In particular, it holds $\varepsilon=1$. By the reverse triangle inequality, we have

$$\left\{t\in I_r:\frac{1}{t}|x_t|\geq|A|+1\right\}\subseteq\left\{t\in I_r:\frac{1}{t}|x_t-A|\geq 1\right\}$$

and

$$\frac{1}{E_r}\left|\left\{t\in I_r:\frac{1}{t}|x_t|\geq|A|+1\right\}\right|\leq\frac{1}{E_r}\left|\left\{t\in I_r:\frac{1}{t}|x_t-A|\geq 1\right\}\right|$$

for all $r \in \mathbb{N}$. Taking limit, the previous inequality and selecting $B = |A| + 1$, we get $x = (x_t) \in SH_\lambda^b$. □

Theorem 5.10 *If a sequence $x = (x_t)$ is bounded, then it is λ—statistically logarithmic bounded.*

Proof Let $x \in \ell_\infty$. If so, there exists such a number $B \geq 0$ that $\frac{|x_t|}{t} \leq |x_t| \leq B$ for every $t \in \mathbb{N}$. Then,

$$\left\{t \in I_r : \frac{1}{t}|x_t| > B\right\} = \emptyset$$

which yields

$$\lim_{r\to\infty} \frac{1}{E_r}\left|\left\{t \in I_r : \frac{1}{t}|x_t| > B\right\}\right| = 0.$$

Thus, $x \in SH_\lambda^b$.

To demonstrate strictness, take $(\lambda_r) = (r)$ and define a sequence $x = (x_t)$ by

$$x_t = \begin{cases} r\left(\sqrt{\log r}\right), & \text{if } t = r \\ 0, & \text{otherwise.} \end{cases} \quad r = 1, 2, \ldots.$$

For $A = 0$ and $(\lambda_r) = (r)$, we have

$$\frac{1}{E_r}\sum_{t\in I_r}\frac{|x_t|}{t} = \frac{\sqrt{\log r}}{\ell_r} \approx \frac{\sqrt{\log r}}{\log r} \to 0 \text{ as } r \to \infty$$

hence $x \in [SH_\lambda] \subseteq [LH_\lambda]$ i.e. $x \in SH_\lambda^b$ but $x \notin \ell_\infty$. □

Theorem 5.11 *We have*

(i) If

$$\lim_{r\to\infty} \inf \frac{E'_r}{E_r} > 0$$

holds then $SH_\eta^b \subseteq SH_\nu^b$,

(ii) If

$$\lim_{r\to\infty} \frac{E'_r}{E_r} = 1$$

holds then $SH_\nu^b = SH_\eta^b$.

Proof (i) Let $\lim_{r\to\infty} \inf \frac{E'_r}{E_r} > 0$ and $x = (x_t) \in SH_\eta^b$. Then there is a $B \geq 0$ such that

$$\lim_{r\to\infty}\frac{1}{E_r'}\left|\left\{t\in I_r':\frac{1}{t}\left|x_t\right|>B\right\}\right|=0.$$

Thus, we can write

$$\frac{1}{E_r}\left|\left\{t\in I_r:\frac{1}{t}\left|x_t\right|>B\right\}\right|\leq\frac{E_r'}{E_r}\frac{1}{E_r'}\left|\left\{t\in I_r':\frac{1}{t}\left|x_t\right|>B\right\}\right|$$

taking limit as r tends to infinity, we get

$$\lim_{r\to\infty}\frac{1}{E_r}\left|\left\{t\in I_r:\frac{1}{t}\left|x_t\right|>B\right\}\right|=0.$$

Hence $x=(x_t)\in SH_\lambda^b$.

(ii) Let $\lim_{r\to\infty}\frac{E_r'}{E_r}=1$. This yields $\lim_{r\to\infty}\inf\frac{E_r'}{E_r}=1>0$. So $SH_\eta^b\subset SH_\nu^b$ by (i).

Let $x=(x_t)\in SH_\nu^b$ be given. In that case, $B\geq 0$ exists such that

$$\lim_{r\to\infty}\frac{1}{E_r}\left|\left\{t\in I_r:\frac{1}{t}\left|x_t\right|>B\right\}\right|=0.$$

Also, we have

$$\left\{t\in I_r':\frac{1}{t}\left|x_t\right|>B\right\}=\left\{t\in I_r'-I_r:\frac{1}{t}\left|x_t\right|>B\right\}\cup\left\{t\in I_r:\frac{1}{t}\left|x_t\right|>B\right\}$$

Then it is obtained that

$$\begin{aligned}\frac{1}{E_r'}\left|\left\{t\in I_r':\frac{1}{t}\left|x_t\right|>B\right\}\right|&=\frac{1}{E_r'}\left|\left\{t\in I_r'-I_r:\frac{1}{t}\left|x_t\right|>B\right\}\right|\\&\quad+\frac{1}{E_r'}\left|\left\{t\in I_r:\frac{1}{t}\left|x_t\right|>B\right\}\right|\\&\leq\frac{E_r'-E_r}{E_r}+\frac{1}{E_r}\left|\left\{t\in I_r:\frac{1}{t}\left|x_t\right|>B\right\}\right|.\end{aligned}$$

for all $r\geq r_0$. Taking limit, we have

$$\frac{1}{E_r'}\left|\left\{t\in I_r':\frac{1}{t}\left|x_t\right|>B\right\}\right|=0,$$

and so $x=(x_t)\in SH_\eta^b$. Hence $SH_\nu^b\subset H_\eta^b$. We have $SH_\nu^b=SH_\eta^b$. □

Theorem 5.11 leads to the following conclusion.

Corollary 5.1 *Let $\lambda = (\lambda_r) \in \Lambda$ such that for all $r \in \mathbb{N}$, $r \geqslant \lambda_r$. Then we have*

(i) Statistically $(H, 1)$—boundedness is included by λ—statistically logarithmic bounded provided that $\lim_{r\to\infty} \inf \frac{E_r}{\ell_r} > 0$.

(ii) Statistically $(H, 1)$—boundedness and λ—statistically logarithmic bounded are equivalent provided that $\lim_{r\to\infty} \inf \frac{E_r}{\ell_r} = 1$.

5.3 (f, λ)—Harmonic Summability

In this section, we define the concepts of (f, λ)—harmonic summability and (f, λ)—statistical logarithmic convergence of sequences and give some inclusion relations between these spaces.

$f : \mathbb{R}^+ \to \mathbb{R}^+$ is said to be a modulus function with f is increasing, f is continuous from the right at 0, $f(x + y) \leq f(x) + f(y)$ for $x, y \geq 0$ and $f(x) = 0$ if and only if $x = 0$ properties.

f—statistical convergence for any unbounded modulus f defined by

$$\lim_{r\to\infty} \frac{1}{f(r)} f\left(|\{k \leq r : |x_k - A| \geq \varepsilon\}|\right) = 0$$

and we write it as $S^f - \lim x_k = A$ or $x_k \to A\left(S^f\right)$ [1].

Let f be a modulus function and $\lambda \in \Lambda$ in the next definitions.

Definition 5.5 If there is a real number A such that

$$\lim_{r\to\infty} \frac{1}{f(E_r)} \sum_{t\in I_r} f\left(\frac{|x_t|}{t}\right) = A,$$

the sequence $x = (x_t)$ is called $H(\lambda, f)$—summable to the number A.

This space shall be referred to as $H(\lambda, f)$. We write $H(\lambda, f) - \lim x_t = A$ or $x_t \to A\,(H(\lambda, f))$ in this instance. In the special case $(\lambda_r) = (r)$, it will be written $H(f)$ rather than $H(\lambda, f)$.

Definition 5.6 If there is a real number A such that

$$\lim_{r\to\infty} \frac{1}{E_r} \sum_{t\in I_r} f\left(\frac{|x_t - A|}{t}\right) = 0,$$

the sequence $x = (x_t)$ is called SH_λ^f—summable to the number A.

This space shall be referred to as SH_{λ}^{f}. We write $SH_{\lambda}^{f} - \lim x_t = A$ or $x_t \to A\left(SH_{\lambda}^{f}\right)$ in this instance. In the special case $(\lambda_r) = (r)$, it will be written SH^{f} rather than SH_{λ}^{f}.

Definition 5.7 If there is a real number A such that

$$\lim_{r\to\infty} \frac{1}{f(E_r)} \sum_{t\in I_r} f\left(\frac{|x_t - A|}{t}\right) = 0,$$

where $I_r = [r - \lambda_r + 1, r]$, the sequence $x = (x_t)$ is called $SH(\lambda, f)$—summable to the number A.

$SH(\lambda, f)$ will be used to identify this space. In this case, we write $SH(\lambda, f) - \lim x_t = L$ or $x_t \to A(SH(\lambda, f))$. In the special case $(\lambda_r) = (r)$ we shall write $SH(f)$ instead of $SH(\lambda, f)$.

Definition 5.8 If there is a real number A such that

$$\lim_{r\to\infty} \frac{1}{f(E_r)} \sum_{t\in I_r} \frac{|x_t - L|}{t} = 0,$$

the sequence $x = (x_t)$ is called $SH_{\lambda,f}$—summable to the number A.

This space shall be referred to as $SH_{\lambda,f}$. We write $SH_{\lambda,f} - \lim x_t = A$ or $x_t \to A\left(SH_{\lambda,f}\right)$ in this instance. In the special case $(\lambda_r) = (r)$, it will be written SH_f rather than $SH_{\lambda,f}$.

Definition 5.9 If there is a real number A such that

$$\lim_{r\to\infty} \frac{1}{f(E_r)} f\left(\left|\left\{i \in I_r : \frac{1}{t}|x_t - A| \geq \varepsilon\right\}\right|\right) = 0,$$

the sequence $x = (x_t)$ is called $LH(\lambda, f)$—convergent to the number A. $LH(\lambda, f)$ will be used to identify this space. We write $LH(\lambda, f) - \lim x_t = L$ or $x_t \to A(LH(\lambda, f))$ in this instance. In the special case $(\lambda_r) = (r)$, it will be written $LH(f)$ rather than $LH(\lambda, f)$.

Let f be an unbounded modulus in the next theorems.

Theorem 5.12 *The classes of sequences $H(\lambda, f)$ and $SH(\lambda, f)$ are linear spaces.*

Proposition 5.1 ([24]) *Let f be a modulus and $0 < \tau < 1$. Then for each $\|m\| \geq \tau$, we have $f(\|m\|) \leq 2f(1)\tau^{-1}\|m\|$.*

Theorem 5.13 $\lambda \in \Lambda$. *If $\lim_{\zeta\to\infty} \inf \frac{f(\zeta)}{\zeta} > 0$, then $SH(\lambda, f) = SH_{\lambda,f}$.*

Proof Let $x \in SH(\lambda, f)$. If $\lim_{\zeta\to\infty} \inf \frac{f(\zeta)}{\zeta} > 0$, in that case, $d > 0$ exists such that $f(\zeta) > d\zeta$ for $\zeta > 0$. Thus,

$$\frac{1}{f(E_r)} \sum_{t\in I_r} f\left(\frac{|x_t - A|}{t}\right) \geq \frac{1}{f(E_r)} \sum_{t\in I_r} \left[d\frac{|x_t - A|}{t}\right] = \frac{d}{f(E_r)} \sum_{t\in I_r} \frac{|x_t - A|}{t},$$

and therefore, $SH(\lambda, f) \subset SH_{\lambda,f}$.

Now let $x \in SH_{\lambda,f}$. Then we have

$$\frac{1}{f(E_r)} \sum_{t\in I_r} \frac{|x_t - A|}{t} \to 0 \text{ as } r \to \infty.$$

Let $0 < \gamma < 1$. So, we get

$$\begin{aligned}\frac{1}{f(E_r)} \sum_{t\in I_r} \frac{|x_t - A|}{t} &\geq \frac{1}{f(E_r)} \sum_{\substack{t\in I_r\\ \frac{|x_t-A|}{t}\geq\gamma}} \frac{|x_t - A|}{t}\\ &\geq \frac{1}{f(E_r)} \sum_{\substack{t\in I_r\\ \frac{|x_t-A|}{t}\geq\gamma}} \left[\frac{f\left(\frac{|x_t-A|}{t}\right)}{2f(1)\gamma^{-1}}\right]\\ &\geq \frac{1}{f(E_r)} \frac{\gamma}{2f(1)} \sum_{t\in I_r} f\left(\frac{|x_t - A|}{t}\right)\end{aligned}$$

by Proposition 5.1. Therefore, $x \in SH(\lambda, f)$.

The equality $SH(\lambda, f) = SH_{\lambda,f}$ cannot be hold if $\lim_{\zeta\to\infty} \inf \frac{f(\zeta)}{\zeta} = 0$ as shown in the following example:

$x = (x_t)$ be defined as

$$x_t = \begin{cases} r(\log r), & \text{if } t = r \\ 0, & \text{otherwise.}\end{cases} \quad r = 1, 2, \ldots.$$

and let $f(x) = \sqrt{x}$.

For $A = 0$ and $(\lambda_r) = (r)$, we have

$$\frac{1}{f(E_r)} \sum_{t\in I_r} f\left(\frac{|x_t|}{t}\right) = \frac{\sqrt{\log r}}{f(\ell_r)} \approx \frac{\sqrt{\log r}}{\log r} \to 0 \text{ as } r \to \infty$$

hence $x \in SH(\lambda, f)$, but

$$\frac{1}{f(E_r)}\sum_{t\in I_r}\frac{|x_t|}{t}=\frac{\log r}{f(\ell_r)}\approx\frac{\log r}{\sqrt{\log r}}\to\infty \text{ as } r\to\infty$$

so, $x \notin SH_{\lambda,f}$. □

Maddox [19] demonstrated the existence of an unbounded modulus f for which d is a positive constant such that $f(uv)\geq df(u)f(v)$, for all $u\geq 0$, $v\geq 0$.

Theorem 5.14 *If* $\lim_{\zeta\to\infty}\frac{f(\zeta)}{\zeta}>0$, *then* $SH_{\lambda}^{f}\subseteq LH(\lambda, f)$.

Proof Let $x\in SH_{\lambda}^{f}$ and $\lim_{\zeta\to\infty}\frac{f(\zeta)}{\zeta}>0$. For $\varepsilon>0$, we have

$$\begin{aligned}\frac{1}{E_r}\sum_{t\in I_r}f\left(\frac{|x_t-A|}{t}\right)&\geq\frac{1}{E_r}f\left(\sum_{t\in I_r}\frac{|x_t-A|}{t}\right)\geq\frac{1}{E_r}f\left(\sum_{\substack{t\in I_r\\ \frac{|x_t-A|}{t}\geq\varepsilon}}\frac{|x_t-A|}{t}\right)\\&\geq\frac{1}{E_r}f\left(\left|\left\{t\in I_r:\frac{|x_t-A|}{t}\geq\varepsilon\right\}\right|\varepsilon\right)\\&\geq\frac{d}{E_r}f\left(\left|\left\{t\in I_r:\frac{|x_t-A|}{t}\geq\varepsilon\right\}\right|\right)f(\varepsilon)\\&=\frac{d}{E_r}\frac{f\left(\left|\left\{t\in I_r:\frac{|x_t-A|}{t}\geq\varepsilon\right\}\right|\right)}{f(E_r)}f(E_r)f(\varepsilon).\end{aligned}$$

Therefore, $SH_{\lambda}^{f}-\lim x_t=A$ implies $LH(\lambda, f)-\lim x_t=A$. □

Theorem 5.15 *We have* $SH(\lambda, f)\subseteq LH(\lambda, f)$.

Proof Let $x\in SH(\lambda, f)$ and $\varepsilon>0$ be given. We may write

$$\begin{aligned}\frac{1}{f(E_r)}\sum_{t\in I_r}f\left(\frac{|x_t-A|}{t}\right)&=\frac{1}{f(E_r)}\left[\sum_{\substack{t\in I_r\\ \frac{|x_t-A|}{t}\geq\varepsilon}}f\left(\frac{|x_t-A|}{t}\right)+\sum_{\substack{t\in I_r\\ \frac{|x_t-A|}{t}<\varepsilon}}f\left(\frac{|x_t-A|}{t}\right)\right]\\&\geq\frac{1}{f(E_r)}\sum_{\substack{t\in I_r\\ \frac{|x_t-A|}{t}\geq\varepsilon}}[f(\varepsilon)]\\&\geq\frac{1}{Hf(E_r)}\left[f\left(\sum_{\substack{t\in I_r\\ \frac{|x_t-A|}{t}\geq\varepsilon}}\varepsilon\right)\right]\end{aligned}$$

$$\geq \frac{1}{Hf(E_r)} f\left(\left|\left\{t \in I_r : \frac{|x_t - A|}{t} \geq \varepsilon\right\}\right| \varepsilon\right)$$
$$\geq \frac{c}{Hf(E_r)} f\left(\left|\left\{t \in I_r : \frac{|x_t - A|}{t} \geq \varepsilon\right\}\right|\right) f(\varepsilon).$$

Hence, $x \in LH(\lambda, f)$. □

Theorem 5.16 *If* $\liminf_r \frac{f(E_r)}{f(\ell_r)} > 1$, *then* $LH(f) \subseteq LH(\lambda, f)$.

Proof If $LH(f) - \lim x_t = A$, then for r that is large enough, we get

$$\frac{1}{f(\ell_r)} f\left(\left|\left\{t \leq r : \frac{|x_t - A|}{t} \geq \varepsilon\right\}\right|\right) \geq \frac{1}{f(\ell_r)} f\left(\left|\left\{t \in I_r : \frac{|x_t - A|}{t} \geq \varepsilon\right\}\right|\right)$$
$$= \frac{f(E_r)}{f(\ell_r)} \frac{1}{f(E_r)} f\left(\left|\left\{t \in I_r : \frac{|x_t - A|}{t} \geq \varepsilon\right\}\right|\right)$$

for every $\varepsilon > 0$. The proof is finished. □

Theorem 5.17 *If* $(x_t) \in AH(f) \cap AH(\lambda, f)$, *then* $AH(f) - \lim x_t = AH(\lambda, f) - \lim x_t$ *such that* $|f(u) - f(v)| = f(|u - v|)$ *for* $u \geq 0$, $v \geq 0$.

Proof Suppose $LH(f) - \lim x_t = A_1$, $AH(\lambda, f) - \lim x_t = A_2$ and $A_1 \neq A_2$. Let $\frac{|A_1 - A_2|}{2} > \varepsilon$. We get

$$\lim_{r\to\infty} \frac{1}{f(\ell_r)} f\left(\left|\left\{t \leq r : \frac{|x_t - A_1|}{t} \geq \varepsilon\right\}\right|\right) = 0,$$

and

$$\lim_{r\to\infty} \frac{1}{f(E_r)} f\left(\left|\left\{t \in I_r : \frac{|x_t - A_2|}{t} \geq \varepsilon\right\}\right|\right) = 0$$

for $\varepsilon > 0$. On the other hand, since $\log r < r$ and $f(\log r) < f(r)$, we have $f(\ell_r) < f(r)$.

We can write

$$\frac{f\left(\left|\left\{t \leq r : \frac{|A_1 - A_2|}{t} \geq 2\varepsilon\right\}\right|\right)}{f(r)} < \frac{f\left(\left|\left\{t \leq r : \frac{|A_1 - A_2|}{t} \geq 2\varepsilon\right\}\right|\right)}{f(\ell_r)}$$

and

$$\frac{f\left(\left|\left\{t \leq r : \frac{|A_1 - A_2|}{t} \geq 2\varepsilon\right\}\right|\right)}{f(r)} \leq \frac{f\left(\left|\left\{t \leq r : \frac{|x_t - A_1|}{t} \geq \varepsilon\right\}\right|\right)}{f(r)} + \frac{f\left(\left|\left\{t \leq r : \frac{|x_t - A_2|}{t} \geq \varepsilon\right\}\right|\right)}{f(r)}.$$

Limiting to $r \to \infty$, we obtain

$$1 \leq 0 + \lim_{r\to\infty} \frac{f\left(\left|\left\{t \leq r : \frac{|x_t - A_2|}{t} \geq \varepsilon\right\}\right|\right)}{f(r)} \leq 1,$$

and so

$$\lim_{r\to\infty} \frac{f\left(\left|\left\{t \leq r : \frac{|x_t - A_2|}{t} \geq \varepsilon\right\}\right|\right)}{f(r)} = 1 \text{ and } \lim_{r\to\infty} \frac{f\left(\left|\left\{t \leq r : \frac{|x_t - A_2|}{t} \geq \varepsilon\right\}\right|\right)}{f(\ell_r)} > 1$$

We consider the subsequence

$$\frac{1}{f(\ell_s)} f\left(\left|\left\{t \leq k_s : \frac{|x_t - A_2|}{t} \geq \varepsilon\right\}\right|\right)$$

of sequence

$$\frac{1}{f(\ell_r)} f\left(\left|\left\{t \leq r : \frac{|x_t - A_2|}{t} \geq \varepsilon\right\}\right|\right).$$

Then

$$\begin{aligned}
\frac{1}{f(\ell_s)} f\left(\left|\left\{t \leq k_s : \frac{|x_t - A_2|}{t} \geq \varepsilon\right\}\right|\right) &= \frac{1}{f(\ell_s)} f\left(\left|\left\{t \in \bigcup_{r=1}^{s} I_r : \frac{|x_t - A_2|}{t} \geq \varepsilon\right\}\right|\right) \\
&= \frac{1}{f(\ell_s)} f\left(\sum_{r=1}^{s} \left|\left\{t \in I_r : \frac{|x_t - A_2|}{t} \geq \varepsilon\right\}\right|\right) \\
&\leq \frac{1}{f(\ell_s)} \sum_{r=1}^{s} f\left(\left|\left\{t \in I_r : \frac{|x_t - A_2|}{t} \geq \varepsilon\right\}\right|\right) \\
&= \frac{1}{f(\ell_s)} \sum_{r=1}^{s} f(E_r) \frac{1}{f(E_r)} f\left(\left|\left\{t \in I_r : \frac{|x_t - A_2|}{t} \geq \varepsilon\right\}\right|\right)
\end{aligned}$$

and

$$\begin{aligned}
\sum_{r=1}^{s} f(E_r) &= f(E_1) + f(E_2) + \cdots + f(E_s) \\
&= f\left(\sum_{t\in I_1} \frac{1}{t}\right) + f\left(\sum_{t\in I_2} \frac{1}{t}\right) + \cdots + f\left(\sum_{t\in I_s} \frac{1}{t}\right)
\end{aligned}$$

$$\leq sf\left(\sum_{t\in I_s}\frac{1}{t}\right)$$
$$= sf\left(\sum_{t=1}^{s}\frac{1}{t}-\sum_{t=1}^{s-\lambda_s}\frac{1}{t}\right)$$
$$\leq sf\left(\sum_{t=1}^{s}\frac{1}{t}\right)=sf\left(\ell_s\right).$$

We have

$$\frac{1}{f\left(\ell_s\right)}f\left(\left|\left\{t\leq k_s:\frac{|x_t-A_2|}{t}\geq\varepsilon\right\}\right|\right)\leq\frac{\sum\limits_{r=1}^{s}f\left(E_r\right)}{\left(\sum\limits_{r=1}^{s}f\left(E_r\right)\right)/s}\frac{1}{f\left(E_r\right)}f\left(\left|\left\{t\in I_r:\frac{|x_t-A_2|}{t}\geq\varepsilon\right\}\right|\right)$$

we get

$$\frac{1}{f\left(\ell_s\right)}f\left(\left|\left\{t\leq k_s:\frac{|x_t-A_2|}{t}\geq\varepsilon\right\}\right|\right)\to 0.$$

But

$$\lim_{r\to\infty}\frac{f\left(\left|\left\{t\leq r:\frac{|x_t-A_2|}{t}\geq\varepsilon\right\}\right|\right)}{f\left(\ell_r\right)}>1.$$

Contradiction. As a result, $A_1 = A_2$. □

We can get the following Corollary 5.2 from Theorem 5.17.

Corollary 5.2 *Let* $\lambda, \mu \in \Lambda$. *If* $(x_t) \in LH(f) \cap (LH(\lambda, f) \cap LH(\mu, f))$, *then* $LH(\lambda, f) - \lim x_t = LH(\mu, f) - \lim x_t$.

Theorem 5.18 $SH(\lambda, f) - \lim x_t = A$ *uniquely.*

Proof Assume that $SH(\lambda, f) - \lim x_t = A_1$ and $SH(\lambda, f) - \lim x_t = A_2$. Then

$$\lim_r\frac{1}{f\left(E_r\right)}\sum_{t\in I_r}f\left(\frac{|x_t-A_1|}{t}\right)=0,$$

and

$$\lim_r\frac{1}{f\left(E_r\right)}\sum_{t\in I_r}f\left(\frac{|x_t-A_2|}{t}\right)=0.$$

We obtain using definition of f,

$$\frac{1}{f(E_r)}\sum_{t\in I_r} f\left(\frac{|A_1-A_2|}{t}\right) \leq \frac{1}{f(E_r)}\left(\sum_{t\in I_r} f\left(\frac{|x_t-A_1|}{t}\right)+\sum_{t\in I_r} f\left(\frac{|x_t-A_2|}{t}\right)\right)$$
$$=\frac{1}{f(E_r)}\sum_{t\in I_r} f\left(\frac{|x_t-A_1|}{t}\right)+\frac{1}{f(E_r)}\sum_{t\in I_r} f\left(\frac{|x_t-A_2|}{t}\right).$$

Hence,

$$\lim_r \frac{1}{f(E_r)}\sum_{t\in I_r} f\left(\frac{|A_1-A_2|}{t}\right)=0.$$

As a result, the limit is unique. □

Theorem 5.19 *If*

$$\liminf_{r\to\infty}\frac{f(E_r)}{f(E_r')}>0,$$

then $SH(\eta, f)\subseteq SH(\nu, f)$.

Proof Let $x\in SH(\eta, f)$. We get

$$\frac{1}{f(E_r')}\sum_{t\in I_r'} f\left(\frac{|x_t-A|}{t}\right)=\frac{1}{f(E_r')}\sum_{t\in I_r'-I_r} f\left(\frac{|x_t-A|}{t}\right)+\frac{1}{f(E_r')}\sum_{t\in I_r} f\left(\frac{|x_t-A|}{t}\right)$$
$$\geq\frac{1}{f(E_r')}\sum_{t\in I_r} f\left(\frac{|x_t-A|}{i}\right)$$
$$\geq\frac{f(E_r)}{f(E_r')}\frac{1}{f(E_r)}\sum_{t\in I_r} f\left(\frac{|x_t-A|}{t}\right).$$

Thus, if $x\in SH(\eta, f)$, then $x\in SH(\nu, f)$. □

References

1. Aizpuru, A., Listán-García, M.C., Rambla-Barreno, F.: Density by moduli and statistical convergence. Quaest. Math. **37**(4), 525–530 (2014)
2. Akbas, K.E., Isik, M.: On asymptotically λ—statistical equivalent sequences of order α in probability. Filomat **34**(13), 4359–4365 (2020)
3. Altin, Y., Et, M., Basarir, M.: On some generalized difference sequences of fuzzy numbers. Kuwait J. Sci. Eng. **34**(1A), 1–14 (2007)
4. Altinok, H., Et, M., Çolak, R.: Some remarks on generalized sequence space of bounded variation of sequences of fuzzy numbers. Iran. J. Fuzzy Syst. **11**(5), 39–46 (2014)

5. Aral, N.D., Şengul Kandemir, H.: $I-$lacunary statistical convergence of order β of difference sequences of fractional order. Facta Univ. Ser. Math. Inform. **36**(1), 43–55 (2021)
6. Cakalli, H.: A variation on statistical ward continuity. Bull. Malays. Math. Sci. Soc. **40**, 1701–1710 (2017)
7. Çakallı, H.: A study on statistical convergence. Funct. Anal. Approx. Comput. **1**(2), 19–24 (2009)
8. Et, M., Cinar, M., Sengul Kandemir, H.: Deferred statistical convergence of order α in metric spaces. AIMS Math. **5(4)**, 3731–3740 (2020)
9. Çolak, R.: Statistical convergence of order α. In: Modern Methods in Analysis and its Applications, pp. 121–129. Anamaya Pub, New Delhi, India (2010)
10. Connor, J.S.: The statistical and strong p-Cesaro convergence of sequences. Analysis **8**, 47–63 (1988)
11. Di Maio, G., Kočinac, L.D.R.: Statistical convergence in topology. Topol. Appl. **156**, 28–45 (2008)
12. Fast, H.: Sur la convergence statistique. Colloq. Math. **2**, 241–244 (1951)
13. Fridy, J.: On statistical convergence. Analysis **5**, 301–313 (1985)
14. Gungor, M., Et, M.: Δ^r-strongly almost summable sequences defined by Orlicz functions. Indian J. Pure Appl. Math. **34**(8), 1141–1151 (2003)
15. Gungor, M., Et, M., Altin, Y.: Strongly (V_σ, λ, q)-summable sequences defined by Orlicz functions. Appl. Math. Comput. **157**(2), 561–571 (2004)
16. Isik, M., Akbas, K.E.: On λ-statistical convergence of order α in probability. J. Inequal. Spec. Funct. **8**(4), 57–64 (2017)
17. Isik, M., Akbas, K.E.: On asymptotically lacunary statistical equivalent sequences of order α in probability. ITM Web Conf. **13**, 01024 (2017). https://doi.org/10.1051/itmconf/20171301024
18. Leindler, L.: Uber die de la Vallee-Pousinsche Summierbarkeit allgemeiner Orthogonalreihen. Acta Math. Acad. Sci. Hungar **16**, 375–387 (1965)
19. Maddox, I.J.: Sequence spaces defined by a modulus. Math. Proc. Camb. Philos. Soc. **100**, 161–166 (1986)
20. Malkowsky, E., Savas, E.: Some λ-sequence spaces defined by a modulus. Arch. Math. **36**(3), 219–228 (2000)
21. Mursaleen, M.: λ-statistical convergence. Math. Slovaca **50**(1), 111–115 (2000)
22. Moricz, F.: Theorems relating to statistical harmonic summability and ordinary convergence of slowly decreasing or oscillating sequences. Analysis **24**, 127–145 (2004)
23. Nuray, F.: Lacunary statistical harmonic summability. J. Appl. Anal. Comput. https://doi.org/10.11948/20210155
24. Pehlivan, S., Fisher, B.: Some sequence spaces defined by a modulus. Math. Slovaca. **45**(3), 275–280 (1995)
25. Salat, T.: On statistically convergent sequences of real numbers. Math. Slovaca **30**, 139–150 (1980)
26. Savas, E., Et, M.: On (Δ_λ^m, I)-statistical convergence of order α. Period. Math. Hungar. **71**, 135–145 (2015)
27. Schoenberg, I.J.: The integrability of certain functions and related summability methods. Am. Math. Monthly **66**, 361–375 (1959)
28. Sezer, S.A.: Statistical harmonic summability of sequences of fuzzy numbers. Soft Comput. (2020). https://doi.org/10.1007/s00500-020-05151-9
29. Steinhaus, H.: Sur la convergence ordinaire et la convergence asymptotique. Colloq. Math. **2**, 73–74 (1951)
30. Şengül, H.: Some Cesàro-type summability spaces defined by a modulus function of order (α, β). Commun. Fac. Sci. Univ. Ank. Sér. A1 Math. Stat. **66**(2), 80–90 (2017)
31. Şengül, H., Et, M.: On lacunary statistical convergence of order α. Acta Math. Sci. Ser. B Engl. Ed. **34**(2), 473–482 (2014)
32. Şengül, H., Et, M., Çakallı, H.: On (f, I)—lacunary statistical convergence of order α of sequences of sets. Bol. Soc. Parana. Mat. **38**(7), 85–97 (2020)

33. Şengül, H., Et, M., Altin, Y.: f—lacunary statistical convergence and strong f—lacunary summability of order α of double sequences. Facta Univ. Ser. Math. Inform. **35**(2), 495–506 (2020)
34. Şengül, H., Et, M.: f—lacunary statistical convergence and strong f—lacunary summability of order α. Filomat **32**(13), 4513–4521 (2018)

Chapter 6
A New Approach on Statistical Relative Modular Deferred Weighted Mean Based on (p, q)-integers

Bidu Bhusan Jena and Susanta Kumar Paikray

Abstract The notion of statistical deferred weighted summability technique has been explored by Srivastava et al. [46] in the recent past. The present article deals with the deferred weighted approach of summability in the modular space for the generalized double sequence of functions associated with (p, q)-integers. In fact, we basically introduce here the concept of relative modular deferred weighted statistical convergence and statistical relative modular deferred weighted summability for a double sequence of functions and build an inclusion connection between them.

Keywords P-convergent · Statistical convergence · Statistical relative modular deferred weighted summability · Relative modular deferred weighted statistical convergence

6.1 Introduction

The gradual improvement in the study of sequence spaces leads to the advancement of the concept of statistical convergence, which is quite more prevailing than the usual convergence. The credit for such development goes to two eminent mathematicians Fast [18] and Steinhaus [48], and this concept makes the convergence analysis much wider. This notion was subsequently extended to the cases of double sequences by Mursaleen and Edely [34]. Nowadays, this potential idea has been applied in numerous disciplines of pure and applied mathematics and analytical statistics as well. In particular, it is very much useful in the study of Machine Learning, Soft Computing, Number Theory, Measure theory, Probability theory, etc. For some latest works, the interested learners may refer [5, 11, 21].

B. B. Jena
Faculty of Science (Mathematics), Sri Sri University, Cuttack 754006, Odisha, India

S. K. Paikray (✉)
Department of Mathematics, Veer Surendra Sai University of Technology, Burla 768018, Odisha, India
e-mail: skpaikray_math@vssut.ac.in

B. Hazarika et al. (eds.), *Advances in Functional Analysis and Fixed-Point Theory*, Industrial and Applied Mathematics,
https://doi.org/10.1007/978-981-99-9207-2_6

Suppose that the set of all real-valued sequences be ω. Recall that, in the sequence space, any subspace of ω is also a sequence space. Let c_0 and c be the corresponding classes of null and convergent sequences. Let the class of all bounded linear spaces be ℓ_∞. For real or complex terms sequence, we have

$$\|\chi\|_\infty = \sup_k |\chi_k| \quad (k \in \mathbb{N}).$$

Note that, the above-specified spaces are all Banach spaces under this usual norm.

The technique of difference sequence space was considered initially by Kızmaz [29] and was subsequently generalized by Et et al. (see [14], [17]) in the sense as follows:

$$X(\Delta^r) = \left\{\chi = (\chi_k) : (\Delta^r \chi_k) \in X\right\},$$

where X may be ℓ_∞, c or c_0 and $r \in \mathbb{N}$, $\Delta^0 \chi = (\chi_k)$ with $\Delta^r \chi = (\Delta^r \chi_k) = (\Delta^{r-1}\chi_k - \Delta^{r-1}\chi_{k+1})$.

The below-mentioned expression is the binomial representation of the generalized difference sequence:

$$\Delta^r \chi_k = \sum_{\upsilon=0}^{r} (-1)^\upsilon \binom{r}{\upsilon} \chi_{k+\upsilon}.$$

The sequence space $\ell_\infty(\Delta^r)$, $c(\Delta^r)$, $c_0\Delta^r$ are Banach spaces normed by

$$\|\chi\|_\Delta = \sum_{\iota=1}^{\mu} |\chi_\iota| + \|\Delta^\mu \chi\|_\infty,$$

for $X = \ell_\infty(\Delta^r),\ c(\Delta^r)$ and $c_0\Delta^r$.

Recalling the work of Pringsheim [41], a double sequence $(\chi_{\mu,\nu})$ is said to be P-convergent to a number ℓ, if for the given $\epsilon > 0$,

$$|\chi_{\mu,\nu} - \ell| < \epsilon, \quad \text{for } \mu, \nu \geqq \nu_0 \in \mathbb{N},$$

and it is denoted by $P - \lim \chi_{\mu,\nu} = \ell$.

Likewise, $(\chi_{\mu,\nu})$ is bounded, if there is a positive constant $\mathscr{K}$ for which $|\chi_{\mu,\nu}| \leqq \mathscr{K}$, for all $\mu, \nu \in \mathbb{N}$. Note that, unlike single sequences case, a double sequence need not be bounded even if it is convergent. Furthermore, $(\chi_{\mu,\nu})$ is non-increasing in *Pringsheim's sense* if, $\chi_{\mu+1,\nu} \leq \chi_{\mu,\nu}$ and $\chi_{\mu,\nu+1} \leqq \chi_{\mu,\nu}$.

Let $\mathscr{G} \subset \mathbb{N} \times \mathbb{N}$ be a two-dimensional set of positive integers, and let $\mathscr{G}(\iota, \tau) = \{(\mu, \nu) : \mu \leqq \iota \text{ and } \nu \leqq \tau\}$. Then, the *double natural or asymptotic density* $\delta(\mathscr{G})$ of $\mathscr{G}$ is defined by

$$\delta(\mathscr{G}) = P - \lim_{\iota,\tau} \frac{1}{\iota, \tau} |\mathscr{G}(\iota, \tau)|,$$

subject to the existence of the limit.

A double real sequence $(\chi_{\mu,\nu})$ is said to be statistically convergent to ℓ in the sense of *Pringsheim*, if for every $\epsilon > 0$,

$$\delta(\mathscr{G}_\epsilon(\iota, \tau)) = 0,$$

such taht

$$\delta(\mathscr{G}_\epsilon(\iota, \tau)) = \{(\mu, \nu), \mu \leqq \iota, \nu \leqq \tau \text{ and } |\chi_{\mu,\nu} - \ell| \geqq \epsilon\}.$$

In this case, we write

$$\text{stat}^2 \lim_{\mu,\nu} \chi_{\mu,\nu} = \ell.$$

Example 6.1 Considering $\chi = (\chi_{\mu,\nu})$ a double sequence of the following form:

$$\chi_{\mu,\nu} = \begin{cases} \sqrt{\nu\mu} \text{ when } \mu = k^2, \ \nu = l^2 \text{ for all } k, l \in \mathbb{N} \\ \frac{1}{\nu\mu} \quad \text{otherwise,} \end{cases}$$

it is easy to see that the sequence $(\chi_{\mu,\nu})$ is not P-convergent in the usual ordinary sense, while the statistical limit of $(\chi_{\mu,\nu})$ is 0.

Let $\mathscr{I} = [0, \infty) \subseteq \mathbb{R}$ be such that υ is the Lebesgue measure on $\mathscr{J}$. Then, by $X(\mathscr{J}^2)$ ($\mathscr{J}^2 = [0, \infty) \times [0, \infty)$), we specify the space of all measurable functions (real-valued) on $\mathscr{J}^2$ subject to the equality almost everywhere. Further, let $C(\mathscr{J}^2)$ and $C^\infty(\mathscr{J}^2)$ be the respective spaces of all continuous and infinitely differentiable functions on $\mathscr{J}^2$.

Next, we recollect that a functional

$$\omega : X(\mathscr{J}^2) \to [0, \infty)$$

is a *modular* on $X(\mathscr{J}^2)$ such that the usual conditions (see [28]) hold.

We further recall the work of Bardaro et al. [6, 7] that:

- a modular ω is called as $\mathscr{N}$*-quasi convex* when there is a constant $\mathscr{N} \geq 1$ for which

$$\omega(\zeta h + \chi h') \leq \mathscr{N}\zeta\omega(\mathscr{N}h) + \mathscr{N}\chi\omega(\mathscr{N}h')$$

holds for every $h, h' \in X(\mathscr{J}^2)$, $\zeta, \chi \geq 0$ with $\zeta + \chi = 1$. Particularly, if $\mathscr{N} = 1$, then ω is called *convex*. Moreover,

- a modular ω is *$\mathscr{N}$-quasi semi-convex*, if there is a constant $\mathscr{N} \geqq 1$ for which

$$\omega(\lambda h) \leqq \mathscr{N}\lambda\omega(\mathscr{N}h)$$

exists for all $h \in X(\mathscr{J}^2)$ and $\lambda \in (0, 1]$.

Of course, it is trivial that every $\mathscr{N}$-quasi semi-convex modular is $\mathscr{N}$-quasi convex.

Now based on the modular ω, we consider some subspaces of $X(\mathscr{J}^2)$ by means as follows:

$$L^{\omega}(\mathscr{J}^2) = \{h \in X(\mathscr{J}^2) : \lim_{\lambda \to 0^+} \omega(\lambda h) = 0\}$$

and

$$E^{\omega}(\mathscr{J}^2) = \{h \in L^{\omega}(\mathscr{J}^2) : \omega(\lambda h) < +\infty \text{ for all } \lambda > 0\}.$$

Here, $L^{\omega}(\mathscr{J}^2)$ is known as the modular space emerged by ω and $E^{\omega}(\mathscr{J}^2)$ is the space of the finite elements of $L^{\omega}(\mathscr{J}^2)$. Further, it is interesting to see that, if ω is $\mathscr{N}$-quasi semi-convex, then the space

$$\{h \in X(\mathscr{J}^2) : \omega(\lambda h) < +\infty \text{ for all } \lambda > 0\}$$

is same as $L^{\omega}(\mathscr{J}^2)$. Moreover, for a *convex modular* ω in $X(\mathscr{J}^2)$, we may identify an *F-norm* of the form:

$$\|h\|_{\omega} = \inf\left\{\lambda > 0 : \omega\left(\frac{h}{\lambda}\right) \leqq 1\right\}.$$

The idea of modular has been useful in various contexts which was studied in [7, 36]. Moore [33] and Chittenden [8] studied relatively uniform convergence of functions sequence defined over the closed intervals $[a, b] \subseteq \mathbb{R}$.

Next, we recollect the definition of uniform convergence compatible with a scale function as follows.

A functions sequence (h_{υ}) defined on an interval $[a, b]$, *converges relatively uniformly* to a limit function $h(\chi)$, if there exists a scale function (non-zero) $\sigma(\chi)$ in $[a, b]$, such that for every $\epsilon > 0$, there is an integer υ_{ϵ} and for every $\upsilon > \upsilon_{\epsilon}$,

$$\left|\frac{h_{\upsilon}(\chi) - h(\chi)}{\sigma(\chi)}\right| \leqq \epsilon$$

uniformly exits for every $\chi \in [a, b] \subseteq \mathbb{R}$.

Example 6.2 For each $\nu \in \mathbb{N}$, let $h_\nu : [0, 1] \to \mathbb{R}$ be given by

$$h_\nu(\chi) = \begin{cases} \frac{\nu\chi}{1+\nu^2\chi^2} & (0 < \chi \leqq 1) \\ 0 & (\chi = 0). \end{cases}$$

The sequence (h_ν) of functions is nither classically nor statistically uniformly convergent in $[0, 1]$; but converges uniformly to $h = 0$ compatible (relative) to a scale function

$$\sigma(\chi) = \begin{cases} \frac{1}{\chi} & (0 < \chi \leqq 1) \\ 0 & (\chi = 0) \end{cases}$$

on $[0, 1]$.
Here, we write

$$h_\nu \rightrightarrows h = 0 \quad ([0, 1]; \sigma)\,.$$

Later, as the q-calculus evolved, many investigators focused on developing additional generalizations of positive linear operators based on q-integers (see [3] and [4]). The (p, q)-analogue of Bernstein operators was recently introduced by Mursaleen et al. [35] in connection with (p, q)-integers. Subsequently, various approximation results for Baskakov operators and Bernstein-Schurer operators were examined for (p, q)-integers by [1].

Recalling some basic facts about (p, q)-integers, we have for any $(\nu \in \mathbb{N})$ the (p, q)-integer $[\nu]_{p,q}$ is specified by

$$[\nu]_{p,q} = \begin{cases} \frac{p^\nu - q^\nu}{p-q} & (\nu \geq 1) \\ 0 & (\nu = 0) \end{cases}$$

where $0 < q < p \leq 1$.

The (p, q)-factorial takes the form

$$[\nu]!_{p,q} = \begin{cases} [1]_{p,q}[2]_{p,q}\cdots[\nu]_{p,q} & (\nu \geq 1) \\ 1 & (\nu = 0). \end{cases}$$

The (p, q)-binomial coefficient is given by

$$\begin{bmatrix} \nu \\ k \end{bmatrix}_{p,q} = \frac{[\nu]!_{p,q}}{[k]!_{p,q}\,[\nu - k]!_{p,q}} \qquad \text{for all } \nu, k \in \mathbb{N} \text{ and } \nu \geq k.$$

Kadak [27] and Paikray et al. [49] explored the weighted statistical convergence based on (p, q)-integers for single sequence.

Let $\chi = (\chi_k)$ be any number sequence, suppose that $0 < q < p \leq 1$ and r be a non-negative integer. The operator

$$\Delta^{[r]}_{p,q} : \omega \to \omega$$

is identified by

$$\Delta^{[r]}_{p,q}(\chi_\nu) = \sum_{\iota=0}^{\infty} (-1)^{\iota} \begin{bmatrix} r \\ \iota \end{bmatrix}_{p,q} \chi_{\nu-\iota}.$$

Next, we wish to present a difference (generalized) operator on double functions sequences $(h_{\mu,\nu}(\chi, \gamma))$ by using (p, q) integers.

Let $(h_{\mu,\nu}(\chi, \gamma))$ be a real-valued double functions sequence defined over a two-dimensional subset $E \subset \mathbb{R}^2$ and suppose that $0 < q < p \leq 1$ and r be a non-negative integer. We present the generalized difference operator of double sequences of functions as follows:

$$\Delta^{\theta,\vartheta}_{p,q}(h_{\mu,\nu}) = \sum_{\iota,\tau=0}^{\infty,\infty} (-1)^{\iota+\tau} \begin{bmatrix} \theta \\ \iota \end{bmatrix}_{p,q} \begin{bmatrix} \vartheta \\ \tau \end{bmatrix}_{p,q} h_{\mu-\iota,\nu-\tau}(\chi, \gamma). \tag{6.1}$$

Throughout, it is assumed that the double series defined in (6.1) converges uniformly in the sense of Pringsheim to a function $h(\chi, \gamma)$ on E^2. For instance,

$$\begin{aligned}
\Delta^{\theta,\vartheta}_{p,q}(h_{\mu,\nu}) &= \sum_{\iota=0}^{2} (-1)^{\iota} \begin{bmatrix} \theta \\ \iota \end{bmatrix}_{p,q} \left\{ \sum_{\tau=0}^{2} (-1)^{\tau} \begin{bmatrix} \vartheta \\ \tau \end{bmatrix}_{p,q} h_{\mu-\iota,\nu-\tau}(\chi, \gamma) \right\} \\
&= \begin{bmatrix} \theta \\ 0 \end{bmatrix}_{p,q} \left\{ \begin{bmatrix} \vartheta \\ 0 \end{bmatrix}_{p,q} h_{\mu,\nu} - \begin{bmatrix} \vartheta \\ 1 \end{bmatrix}_{p,q} h_{\mu,\nu-1} + \begin{bmatrix} \vartheta \\ 2 \end{bmatrix}_{p,q} h_{\mu,\nu-2} \right\} \\
&\quad - \begin{bmatrix} \theta \\ 1 \end{bmatrix}_{p,q} \left\{ \begin{bmatrix} \vartheta \\ 0 \end{bmatrix}_{p,q} h_{\mu-1,\nu} - \begin{bmatrix} \vartheta \\ 1 \end{bmatrix}_{p,q} h_{\mu-1,\nu-1} + \begin{bmatrix} \vartheta \\ 2 \end{bmatrix}_{p,q} h_{\mu-1,\nu-2} \right\} \\
&\quad + \begin{bmatrix} \theta \\ 2 \end{bmatrix}_{p,q} \left\{ \begin{bmatrix} \vartheta \\ 0 \end{bmatrix}_{p,q} h_{\mu-2,\nu} - \begin{bmatrix} \vartheta \\ 1 \end{bmatrix}_{p,q} h_{\mu-2,\nu-1} + \begin{bmatrix} \vartheta \\ 2 \end{bmatrix}_{p,q} h_{\mu-2,\nu-2} \right\} \\
&= h_{\mu,\nu} - \left(\frac{p^{\vartheta} - q^{\vartheta}}{p - q} \right) h_{\mu,\nu-1} + \left(\frac{p^{\vartheta} - q^{\vartheta}}{p - q} \right) \left(\frac{p^{\vartheta-1} - q^{\vartheta-1}}{p^2 - q^2} \right) h_{\mu,\nu-2}
\end{aligned}$$

$$
\begin{aligned}
&-\left(\frac{p^\theta - q^\theta}{p-q}\right) h_{\mu-1,\nu} + \left(\frac{(p^\theta - q^\theta)(p^\vartheta - q^\vartheta)}{(p-q)^2}\right) h_{\mu-1,\nu-1} \\
&-\left(\frac{(p^\theta - q^\theta)(p^\vartheta - q^\vartheta)(p^{\vartheta-1} - q^{\vartheta-1})}{(p-q)^3(p+q)}\right) h_{\mu-1,\nu-2} \\
&+\left(\frac{(p^\theta - q^\theta)(p^{\theta-1} - q^{\theta-1})}{(p-q)^2(p+q)}\right) h_{\mu-2,\nu} \\
&-\left(\frac{(p^\theta - q^\theta)(p^{\theta-1} - q^{\theta-1})(p^\vartheta - q^\vartheta)}{(p-q)^3(p+q)}\right) h_{\mu-2,\nu-1} \\
&+\left(\frac{(p^\theta - q^\theta)(p^{\theta-1} - q^{\theta-1})}{(p-q)^2(p+q)}\right)\left(\frac{(p^\vartheta - q^\vartheta)(p^{\vartheta-1} - q^{\vartheta-1})}{(p-q)^2(p+q)}\right) h_{\mu-2,\nu-2}.
\end{aligned}
$$

It is clear that, the operator defined in (6.1) is linear and generates the μ, ν-th order difference functions sequences. Taking $\theta = \vartheta = 1$ and $0 < q < p \leq 1$, the operator (difference) defined by (6.1) reduces to ordinary double difference operator $\Delta^{1,1}(h_{\mu,\nu})$ defined by

$$\Delta^{1,1}(h_{\mu,\nu}) = h_{\mu,\nu} - h_{\mu,\nu-1} - h_{\mu-1,\nu} + h_{\mu-1,\nu-1} \quad (\forall\ \chi, \gamma \in E^2).$$

We now present an example associated with the difference operator $\Delta^{\theta,\vartheta}_{p,q}(h_{\mu,\nu})$.

Example 6.3 Consider a sequence of functions with domain $E \subseteq R^2$ by $(h_{\mu,\nu}(\chi,\gamma)) = \mu\chi^2 + \nu\gamma^2$ $(\mu, \nu \in \mathbb{N})$. Let $\theta = \vartheta = 2$ and $0 < q < p \leq 1$. Then, clearly the sequence $(h_{\mu,\nu})$ is not P-convergent in the ordinary sense. On the other side, for the operator $\Delta^{2,2}_{p,q}(h_{\mu,\nu})$, we have

$$
\begin{aligned}
\Delta^{2,2}_{p,q}(h_{\mu,\nu}) &= h_{\mu,\nu} - [2]_{p,q} h_{\mu,\nu-1} + [2]_{p,q}\frac{1}{p+q} h_{\mu,\nu-2} - [2]_{p,q} h_{\mu-1,\nu} \\
&\quad + [2]_{p,q}[2]_{p,q} h_{\mu-1,\nu-1} - [2]_{p,q}[2]_{p,q}\frac{1}{p+q} h_{\mu-1,\nu-2} \\
&\quad + [2]_{p,q}\frac{1}{p+q} h_{\mu-2,\nu} - [2]_{p,q}[2]_{p,q}\frac{1}{p+q} h_{\mu-2,\nu-1} \\
&\quad + [2]_{p,q}\frac{1}{p+q}[2]_{p,q}\frac{1}{p+q} h_{\mu-2,\nu-2} \\
&= (p+q)\{(\chi^2 + 3\gamma^2) - (p+q)(\chi^2 + \gamma^2)\}.
\end{aligned}
$$

Evidently, depending on the values of p and q, the difference functions sequence $\Delta^{2,2}_{p,q}(h_{\mu,\nu})$ has varying limits. This is basically due to the definition of (p, q)-integers. However, in view of obtaining a criterion for P-convergence, we must have to come out this difficult situation. Certainly, there are two aspects to get rid of these difficulties. One aspect is by taking $p = q = 1$ and consequently the operator reduces to the as usual difference sequence, and the second aspect is to replace by double sequences $p = p_{\mu,\nu}$ and $q = q_{\mu,\nu}$ such that

$$P - \lim_{\mu,\nu} q_{\mu,\nu} = \ell_1 \quad \text{and}$$
$$P - \lim_{\mu,\nu} p_{\mu,\nu} = \ell_2 \; (0 < \ell_1, \ell_2 \leq 1) \quad \text{and} \quad (0 < q_{\mu,\nu} < p_{\mu,\nu} \leq 1). \tag{6.2}$$

Afterwards, the difference sequence of functions $\Delta^{2,2}_{p,q}(h_{\mu,\nu})$ converges to the value $3 - (\vartheta^2 + \theta\vartheta + \theta^2)$. Thus, if one takes $q_\nu = \left(\frac{\nu+1}{\nu+1+s}\right) < \left(\frac{\nu+1}{\nu+1+t}\right) = p_\nu$, $0 < q_\nu < p_\nu \leq 1\,(s > t > 0)$, then $\lim_\nu q_\nu = 1 = \lim_\nu p_\nu$, and therefore, $\Delta^{[3]}_{p,q}(\chi_\nu) \to 0 \;(\nu \to \infty)$.

Remark 6.1 If $r = 1$, $\lim_n q_\nu = 1$ and $\lim_\nu p_\nu = 1$, then the operator $\Delta^{[r]}_{p,q}$ is same as the $\Delta^{[1]}$. Moreover, for $r = 0$, $\lim_\nu q_\nu = 1$ and $\lim_\nu p_\nu = 1$, the operator (difference) $\Delta^{[r]}_{p,q}$ recedes to the general sequence (χ_ν).

A finding on statistical approximation by sequences (double) of positive linear operators on modular space was produced by Orhan and Demirci [37] in 2015. Later, Demirci and Burccak [9] proposed the concept of positive linear operators achieving A-statistical relative modular convergence. Recent findings on statistical relative approximation on modular spaces were established by Demirci and Orhan [10]. Deferred weighted statistical convergence was used by Srivastava et al. [46] quite recently to obtain certain approximation results on the Banach space. Again, Srivastava et al. [47] further introduced deferred weighted equi-statistical convergence to prove certain approximation results. For certain related current works in this direction, see [22–24, 31, 38–40, 42–45].

The above-mentioned works, motivated us to introduce here the concept of relative modular deferred weighted statistical convergence and statistical relative modular deferred weighted summability for a double sequence of functions and build an inclusion connection between them.

6.2 Relative Modular Deferred Weighted Mean

In the year 1932, Agnew (see below [2]) defined deferred Cesàro mean of sequences of real numbers such as:

$$(D_{a,b}\chi)_\nu = \frac{1}{b_\nu - a_\nu} \sum_{k=a_\nu+1}^{b_\nu} \chi_k,$$

where (a_ν) and (b_ν) are sequences of integers (non-negative) for which

$$a_\nu < b_\nu \;(\nu \in \mathbb{N}) \quad \text{and} \quad \lim_{\nu\to\infty} b_\nu = +\infty.$$

In the year 2016, üçükaslan and Yılmaztürk (see [30]) proposed the idea of deferred statistical convergence. Accordingly, Let (a_ν) and (b_ν) be two non-negative integers sequence, a sequence (χ_k) is deferred statistically convergent to L if, for every $\epsilon > 0$,

$$\frac{1}{b_\nu - a_\nu}|\{a_\nu < k \leqq b_\nu \ : \ |\chi_k - L| \geqq \epsilon\}| = 0.$$

After then several works are carried out with regard to this concept (see below [12, 20, 21, 25]).

Recently, in the year 2018, Srivastava et al. [46] explored the class of weighted statistical convergence connecting to some Korovkin-type approximation theorems. Subsequently, Mishra et al. [32] considered (p, q)-integers and demonstrated a new class of statistical deferred weighted A-summability and proved some associated approximation theorems. Very recently, Et et al. [16] established Δ^{μ}–weighted statistical convergence, and later on in the year 2022, Et et al. [15] also demonstrated the generalized deferred statistical convergence. Again, in the same year, Et et al. [13] established a result on deferred statistical boundedness of order θ.

From the above literature, we now define the new deferred weighted summability mean $(\mathfrak{D}_{\mathfrak{N}}\,(h_{\mu,\nu}))$ for the double sequence $(h_{\mu,\nu})$ of functions under (p, q)-integers,

$$\mathfrak{D}_{\mathfrak{N}}\left(\Delta_{p,q}^{\theta,\vartheta}(h_{\mu,\nu})\right) = \frac{1}{\mathscr{P}_{\mu}\mathscr{Q}_{\nu}} \sum_{u,v=a_\nu+1}^{b_\mu,b_\nu} t_u s_v(1 + r_1^u r_2^v)\Delta_{p,q}^{\theta,\vartheta}(h_{\mu,\nu})(\chi, \gamma), \quad (0 < r_1, r_2 < 1) \tag{6.3}$$

where (s_ν) and (t_ν) are the non-negative real number sequences satisfying

$$\mathscr{P}_{\nu} = \sum_{v=a_\nu+1}^{b_\nu} s_v \text{ and } \mathscr{Q}_{\mu} = \sum_{u=a_\nu+1}^{b_\mu} t_v.$$

Definition 6.1 Let (a_ν) and (b_ν) be non-negative integers sequence and let θ and ϑ be two non-negative integers. Also, let $p = (p_{\mu,\nu})$ and $q = (q_{\mu,\nu})$ be two double sequences of real numbers satisfying the property (6.2). A double sequence $(h_{\mu,\nu})$ of functions in $L^{\omega}(\mathscr{J}^2)$ is *relatively modularly deferred weighted* $(\mathfrak{D}_{\mathfrak{N}}\,(h_{\mu,\nu}))$-*summable* to a function h on $L^{\omega}(\mathscr{J}^2)$ under the difference operator (generalized) based on (p, q)-integers iff there is a scale function (non-negative) $\sigma \in X(\mathscr{J}^2)$ for which

$$P - \lim_{\mu,\nu\to\infty} \omega\left(\lambda\left(\frac{\mathfrak{D}_{\mathfrak{N}}\,(\Delta_{p,q}^{\theta,\vartheta}(h_{\mu,\nu})) - h}{\sigma}\right)\right) = 0 \text{ for certain } \lambda_0 > 0.$$

In this case, we write

$$\mathfrak{D}_{\mathfrak{N}} \lim_{\mu,\nu} \left\| \frac{(\Delta^{\theta,\vartheta}_{p,q}(h_{\mu,\nu})) - h}{\sigma} \right\|_{\omega} = 0 \text{ for some } \lambda_0 > 0.$$

Definition 6.2 Let (a_ν) and (b_ν) be non-negative integers sequence and let θ and ϑ be two integers (non-negative). Also let $p = (p_{\mu,\nu})$ and $q = (q_{\mu,\nu})$ be two double sequences of real numbers satisfying the property (6.2). A functions sequence $(h_{\mu,\nu})$ in $L^{\omega}(\mathscr{J}^2)$ is F-norm relatively deferred weighted summable (or strongly relatively deferred weighted summable) to h on $L^{\omega}(\mathscr{J}^2)$ under the difference operator (generalized) based on (p, q)-integers iff

$$P - \lim_{\mu,\nu\to\infty} \omega \left(\lambda \left(\frac{\mathfrak{D}_{\mathfrak{N}} \, (\Delta^{\theta,\vartheta}_{p,q}(h_{\mu,\nu})) - h}{\sigma} \right) \right) = 0 \text{ for certain } \lambda > 0.$$

In this case, we write

$$\mathscr{F} - \mathfrak{D}_{\mathfrak{N}} \lim_{\mu,\nu} \left\| \frac{(\Delta^{\theta,\vartheta}_{p,q}(h_{\mu,\nu})) - h}{\sigma} \right\|_{\omega} = 0 \text{ for some } \lambda_0 > 0.$$

It is instantly seen that, Definition 6.1 and Definition 6.2 are the same subject to the existence of the modular ω under the Δ_2-condition. That is to say that, there is a constant $\mathfrak{M} > 0$ for which $\omega(2h) \leq \mathfrak{M}\omega(h)$ for every $h \in X(\mathscr{J}^2)$. Precisely, relatively strong summability of the double sequence $(h_{\mu,\nu})$ to h under the difference operator (generalized) based on (p, q)-integers is identical to the condition

$$P - \lim_{\mu,\nu} \omega \left(2^{\nu}\lambda \left(\frac{\mathfrak{D}_{\mathfrak{N}} \, (\Delta^{\theta,\vartheta}_{p,q}(h_{\mu,\nu})) - h}{\sigma} \right) \right) = 0$$

for all $\nu \in \mathbb{N}$ and certain $\lambda > 0$. Thus, if $(h_{\mu,\nu})$ is relatively modularly deferred weighted $(\mathfrak{D}_{\mathfrak{N}} \, (h_{\mu,\nu}))$-summable to h under the generalized difference operator based on (p, q)-integers, then by Definition 1, there exists a $\lambda > 0$ such that

$$P - \lim_{\mu,\nu} \omega \left(\lambda \left(\frac{\mathfrak{D}_{\mathfrak{N}} \, (\Delta^{\theta,\vartheta}_{p,q}(h_{\mu,\nu})) - h}{\sigma} \right) \right) = 0.$$

Clearly, under Δ_2-condition, it implies that

$$\omega \left(2^{\nu}\lambda \left(\frac{\mathfrak{D}_{\mathfrak{N}} \, (\Delta^{\theta,\vartheta}_{p,q}(h_{\mu,\nu})) - h}{\sigma} \right) \right) \leqq \mathfrak{M}^{\nu} \omega \left(\lambda \left(\frac{\mathfrak{D}_{\mathfrak{N}} \, (\Delta^{\theta,\vartheta}_{p,q}(h_{\mu,\nu})) - h}{\sigma} \right) \right).$$

Therefore, we obtain

$$P - \lim_{\mu,\nu} \omega \left(2^{\nu}\lambda \left(\frac{\mathfrak{D}_{\mathfrak{N}} \, (\Delta_{p,q}^{\theta,\vartheta}(h_{\mu,\nu})) - h}{\sigma} \right) \right) = 0.$$

Definition 6.3 Let (a_ν) and (b_ν) be non-negative integers sequence and let θ and ϑ be two non-negative integers. Also, let $p = (p_{\mu,\nu})$ and $q = (q_{\mu,\nu})$ be two double sequences of real numbers satisfying the property (6.2). A sequence (double) $(h_{\mu,\nu})$ of functions in $L^{\omega}(\mathscr{J}^2)$ is modular relative deferred weighted $(\mathfrak{D}_{\mathfrak{N}}(h_{\mu,\nu}))$ statistical convergent to a function $h \in L^{\omega}(\mathscr{J}^2)$ under the generalized difference operator based on (p, q)-integers, if there exists a scale function (non-zero) $\sigma \in X(\mathscr{J}^2)$ for which, for each $\epsilon > 0$, the set

$$P - \lim_{\mu,\nu} \frac{1}{\mathscr{P}_\mu \mathscr{Q}_\nu} \Bigg\{ (u, v) : u \leqq \mathscr{P}_\mu, \, v \leqq \mathscr{Q}_\mu \text{ and } \omega \left(\lambda_0 \left(\frac{t_u s_v (1 + r_1^u r_2^v) |(\Delta_{p,q}^{\theta,\vartheta}(h_{u,v})) - h|}{\sigma} \right) \right) \geqq \epsilon \Bigg\}$$

ensures relative deferred weighted density zero, that means,

$$P - \lim_{\mu,\nu} \frac{1}{\mathscr{P}_\mu \mathscr{Q}_\nu} \left| \Bigg\{ (u, v) : u \leqq \mathscr{P}_\mu, \, v \leqq \mathscr{Q}_\mu \text{ and } \omega \left(\lambda_0 \left(\frac{t_u s_v (1 + r_1^u r_2^v) |(\Delta_{p,q}^{\theta,\vartheta}(h_{u,v})) - h|}{\sigma} \right) \right) \geqq \epsilon \Bigg\} \right| = 0$$

for certain $\lambda > 0$.
In this case, we write

$$\mathrm{stat}_{\mathfrak{D}_{\mathfrak{N}}} \lim_{\mu,\nu} \left\| \frac{(\Delta_{p,q}^{\theta,\vartheta}(h_{\mu,\nu})) - h}{\sigma} \right\|_{\omega} = 0.$$

Moreover, $(h_{\mu,\nu})$ is r F-norm elatively deferred weighted $(\mathfrak{D}_{\mathfrak{N}} \, (h_{\mu,\nu}))$ statistical convergent (or strongly relatively deferred weighted $(\mathfrak{D}_{\mathfrak{N}} \, (h_{\mu,\nu}))$- statistical convergent) to a function $h \in X(\mathscr{J}^2)$ under the difference operator (generalized) based on (p, q)-integers iff

$$P - \lim_{\mu,\nu} \frac{1}{\mathscr{P}_\mu \mathscr{Q}_\nu} \left| \Bigg\{ (u, v) : u \leqq \mathscr{P}_\mu, \, v \leqq \mathscr{Q}_\mu \text{ and } \omega \left(\lambda_0 \left(\frac{t_u s_v (1 + r_1^u r_2^v) |(\Delta_{p,q}^{\theta,\vartheta}(h_{u,v})) - h|}{\sigma} \right) \right) \geqq \epsilon \Bigg\} \right| = 0$$

for certain $\lambda > 0$, where $\sigma \in X(\mathscr{J}^2)$ is a scale function (non-zero) and $\epsilon > 0$ (arbitrarily small).
We write here

$$\mathscr{F} - \text{stat}_{\mathfrak{D}_{\mathfrak{N}}} \lim_{\mu,\nu} \left\| \frac{(\Delta_{p,q}^{\theta,\vartheta}(h_{u,v})) - h}{\sigma} \right\|_{\omega} = 0.$$

Definition 6.4 Let (a_ν) and (b_ν) be sequences of non-negative integers and let θ and ϑ be two integers (non-negative). Let $p = (p_{\mu,\nu})$ and $q = (q_{\mu,\nu})$ be the two double sequences of real numbers satisfying the property (6.2). A double sequence $(h_{\mu,\nu})$ of functions in $L^{\omega}(\mathscr{J}^2)$ is statistical relative modular deferred weighted ($\mathfrak{D}_{\mathfrak{N}}$ $(h_{\mu,\nu})$)-summable to a function $h \in L^{\omega}(\mathscr{J}^2)$ under the generalized difference operator based on (p, q)-integers, if there is a scale function (non-zero) $\sigma \in X(\mathscr{J}^2)$ for which, for each $\epsilon > 0$, the set

$$P - \lim_{\mu,\nu} \frac{1}{\mu,\nu} \Bigg\{ (u, v) : u \leqq \mu, v \leqq \nu \text{ and } \omega\left(\lambda_0 \left(\frac{\mathfrak{D}_{\mathfrak{N}} \, (\Delta_{p,q}^{\theta,\vartheta}(h_{\mu,\nu})) - h}{\sigma}\right)\right) \geqq \epsilon \Bigg\} \text{ for some } \lambda_0 > 0$$

ensures relative deferred weighted density zero, that means,

$$P - \lim_{\mu,\nu} \frac{1}{\mu\nu} \left| \Bigg\{ (u, v) : u \leqq \mu, v \leqq \nu \text{ and } \omega\left(\lambda_0 \left(\frac{\mathfrak{D}_{\mathfrak{N}}(\Delta_{p,q}^{\theta,\vartheta}(h_{\mu,\nu})) - h}{\sigma}\right)\right) \geqq \epsilon \Bigg\} \right| = 0 \text{ for some } \lambda_0 > 0.$$

, we write

$$\mathfrak{D}_{\mathfrak{N}} \text{ stat } \lim_{\mu,\nu} \left\| \frac{(\Delta_{p,q}^{\theta,\vartheta}(h_{u,v})) - h}{\sigma} \right\|_{\omega} = 0.$$

Furthermore, $(h_{\mu,\nu})$ is statistical relative F-norm (locally convex) deferred weighted ($\mathfrak{D}_{\mathfrak{N}}$ $(h_{\mu,\nu})$)-summable (or statistical relative strong deferred weighted ($\mathfrak{D}_{\mathfrak{N}}$ $(h_{\mu,\nu})$)-summable) to a function $h \in X(\mathscr{J}^2)$ under the difference operator (generalized) based on (p, q)-integers iff

$$P - \lim_{\mu,\nu} \frac{1}{\mu,\nu} \left| \Bigg\{ (u, v) : u \leqq \mu, v \leqq \nu \text{ and } \omega\left(\lambda_0 \left(\frac{\mathfrak{D}_{\mathfrak{N}} \, (\Delta_{p,q}^{\theta,\vartheta}(h_{\mu,\nu})) - h}{\sigma}\right)\right) \geqq \epsilon \Bigg\} \right| = 0 \text{ for certain } \lambda > 0,$$

where $\sigma \in X(\mathscr{J}^2)$ is a scale function (non-zero) and $\epsilon > 0$ (arbitrarily small).

In this case, we write

$$\mathscr{F} - \mathfrak{D}_{\mathfrak{N}} \text{ stat } \lim_{\mu,\nu} \left\| \frac{(\Delta_{p,q}^{\theta,\vartheta}(h_{\mu,\nu})) - h}{\sigma} \right\|_{\omega} = 0.$$

Remark 6.2 If we substitute $a_\nu = 0, b_\nu = \nu,\ b_\mu = \mu$ and $t_\mu = s_\nu = 1$ in the Definition 3, then it successfully reduces to relative modular statistical convergence (see [10]).

Next, we need the similar assumptions as in [9] on a modular space for our present study.

Based upon the above assumptions, it is observed that if a modular ω is monotone and finite, then we have $C(\mathscr{J}^2) \subset L^\omega(\mathscr{J}^2)$. Furthermore, if ω is monotone and strongly finite, then $C(\mathscr{J}^2) \subset E^\omega(\mathscr{J}^2)$. Also, if ω is monotone, absolute finite and absolute continuous, then $\overline{C^\infty(\mathscr{J}^2)} = L^\omega(\mathscr{J}^2)$, where closure $\overline{C^\infty(\mathscr{J}^2)}$ is compact with respect to the modular space.

Now we propose the following inclusion relations connecting relative deferred weighted statistical convergence and statistical relative deferred weighted summability by using difference operator (generalized) based on (p, q)-integers in a modular space.

Theorem 6.1 *Let (a_ν) and (b_ν) be sequences of non-negative integers and let θ and ϑ be non-negative integers. Also, let $p = (p_{\mu,\nu})$ and $q = (q_{\mu,\nu})$ be the two sequences of real numbers satisfying the property (6.2). Let ω be a monotone, strongly finite and $\mathscr{N}$-quasi convex modular on $L^\omega(\mathscr{J}^2)$. If a double sequence $(h_{\mu,\nu})$ of functions belongs to $L^\omega(\mathscr{J}^2)$ is bounded and relatively modular deferred weighted statistically convergent to the function $h \in L^\omega(\mathscr{J}^2)$ based on (p, q)-integers, then it is statistically relative modular deferred weighted summable to the function h based on (p, q)-integers, but the converse statement is not usually true.*

Proof Assume that $\Delta_{p,q}^{\theta,\vartheta}(h_{\mu,\nu}) \in L^\omega(\mathscr{J}^2) \cap \ell_\infty$. We set

$$\mathscr{G}_\epsilon = \Bigg\{(u, v) : u \leqq \mu, v \leqq \nu$$
$$\text{and } \omega\left(\lambda_0\left(\frac{\Delta_{p,q}^{\theta,\vartheta}(h_{u,v}) - h}{\sigma}\right)\right) \geqq \epsilon \text{ for some } \lambda_0 > 0\Bigg\}$$

and

$$\mathscr{G}_\epsilon^c = \Bigg\{(u, v) : u \leqq \mu, v \leqq \nu$$
$$\text{and } \omega\left(\lambda_0\left(\frac{\Delta_{p,q}^{\theta,\vartheta}(h_{u,v}) - h}{\sigma}\right)\right) > \epsilon \text{ for some } \lambda_0 > 0\Bigg\}.$$

From our proposed mean, we have

$$P-\lim_{u,v}\frac{1}{\mathscr{P}_{\mu}\mathscr{Q}_{\nu}}\sum_{u,v=a_{\nu}+1}^{b_{\mu},b_{\nu}}t_u s_v(1+r_1^u r_2^v)=0. \tag{6.4}$$

Thus, we obtain

$$\begin{aligned}
&\omega\left(\lambda_0\left(\mathfrak{D}_{\mathfrak{N}}\left(\frac{\Delta^{\theta,\vartheta}_{p,q}(h_{\mu,\nu})-h}{\sigma}\right)\right)\right)\\
&=\omega\left(\lambda_0\left(\frac{1}{\mathscr{P}_{\mu}\mathscr{Q}_{\nu}}\sum_{u,v=a_{\nu}+1}^{b_{\mu},b_{\nu}}t_u s_v(1+r_1^u r_2^v)\left(\frac{\Delta^{\theta,\vartheta}_{p,q}(h_{u,v})-h}{\sigma}\right)\right)\right)\\
&\leqq\omega\Bigg(\frac{\lambda_0}{\mathscr{P}_{\mu}\mathscr{Q}_{\nu}}\sum_{\substack{u,v=a_{\nu}+1,\\(u,v)\in\mathscr{G}_{\epsilon}}}^{b_{\mu},b_{\nu}}t_u s_v(1+r_1^u r_2^v)|\Delta^{\theta,\vartheta}_{p,q}(h_{u,v})-h|\\
&\quad+\frac{\lambda_0}{\mathscr{P}_{\mu}\mathscr{Q}_{\nu}}\sum_{\substack{u=0,v=b_{\nu}+1,\\(u,v)\in\mathscr{G}_{\epsilon}}}^{b_{\mu},\infty}t_u s_v(1+r_1^u r_2^v)\left|\frac{\Delta^{\theta,\vartheta}_{p,q}(h_{u,v})-h}{\sigma}\right|\\
&\quad+\frac{\lambda_0}{\mathscr{P}_{\mu}\mathscr{Q}_{\nu}}\sum_{\substack{u=b_{\mu}+1,v=0,\\(u,v)\in\mathscr{G}_{\epsilon}}}^{\infty,b_{\nu}}t_u s_v(1+r_1^u r_2^v)\left|\frac{\Delta^{\theta,\vartheta}_{p,q}(h_{u,v})-h}{\sigma}\right|\\
&\quad+\frac{\lambda_0}{\mathscr{P}_{\mu}\mathscr{Q}_{\nu}}\sum_{\substack{u=b_{\mu}+1,v=b_{\nu}+1,\\(u,v)\in\mathscr{G}_{\epsilon}}}^{\infty,\infty}t_u s_v(1+r_1^u r_2^v)\left|\frac{\Delta^{\theta,\vartheta}_{p,q}(h_{u,v})-h}{\sigma}\right|\\
&\quad+\omega\Bigg(\frac{\lambda_0}{\mathscr{P}_{\mu}\mathscr{Q}_{\nu}}\sum_{\substack{u,v=a_{\nu}+1,\\(u,v)\in\mathscr{G}_{\epsilon}^c}}^{b_{\mu},b_{\nu}}t_u s_v(1+r_1^u r_2^v)\left|\frac{\Delta^{\theta,\vartheta}_{p,q}(h_{u,v})-h}{\sigma}\right|\\
&\quad+\frac{\lambda_0}{\mathscr{P}_{\mu}\mathscr{Q}_{\nu}}\sum_{\substack{u=0,v=b_{\nu}+1,\\(u,v)\in\mathscr{G}_{\epsilon}^c}}^{b_{\mu},\infty}t_u s_v(1+r_1^u r_2^v)\left|\frac{\Delta^{\theta,\vartheta}_{p,q}(h_{u,v})-h}{\sigma}\right|\\
&\quad+\frac{\lambda_0}{\mathscr{P}_{\mu}\mathscr{Q}_{\nu}}\sum_{\substack{u=b_{\mu}+1,v=0;\\(u,v)\in\mathscr{G}_{\epsilon}^c}}^{\infty,b_{\nu}}t_u s_v(1+r_1^u r_2^v)\left|\frac{\Delta^{\theta,\vartheta}_{p,q}(h_{u,v})-h}{\sigma}\right|\\
&\quad+\frac{\lambda_0}{\mathscr{P}_{\mu}\mathscr{Q}_{\nu}}\sum_{\substack{u=b_{\mu}+1,v=b_{\nu}+1\\(u,v)\in\mathscr{G}_{\epsilon}^c}}^{\infty,\infty}t_u s_v(1+r_1^u r_2^v)\left|\frac{\Delta^{\theta,\vartheta}_{p,q}(h_{u,v})-h}{\sigma}\right|
\end{aligned}$$

$$+\mathscr{K}\left|\frac{1}{\mathscr{P}_\mu\mathscr{Q}_\nu}\sum_{u,v=a_\nu+1}^{\infty,\infty} t_u s_v(1+r_1^u r_2^v)-1\right|\Bigg),$$

where $\mathscr{K}=\sup_{\chi,\gamma}|\Delta_{p,q}^{\theta,\vartheta}h(\chi,\gamma)|$. Further, ω being monotone, strongly finite and $\mathscr{N}$-quasi convex modular on $L^\omega(\mathscr{J}^2)$, it follows that

$$\begin{aligned}
&\omega\left(\lambda_0\left(\mathfrak{D}_{\mathfrak{N}}\left(\frac{\Delta_{p,q}^{\theta,\vartheta}(h_{\mu,\nu})-h}{\sigma}\right)\right)\right)\\
&\leqq 3\omega\left(\frac{9\lambda_0|\mathscr{G}_\epsilon|G}{\mathscr{P}_\mu\mathscr{Q}_\nu}\sum_{\substack{u,v=a_\nu+1,\\(u,v)\in\mathscr{G}_\epsilon}}^{b_\mu,b_\nu} t_u s_v(1+r_1^u r_2^v)\right)\\
&+\epsilon\omega\left(\frac{9\lambda_0|\mathscr{G}_\epsilon|}{\mathscr{P}_\mu\mathscr{Q}_\nu}\sum_{\substack{u,v=a_\nu+1,\\(u,v)\in\mathscr{G}_\epsilon}}^{b_\mu,b_\nu} t_u s_v(1+r_1^u r_2^v)\right)\\
&+\omega\left(\frac{9\lambda_0 G b_\mu b_\nu}{\mathscr{P}_\mu\mathscr{Q}_\nu}\sum_{u,v=a_\nu+1}^{b_\mu,b_\nu} t_u s_v(1+r_1^u r_2^v)\right)\\
&+\omega\left(\frac{9\lambda_0 G b_\mu}{\mathscr{P}_\mu\mathscr{Q}_\nu}\sum_{u=0,v=a_\nu+1}^{b_\mu,\infty} t_u s_v(1+r_1^u r_2^v)\right)\\
&+\omega\left(\frac{9\lambda_0 G b_\nu}{\mathscr{P}_\mu\mathscr{Q}_\nu}\sum_{u=a_\nu+1,v=0}^{\infty,b_\mu} t_u s_v(1+r_1^u r_2^v)\right)\\
&+\epsilon\omega\left(\frac{9\lambda_0}{\mathscr{P}_\mu\mathscr{Q}_\nu}\sum_{u,v=a_\nu+1}^{\infty,\infty} t_u s_v(1+r_1^u r_2^v)\right)\\
&+\omega\left(\frac{9\lambda_0\mathscr{K}}{\mathscr{P}_\mu\mathscr{Q}_\nu}\sum_{u,v=a_\nu+1}^{\infty,\infty} t_u s_v(1+r_1^u r_2^v)-1\right),
\end{aligned}$$

where $G=\max\left|\frac{\Delta_{p,q}^{\theta,\vartheta}h_{u,v}-h(\chi,\gamma)}{\sigma}\right|$ for all $u,v\in\mathbb{N}$ and $(\chi,\gamma)\in\mathscr{J}^2$. Now considering the P-limit as $\mu,\nu\to\infty$ to the last inequality, and using (6.4) and under the conditions of regularity of deferred weighted mean, we certainly obtain

$$P-\lim_{\mu,\nu}\omega\left(\lambda_0\left(\frac{\mathfrak{D}_{\mathfrak{N}}\,(\Delta_{p,q}^{\theta,\vartheta}(h_{\mu,\nu}))-h}{\sigma}\right)\right)=0.$$

This means that $(h_{\mu,\nu})$ is relatively modular deferred weighted $\mathfrak{D}_{\mathfrak{N}}\,(h_{\mu,\nu})$-summable to the function h based on (p,q)-integers. Hence,

$$P-\lim_{\mu,\nu}\frac{1}{\mu,\nu}\left|\left\{(u,v):u\leqq\mu,v\leqq\mu\right.\right.$$
$$\left.\left.\text{and }\omega\left(\lambda_0\left(\frac{\mathfrak{D}_{\mathfrak{N}}\left(\Delta_{p,q}^{\theta,\vartheta}(h_{\mu,\nu})\right)-h}{\sigma}\right)\right)\geqq\epsilon\right\}\right|=0\ \text{ for some }\lambda_0>0.$$

For the converse statement, we illustrate the following Example.

Example 6.4 Let $\mathscr{J}=[0,1]$ and assume that $\varphi:[0,\infty)\rightarrow[0,\infty)$ is a continuous function and $\varphi(0)=0$, $\varphi(u)>0$, $u>0$ with $\lim_{u\to\infty}\varphi(u)=\infty$.
Then, for every real-valued function (measurable) $h\in X(\mathscr{J}^2)$, we may consider the functional ω^{φ} on $X(\mathscr{J}^2)$ of the form

$$\omega^{\varphi}(h)=\int_0^1\int_0^1\varphi(|h_{\mu,\nu}(\chi,\gamma)|)d\chi d\gamma\quad h\in X(\mathscr{J}^2).$$

In this case, φ being convex, then so also ω^{φ} is convex modular on $X(\mathscr{J}^2)$, which evidently satisfies all the assumptions listed above in this section. Next, considering the Orlicz space generated by φ in the following form:

$$L_{\varphi}^{\omega}(\mathscr{J}^2)=\{h\in X(\mathscr{J}^2):\omega^{\varphi}(\lambda(h))<+\infty\ \text{ for some }\lambda>0\}.$$

Let $(h_{\mu,\nu})$ be a sequence (double) of functions. For each $\mu,\nu\in\mathbb{N}$, define $h_{\mu,\nu}$: $[0,1]\times[0,1]\rightarrow\mathbb{R}$ by,

$$h_{\mu,\nu}(\chi)=\begin{cases}1, & (\mu,\nu)\in U\times U\text{ and }(\chi,\gamma)\in(0,\frac{1}{\mu}]\times(0,\frac{1}{\nu}],\\[2ex] 0, & \{(\mu,\nu)\in V\times V\text{ and }(\chi,\gamma)\in(\frac{1}{\mu},0]\times(\frac{1}{\nu},1];\\ & (\mu,\nu)\in U\times V\text{ or }(\mu,\nu)\in V\times U\text{ or }(\chi,\gamma)\in(0,0)\},\end{cases}$$

where U and V are the corrending set of odd and even numbers. Since

$$\omega\lambda(\mathfrak{D}_{\mathfrak{N}}(h_{\mu,\nu}))=\omega\left(\frac{\lambda_0}{\mathscr{P}_{\mu}\mathscr{Q}_n}\sum_{u,v=a_n+1}^{b_{\mu},b_n}t_u s_v(1+r_1^u r_2^v)\right)$$

$$=\lambda_0\begin{cases}\int_0^{1/b_{\mu}}\int_0^{1/b_{\nu}}d\chi d\gamma, & (\mu,\nu)\in U\times U\text{ and }(\chi,\gamma)\in(0,\frac{1}{\mu}]\times(0,\frac{1}{\nu}],\\[2ex] 0, & \{(\mu,\nu)\in V\times V\text{ and }(\chi,\gamma)\in(\frac{1}{\mu},0]\times(\frac{1}{\nu},1];\\[2ex] & (\mu,\nu)\in U\times V\text{ or }(\mu,\nu)\in V\times U\text{ or }(\chi,\gamma)\in(0,0)\},\end{cases}$$

then $(h_{\mu,\nu})$ is modularly relatively deferred weighted summable to $h = 0$, if there exists a scale function (non-zero) $\sigma(\chi, \gamma)$, such that

$$\sigma(\chi, \gamma) = \begin{cases} 1, & (\chi, \gamma) = (0, 0) \\ \frac{1}{\chi\gamma}, & (\chi, \gamma) \in (0, 1] \times (0, 1]. \end{cases}$$

That is,

$$P - \lim_{\mu,\nu} \omega\left(\lambda_0 \left(\frac{\mathfrak{D}_{\mathfrak{N}}(h_{\mu,\nu}) - h}{\sigma}\right)\right) = 0 \text{ for certain } \lambda_0 > 0.$$

Hence, it implies that

$$P - \lim_{\mu,\nu} \frac{1}{\mu, \nu} \left|\left\{(u, v) : u \leqq \mu, v \leqq \nu \right.\right.$$
$$\left.\left. \text{and } \omega\left(\lambda_0 \left(\frac{\mathfrak{D}_{\mathfrak{N}}(h_{\mu,\nu}) - h}{\sigma}\right)\right) \geqq \epsilon\right\}\right| = 0 \text{ for some } \lambda_0 > 0.$$

On the other side, it is not modularly relatively deferred weighted statistical convergent to the function $h = 0$, that means

$$P - \lim_{\mu,\nu} \frac{1}{\mathscr{P}_\mu \mathscr{S}_\nu} \left|\left\{(u, v) : u \leqq \mathscr{P}_\mu, v \leqq \mathscr{Q}_\mu \right.\right.$$
$$\left.\left. \text{and } \omega\left(\lambda_0 \left(\frac{t_u s_v \mathfrak{D}_{\mathfrak{N}} |h_{u,v} - h|}{\sigma}\right)\right) \geqq \epsilon\right\}\right| \neq 0$$

for some $\lambda_0 > 0$.

6.3 Concluding Remarks

Through this study, we have preluded the conception deferred weighted summability mean in the modular space for the generalized double sequences of functions associated with (p, q)-integers. Next, we have established an inclusion relation by connecting relative modular deferred weighted statistical convergence and statistical relative modular deferred weighted summability for a double sequence of functions in association with the (p, q)-integers in this space. Moreover, we have considered certain illustrative examples in support of our findings.

Many researchers have considered different summability means on the sequence spaces to prove several approximation results. A list of some articles has been mentioned in the references. Further, combining the existing ideas and direction of the sequence spaces associated with our proposed mean, many new Korvokin-type approximation theorems can be proved under different settings of algebraic and trigonometric functions.

Influenced by recently published articles by Jena et al. [19, 20], we extract the cognizance of the interested learners concerning the possibilities of establishing some Korvokin-type approximation theorems via deferred Cesàro mean in the usual sequence space as well as the probability space. Also, in view of the latest result of Hazarika et al. [26], the consciousness of the curious readers is drawn out for future researches pertaining to fuzzy approximation theorems.

References

1. Acar, T., Aral, A., Mohiuddine, S.A.: On Kantorovich modifications of (p, q)-Baskakov operators. J. Inequal. Appl. **2016**(98), 1–16 (2016)
2. Agnew, R.P.: On deferred Cesàro means. Ann. Math. **33**, 413–421 (1932)
3. Agrawal, P.N., Kumar, A.S., Sinha, T.A.K.: Stancu-type generalization of modified Schurer operators based on q-integers. Appl. Math. Comput. **226**, 765–776 (2014)
4. Aral, A., Gupta, V.: On q-analogue of Stancu-beta operators. Appl. Math. Lett. **25**, 67–71 (2012)
5. Braha, N.L.: Geometric properties of the second-order Cesàro spaces. Banach J. Math. Anal. **10**, 1–14 (2016)
6. Bardaro, C., Mantellini, I.: A Korovkin theorem in multivariate modular function spaces. J. Funct. Spaces Appl. **7**, 105–120 (2009)
7. Bardaro, C., Musielak, J., Vinti, G.: Nonlinear Integral Operators and Applications, de Gruyter Series in Nonlinear Analysis and Appl, vol. 9. Walter de Gruyter Publ, Berlin (2003)
8. Chittenden, E.W.: On the limit functions of sequences of continuous functions converging relatively uniformly. Trans. AMS **20**, 179–184 (1919)
9. Demirci, K., Burçak, K.: A-statistical relative modular convergence of positive linear operators. Positivity **21**, 847–863 (2016)
10. Demirci, K., Orhan, S.: Statistical relative approximation on modular spaces. Results Math. **71**, 1167–1184 (2017)
11. Dutta, H., Paikray, S.K., Jena, B.B.: On statistical deferred Cesàro summability, in *Current Trends in Mathematical Analysis and Its Interdisciplinary Applications*, eds. by Dutta, H., Kocinac, L.D.R., Srivastava, H.M. (Birkhäuser, Cham, 2019), pp. 885–909
12. Et, M., Baliarsingh, P., Kandemir, H.S., Küçükaslan, M.: On μ-deferred statistical convergence and strongly deferred summable functions. Rev. R. Acad. Cienc. Exactas Fís. Nat. Ser. A Mat. (RACSAM) **115**, 1–14 (2021)
13. Et, M., Bhardwaj, V.K., Gupta, S.: On deferred statistical boundedness of order α. Commun. Stat. Theory. Methods **51**, 8786–8798 (2022)
14. Et, M., Çolak, R.: On some generalized difference sequence spaces. Soochow J. Math. **21**, 377–386 (1995)
15. Et, M., Kandemir, H.S., Cinar, M.: Generalized deferred statistical convergence, in *Approximation Theory, Sequence Spaces and Applications* eds. by Mohiuddine, S.A., Hazarika, B., Nashine, H.K (Springer Nature Singapore Pte Ltd., 2022), pp. 35–52
16. Et, M., Kandemir, H.S., Çakallı, H.: Δ^m-weighted statistical convergence. AIP Conf. Proceed. **2334**(040005), 1–5 (2021)
17. Et, M., Nuray, F.: Δ^m statistical convergence. Indian J. Pure Appl. Math. **32**, 961–969 (2001)
18. Fast, H.: Sur la convergence statistique. Colloq. Math. **2**, 241–244 (1951)
19. Jena, B.B., Paikray, S.K., Mursaleen, M.: Uniform convergence of Fourier series via deferred Cesàro mean and its applications. Math. Methods Appl. Sci. **41**, 5286–5298 (2023)

20. Jena, B.B., Paikray, S.K., Dutta, H.: Statistically Riemann integrable and summable sequence of functions via deferred Cesàro mean. Bull. Iran. Math. Soc. **48**, 1293–1309 (2022)
21. Jena, B.B., Paikray, S.K.: Product of deferred Cesàro and deferred weighted statistical probability convergence and its applications to Korovkin-type theorems. Univ. Sci. **25**, 409–433 (2020)
22. Jena, B.B., Paikray, S.K.: Product of statistical probability convergence and its applications to Korovkin-type theorem. Miskolc. Math. Notes **20**, 969–984 (2019)
23. Jena, B.B., Paikray, S.K., Dutta, H.: On various new concepts of statistical convergence for sequences of random variables via deferred Cesàro mean. J. Math. Anal. Appl. **487**(123950), 1–18 (2020)
24. Jena, B.B., Paikray, S.K., Mohiuddine, S.A., Mishra, V.N.: Relatively equi-statistical convergence via deferred Nörlund mean based on difference operator of fractional-order and related approximation theorems. AIMS Math. **5**, 650–672 (2020)
25. Jena, B.B., Paikray, S.K., Misra, U.K.: Statistical deferred Cesàro summability and its applications to approximation theorems. Filomat **32**, 2307–2319 (2018)
26. Hazarika, B., Alotaibi, A., Mohiuddine, S.A.: Statistical convergence in measure for double sequences of fuzzy-valued functions. Soft Comput. **24**, 6613–6622 (2020)
27. Kadak, U.: On weighted statistical convergence based on (p, q)-integers and related approximation theorems for functions of two variables. J. Math. Anal. Appl. **443**, 752–764 (2016)
28. Kadak, U.: Modularly weighted four dimensional matrix summability with application to Korovkin type approximation theorem. J. Math. Anal. Appl. **468**, 38–63 (2018)
29. Kızmaz, H.: On certain sequence spaces. Canad. Math. Bull. **24**, 169–176 (1981)
30. Küçükaslan, M., Yılmaztürk, M.: On deferred statistical convergence of sequences. Kyungpook Math. J. **56**, 357–366 (2016)
31. Marouf, M.S.: Asymptotic equivalence and summability. Int. J. Math. Math. Sci. **16**, 755–762 (1993)
32. Mishra, L.N., Patro, M., Paikray, S.K., Jena, B.B.: A certain class of statistical deferred weighted A-summability based on (p, q)-integers and associated approximation theorems. Appl. Appl. Math. **14**, 716–740 (2019)
33. Moore, E.H.: An Introduction to a Form of General Analysis, the New Haven Mathematical Colloquium. Yale University Press, New Haven (1910)
34. Mursaleen, M., Edely, O.H.H.: Statistical convergence of double sequences. J. Math. Anal. Appl. **288**, 223–231 (2003)
35. Mursaleen, M., Nasiuzzaman, Md., Nurgali, A.: Some approximation results on Bernstein-Schurer operators defined by (p, q)-integers. J. Inequal. Appl. **2015**(249), 1–12 (2015)
36. Musielak, J.: Orlicz Spaces and Modular Spaces. Lecture Notes in Mathematics, vol. 1034. Springer-Verlag, Berlin (1983)
37. Orhan, S., Demirci, K.: Statistical approximation by double sequences of positive linear operators on modular spaces. Positivity **19**, 23–36 (2015)
38. Patterson, R.F.: On asymptotically statistically equivalent sequences. Demonstratio Math. **36**, 149–153 (2003)
39. Paikray, S.K., Jena, B.B., Misra, U.K.: Statistical deferred Cesàro summability mean based on (p, q)-integers with application to approximation theorems, in *Advances in Summability and Approximation Theory*, eds. by Mohiuddine, S.A., Acar, T (Springer Nature Singapore Pte Ltd., 2019), pp. 203–222
40. Parida, P., Paikray, S.K., Jena, B.B.: Generalized deferred Cesàro equi-statistical convergence and analogous approximation theorems. Proyecciones J. Math. **39**, 307–331 (2020)
41. Pringsheim, A.: Zur theorie der zweifach unendlichen Zahlenfolgen. Math. Ann. **53**, 289–321 (1900)
42. Srivastava, H.M., Jena, B.B., Paikray, S.K.: A certain class of statistical probability convergence and its applications to approximation theorems. Appl. Anal. Discrete Math. **14**, 579–598 (2020)
43. Srivastava, H.M., Jena, B.B., Paikray, S.K.: Statistical deferred Nörlund summability and Korovkin-type approximation theorem. Mathematics **8**(636), 1–11 (2020)

44. Srivastava, H.M., Jena, B.B., Paikray, S.K.: Statistical probability convergence via the deferred Nörlund mean and its applications to approximation theorems. Rev. Real Acad. Cienc. Exactas Fís. Natur. Ser. A Mat. (RACSAM) **114**(144), 1–14 (2020)
45. Srivastava, H.M., Jena, B.B., Paikray, S.K.: Deferred Cesàro statistical probability convergence and its applications to approximation theorems. J. Nonlinear Convex Anal. **20**, 1777–1792 (2019)
46. Srivastava, H.M., Jena, B.B., Paikray, S.K., Misra, U.K.: A certain class of weighted statistical convergence and associated Korovkin type approximation theorems for trigonometric functions. Math. Methods Appl. Sci. **41**, 671–683 (2018)
47. Srivastava, H.M., Jena, B.B., Paikray, S.K., Misra, U.K.: Generalized equi-statistical convergence of the deferred Nörlund summability and its applications to associated approximation theorems. Rev. R. Acad. Cienc. Exactas Fís. Nat. Ser. A Math. (RACSAM) **112**, 1487–1501 (2017)
48. Steinhaus, H.: Sur la convergence ordinaire et la convergence asymptotique. Colloq. Math. **2**, 73–74 (1951)
49. Zraiqat, A., Paikray, S.K., Dutta, H.: A certain class of deferred weighted statistical B-summability involving (p, q)-integers and analogous approximation. Filomat **33**, 1425–1444 (2019)

Chapter 7
Sharper Bounds for Haar Wavelet Coefficients Corresponding to Continuous Real Valued Function Defined on Unit Interval

A. Padmanabha Reddy

Abstract Haar wavelet is the only wavelet family with an explicit expression among all Daubechies wavelet families. This beneficial capability is exploited here to compute some bounds for Haar wavelet (detail) coefficients corresponding to real-valued continuous function on [0,1]. Obtained sharper bounds for the said coefficients at higher resolutions are helpful to study the degree of closeness to zero. We hypothesize that when there are any abrupt changes in the size of detail coefficients, the undererlying function has discontinuity and this discontinuity could be located in a small interval. Location of this kind of interval using Haar wavelet provides the smallest interval among all the Daubechies wavelet families for the given resolution.

Keywords Haar wavelet coefficient · Cauchy–Schwarz inequality · Oscillation of a function · The mean value theorem

7.1 Introduction

Haar wavelet is the only wavelet having analytic expression with compact support which were introduced by Alfred Haar [5] and treated as basis for other family of wavelets. As a consequence of Haar multiresolution analysis on $I = [0, 1]$, we have two sequences of subspaces namely Haar multiresolution approximation spaces $\{V_j\}_{j\geq 0}$ and Haar wavelet spaces $\{W_j\}_{j\geq 0}$ of $L^2([0, 1])$ with the following properties [4, 14]:

i. The Haar space V_j is the set of all functions which are constant on subintervals $I_{j,n} = [\frac{n}{2^j}, \frac{n+1}{2^j})$, for $n = 0, 1, \ldots, 2^j - 1$ of I with dimension 2^j and $V_j \subset V_{j+1}$.

Dedicated to Prof. N. M. Bujurke, INSA Distinguished Professor, Department of Mathematics, K. U. Dharwad, INDIA on the occasion of his 81st birthday anniversary.

A. P. Reddy (✉)
Department of Studies in Mathematics, Vijayanagara Sri Krishnadevara University, Bellary 583105, KA, India
e-mail: apreddy@vskub.ac.in

B. Hazarika et al. (eds.), *Advances in Functional Analysis and Fixed-Point Theory*, Industrial and Applied Mathematics,
https://doi.org/10.1007/978-981-99-9207-2_7

The set $\{\phi_{j,n} = \phi(2^j t - n)\}_{n=0}^{2^j-1}$ is an orthogonal basis for V_j, where ϕ is called Haar scaling function or Father wavelet and is defined as

$$\phi(t) = \begin{cases} 1, & t \in [0, 1) \\ 0, & otherwise. \end{cases}$$

ii. The Haar wavelet space W_j is defined as the orthogonal complement of V_j in V_{j+1} i.e.,

$$V_{j+1} = V_j \oplus W_j.$$

The set of daughter wavelets $\{\psi_{j,n}(t) = \psi(2^j t - n)\}_{n=0}^{2^j-1}$ forms an orthogonal basis for W_j, where ψ is called Haar wavelet or Mother wavelet having least compact support among all higher order Daubechies mother wavelets and is defined by

$$\psi(t) = \begin{cases} 1, & t \in [0, 0.5) \\ -1, & t \in [0.5, 1) \\ 0, & otherwise. \end{cases}$$

Further for a given natural number J, we have the following decomposition:

$$V_J = V_0 \oplus W_0 \oplus W_1 \oplus \cdots \oplus W_{J-1}.$$

Therefore, if $f \in L^2([0, 1))$, then it has the following representation in V_J i.e.,

$$f(x) \approx w_J(x) = s_{0,0}\phi(x) + \sum_{j=0}^{J-1} \sum_{k=0}^{2^j-1} d_{j,k}\psi(2^j x - k). \tag{7.1}$$

Where $s_{0,0} = \int_0^1 f(t)dt$ and $d_{j,k} = 2^j \int_0^1 f(t)\psi_{j,k}(t)dt = 2^j \int_0^1 f(t)\psi(2^j t - k)dt$. Here the coefficients: $\{s_{0,0}\}$ and $\{d_{0,0}, d_{1,0}, d_{1,1}, \ldots, d_{J-1,0}, \ldots, d_{J-1,2^{J-1}-1}\}$ are called Haar scaling and Haar wavelet (detail) coefficients respectively.

Discretization of Haar wavelets and their integrals at collocation points i.e., Haar wavelets discretization methods (HWDM) plays a prominent role in solving various challenging equations in science and engineering [1, 2, 6, 7]. Some interesting properties associated with Haar wavelet matrix in the context of linear algebra are also studied [13, 15]. Further, convergence issues associated with HWDM for differential equations are resolved with the help of size of detail coefficients corresponding to bounded differentiable functions on compact intervals of the real line [8]. This paper concentrates on development and application of bounds for detail coefficients corresponding to real-valued continuous functions on compact intervals of the real line.

The rest of the paper is organised as follows: Sect. 7.2 contains the importance and the need of sharper bounds for detail coefficients. Sharper bounds are developed for the class of said functions in Sect. 7.3. Validation, comparison and application of various inequalities are illustrated on four test functions respectively in Sect. 7.4. Conclusion and future work is depicted in final section.

7.2 Role of Haar Wavelet Coefficients

From Eq. (7.1), we have

$$d_{j,k} = 2^j \left\{ \int_{\frac{k}{2^j}}^{\frac{2k+1}{2^{j+1}}} f(t)dt - \int_{\frac{2k+1}{2^{j+1}}}^{\frac{k+1}{2^j}} f(t)dt \right\}.$$

If f is continuous on I and length of the subinterval $I_{j,n}$ is sufficiently small (j is sufficiently large and fix), then its Haar wavelet coefficient is close to zero. This feature $d_{j,k} \approx 0$ make good use of a resource in various fields such as data compression, denoising of signal, adaptive techniques in numerical analysis and design of preconditioners [10–12]. Overall the size of $d_{j,k}$ play a crucial role in the analysis and applications. In general, rate of closeness to zero ($d_{j,k} \approx 0$) dependence on smoothness of f over subintervals of [0,1] and resolution level j.

Here we derive bounds for detail coefficients using celebrated Cauchy–Schwarz inequality in analysis. For $f \in L^2(I)$, then by the Cauchy–Schwarz inequality, we have the following one of upper bounds for detail coefficients:

$$|d_{j,k}| \leq 2^{j/2}||f||, \tag{7.2}$$

for all $j = 0, 1, \ldots, J-1$ and $k = 0, 1, \ldots, 2^j - 1$.

Majak et al. [8] have estimated the bound for the Haar wavelet coefficients at each level(j) corresponding to the functions with first derivative bounded on I.

$$|d_{j,k}| \leq \frac{\eta}{2^{j+1}}, \quad \text{for} \quad j = 0, 1, \ldots, J-1 \quad \text{and} \quad k = 0, 1, \ldots, 2^j - 1, \tag{7.3}$$

where $|f'(t)| \leq \eta, \forall t \in [0, 1]$.

This bound is very helpful to decide the degree of closeness to zero for sufficiently large j compare to Cauchy–Schwarz based bound.

Author noticed that the behavior of function f over subinterval ($\bar{I}_{j,k} = [\frac{k}{2^j}, \frac{k+1}{2^j}]$) of I is not taken into account in the derivation of above two bounds. Better bounds can be obtained if we consider the oscillation of bounded function over $\bar{I}_{j,k}$ and some properties of continuous function on compact subset of the real line.

7.3 Sharper Bounds for Haar Wavelet Coefficients

The main goal of this article is to investigate bounds for the Haar wavelet coefficients at different resolutions corresponding to continuous function and compare with the existing results [8].

Definition 7.1 The oscillation of a bounded function f on an $[a, b]$ is the *supremum* of the set $\{|f(t_1) - f(t_2)| : t_1, t_2 \in [a, b]\}$ of numbers and is denoted by $W(f; [a, b])$ [3].

Theorem 7.1 $W(f; [a, b]) = sup\{f(t) : t \in [a, b]\} - inf\{f(t) : t \in [a, b]\}$ *[9].*

Theorem 7.2 *Let* $J \in \mathbb{N}$, $f \in C([0,1])$, *then its Haar detail (MRA) coefficients in (1.1) i.e.,* $d_{j,k}(0 \le j \le J-1, \quad 0 \le k \le 2^j - 1)$ *satisfy the following inequalities:*

i. $|d_{j,k}| \le \frac{(M_{j,k} - m_{j,k})}{2} = \frac{W(f; \bar{I}_{j,k})}{2}$;

ii. $\frac{m'_{j,k} - M''_{j,k}}{2} \le d_{j,k} \le \frac{M'_{j,k} - m''_{j,k}}{2}$.

Where $M_{j,k} = sup\left\{f(x) : x \in \left[\frac{k}{2^j}, \frac{k+1}{2^j}\right]\right\}$, $m_{j,k} = inf\left\{f(x) : x \in \left[\frac{k}{2^j}, \frac{k+1}{2^j}\right]\right\}$,

$M'_{j,k} = sup\left\{f(x) : x \in \left[\frac{k}{2^j}, \frac{2k+1}{2^{j+1}}\right]\right\}$, $m'_{j,k} = inf\left\{f(x) : x \in \left[\frac{k}{2^j}, \frac{2k+1}{2^{j+1}}\right]\right\}$ *and*

$M''_{j,k} = sup\left\{f(x) : x \in \left[\frac{2k+1}{2^{j+1}}, \frac{k+1}{2^j}\right]\right\}$, $m''_{j,k} = inf\left\{f(x) : x \in \left[\frac{2k+1}{2^{j+1}}, \frac{k+1}{2^j}\right]\right\}$.

Proof Let $j \in \{0, 1, ..., J-1\}$ and $k \in \{0, 1, ..., 2^j - 1\}$. Suppose $M_{j,k}, m_{j,k}, M'_{j,k}, m'_{j,k}, M''_{j,k}$ and $m''_{j,k}$ are the real numbers defined as in the hypothesis.
Here,

$$d_{j,k} = 2^j \left\{ \int_{\frac{k}{2^j}}^{\frac{2k+1}{2^{j+1}}} f(t)dt - \int_{\frac{2k+1}{2^{j+1}}}^{\frac{k+1}{2^j}} f(t)dt \right\},$$

by the mean value theorem for definite integrals, there exist $e_1 \in \left[\frac{k}{2^j}, \frac{2k+1}{2^{j+1}}\right]$, $e_2 \in \left[\frac{2k+1}{2^{j+1}}, \frac{k+1}{2^j}\right]$ such that

$$d_{j,k} = 2^j \left\{ \frac{1}{2^{j+1}} f(e_1) - \frac{1}{2^{j+1}} f(e_2) \right\}.$$

Hence, $2|d_{j,k}| = |f(e_1) - f(e_2)|$ is less than or equal to oscillation of f on the subinterval $[\frac{k}{2^j}, \frac{k+1}{2^j}]$. Therefore, $|d_{j,k}| \leq \frac{(M_{j,k}-m_{j,k})}{2} = \frac{W(f;\bar{I}_{j,k})}{2}$.

Again,

$$d_{j,k} = 2^j \left\{ \int_{\frac{k}{2^j}}^{\frac{2k+1}{2^{j+1}}} f(t)dt - \int_{\frac{2k+1}{2^{j+1}}}^{\frac{k+1}{2^j}} f(t)dt \right\} \geq 2^j m'_{j,k} \frac{1}{2^{j+1}} - 2^j M''_{j,k} \frac{1}{2^{j+1}} = \frac{m'_{j,k} - M''_{j,k}}{2}.$$

Similarly,

$$d_{j,k} = 2^j \left\{ \int_{\frac{k}{2^j}}^{\frac{2k+1}{2^{j+1}}} f(t)dt - \int_{\frac{2k+1}{2^{j+1}}}^{\frac{k+1}{2^j}} f(t)dt \right\} \leq 2^j M'_{j,k} \frac{1}{2^{j+1}} - 2^j m''_{j,k} \frac{1}{2^{j+1}} = \frac{M'_{j,k} - m''_{j,k}}{2}.$$

□

Corollary 7.1 *Let* $f : [0, 1] \rightarrow \mathbb{R}$ *be a bounded differentiable function i.e.,* $|f'(t)| \leq \eta$, $\forall\, t \in [0, 1]$ *for some* $\eta > 0$, *then* $|d_{j,k}| \leq \frac{W(f;\bar{I}_{j,k})}{2} \leq \frac{\eta}{2^{j+1}}$, $j = 0, 1, \ldots, J-1$ *and* $k = 0, 1, \ldots, 2^j - 1$.

Proof Follows from the fact that $\frac{\eta}{2^{j+1}}$ is one of upper bounds for the set $\left\{ |f(t_1) - f(t_2)| : t_1, t_2 \in \left[\frac{k}{2^j}, \frac{k+1}{2^j}\right] \right\}$. □

Corollary 1 meets the goal of this article. In the next section, we concentrate on the application of derived bounds.

7.4 Results and Discussion

Four test functions are considered to demonstrate the validation, comparison and applicability of obtained bounds for few levels i.e., $j = 0, 1, 2$ and 3. Computations are done with the help of Matlab .

Example 7.1 Let us consider the function e^x on [0, 1]. Since e^x is monotone, parameters appear in various bounds are easy to compute. Bounds for this example are recorded in Table 7.1.

Example 7.2 We consider a function which is non-monotone, continuous on [0, 1] and having unbounded derivative on (0, 1). Detail coefficients and their bounds are listed for the following example in Table 7.2.

$$f(x) = \begin{cases} x\sin(\frac{1}{x^4}), & if \quad x \neq 0 \\ 0, & if \quad x = 0. \end{cases}$$

Table 7.1 Bounds (col. 2, 4 and 6) for Haar wavelet coefficients corresponding to e^x

Parameters	Lower bound $\frac{m'_{j,k}-M''_{j,k}}{2}$	Haar wavelet coefficients $d_{j,k}$	Upper bound $\frac{M'_{j,k}-m''_{j,k}}{2}$	$\|d_{j,k}\|$	$\frac{W(f;\bar{I}_{j,k})}{2}$	$\frac{\eta}{2^{j+1}}$	$2^{\frac{j}{2}}\|\|f\|\|$
$j=0$; $k=0$.	−0.8591	−0.4208	0	0.4208	0.8591	$\frac{e}{2}=$ 1.3591	1.7873
$j=1$; $k=0,1$.	−0.3244	−0.1613	0	0.1613	0.3244	$\frac{e}{4}=$ 0.6796	2.5276
	−0.5348	−0.2660	0	0.2660	0.5348		
	−0.1420	−0.0709	0	0.0709	0.1420		
$j=2$; $k=$ 0, 1, 2, 3 .	−0.1823	−0.0911	0	0.0911	0.1823	$\frac{e}{8}=$ 0.3398	3.5746
	−0.2341	−0.1169	0	0.1169	0.2341		
	−0.3006	−0.1501	0	0.1501	0.3006		
	−0.0666	−0.0333	0	0.0333	0.0666		
	−0.0754	−0.0377	0	0.0377	0.0754		
$j=3$; $k=$ 0, 1, ..., 7.	−0.0855	−0.0427	0	0.0427	0.0855		
	−0.0969	−0.0484	0	0.0484	0.0969		
	−0.1098	−0.0549	0	0.0549	0.1098	$\frac{e}{16}=$ 0.1699	5.0552
	−0.1244	−0.0622	0	0.0622	0.1244		
	−0.1409	−0.0704	0	0.0704	0.1409		
	−0.1597	−0.0798	0	0.0798	0.1597		

Example 7.3 Consider the truncated signum function having jump discontinuity at 0.26.

$$f(x) = \chi_{[0,1]}(x)sgn(x-0.26) = \begin{cases} -1, & if \quad x \in [0, 0.26) \\ 1, & if \quad x \in (0.26, 1] \\ 0, & otherwise. \end{cases}$$

Clearly, f is constant on dyadic subintervals except on the subintervals of $[0, 1]$: $[0, 1]$, $[0, \frac{1}{2}]$ and $[\frac{1}{4}, \frac{2^{j-2}+1}{2^j}]$ for $j \geq 2$. Hence the size of the Haar detail coefficients corresponding to the Haar wavelets with mentioned compact support calculated as

$$d_{0,0} = -0.52, d_{1,0} = -0.96, d_{j,2^{j-2}} = -\frac{2^{j+1}}{100}, \text{ for } j \geq 2.$$

Table 7.2 Bounds (col. 2, 4 and 6) for Haar wavelet coefficients corresponding to Example 7.2

Parameters	Lower bound $\frac{m'_{j,k}-M''_{j,k}}{2}$	Haar wavelet coefficients $d_{j,k}$	Upper bound $\frac{M'_{j,k}-m''_{j,k}}{2}$	$\lvert d_{j,k}\rvert$	$\frac{W(f;\bar{I}_{j,k})}{2}$
$j=0; k=0.$	−0.5447	−0.1505	0.4948	0.1505	0.5441
$j=1;$ $k=0,1.$	−0.2111	0.0076	0.2730	0.0076	0.2466
	−0.7928	−0.4230	0.3057	0.4230	0.7928
$j=2;$ $k=0,1,2,3.$	−0.1125	0.0000	0.1204	0.0000	0.1157
	−0.3758	0.0205	0.3758	0.0205	0.3684
	−0.2672	0.2733	0.6387	0.2733	0.6387
	−0.4597	−0.1701	0.0127	0.1701	0.4598
$j=3; k=$ $0,1,2,...,7.$	−0.0550	−0.0000	0.0547	0.0000	0.0404
	−0.1889	0.0000	0.1766	0.0000	0.1926
	−0.3118	−0.0068	0.3114	0.0068	0.3160
	−0.4492	0.0107	0.4267	0.0107	0.4356
	−0.5735	−0.1960	0.4076	0.1960	0.5735
	−0.3325	−0.0157	0.4171	0.0157	0.3324
	−0.4403	−0.2174	−0.0003	0.2174	0.4402
	−0.0174	−0.0103	0.0322	0.0103	0.0322

f is constant on subintervals: $[\frac{1}{2}, 1]$ and $\left[\frac{k}{2^j}, \frac{k+1}{2^j}\right]$ for $j \geq 2$ and $k \neq 2^{j-2}$. Hence $d_{1,1} = 0$, $d_{j,k} = 0$, for $j \geq 2$ and $k \in \{0, 1, 2, ..., 2^j - 1\} - \{2^{j-2}\}$ follows from the fact that area of each Haar wavelet vanish over its compact support.

At each level j the absolute value of the Haar detail coefficient corresponding to the wavelet whose compact support contains the point 0.26 is larger compare to others. In other words only one detail coefficient is nonzero(abrupt changes) for each level j. This information directs us to locate the subinterval containing the point of discontinuity of f, for larger values of j, length of the said subintervals decreases.

Example 7.4 This example concentrates on discrete signals(functions) and is different from the above three examples. Here we consider the 1024 ($J = 10$) samples of a real-valued function f defined over [0, 1]. Its 512 ($j = 9$) detail coefficients are computed using one level fast wavelet transform [14] and the same is shown in the following stem plot (Fig. 7.1).

In the above stem plot all the detail coefficients are closer to zero except one coefficient. Corresponding larger size detail coefficient ($d_{9,133}$) locates the discontinuity of the underlying function in the interval $\left([\frac{133}{512}, \frac{134}{512}]\right)$ of length $\frac{1}{512}$.

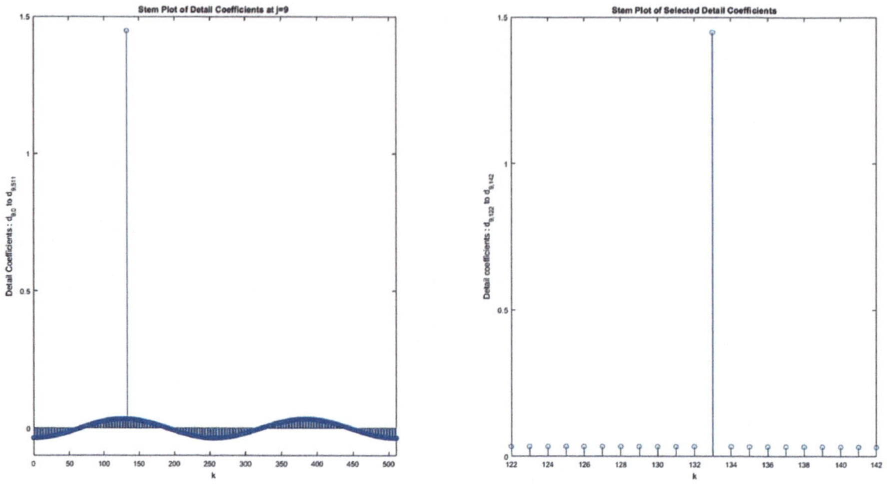

Fig. 7.1 Stem plot of: **a** Haar detial coefficients- $d_{9,k}$; **b** Selected coefficients-$d_{9,122}$ to $d_{9,142}$

Now, we will reveal that the above 1024 samples are from the evaluation of $f(x) = (4sin(4\pi x) - sgn(x - 0.26))\chi_{[0,1]}(x)$ at $\{k/2^{10} : k = 0, 1, \ldots, 2^{10} - 1\}$ results into the member of the Haar space V_{10} as a constant function on subintervals $I_{10,n} = [\frac{n}{2^{10}}, \frac{n+1}{2^{10}})$ of [0, 1], for $n = 0, 1, \ldots, 2^{10} - 1$.

7.5 Conclusion and Future Work

We derived the expressions for lower and upper bounds for each Haar wavelet coefficient $d_{j,k}$ and further we have shown that its absolute value is less than or equal to half of the oscillation of underlying continuous function on $\bar{I}_{j,k}$(compact support of the corresponding Haar wavelet,$\{\psi_{j,n}(.)\}$). Oscillation based bound is sharper than bound existing in the literature and the same is concluded, validated in corollary 3.1, first two test examples respectively. Further, a larger Haar detail coefficient has insights/possesses information that: whenever the detail coefficient is significantly larger compared to its neighbour, it indicates that the considered function has discontinuity(s) in the interval of length $1/2^j$. Last two examples are a testimony to the discussed insights and hence application of derived inequalities.

Designed bounds play a crucial role in approximating the threshold value at each level. Once, if we know the threshold value it is an easy task to construct wavelet based preconditioners for the linear system of equations and sparse representation of functions with respect to Haar wavelet basis. Investigation of threshold value(s) for the said purposes and detection of other types of discontinuities in the absence of noise treated as a future work.

References

1. Awati, V.B., Kumar, N.M., Wakif, A.: Haar wavelet scrutinization of heat and mass transfer features during the convective boundary layer flow of a nanofluid moving over a nonlinearly stretching sheet. Partial Diff. Equ. Appl. Math. **4**, 100192 (2021)
2. Awati, V.B., Kumar, N.M.: Analysis of forced convection boundary layer flow and heat transfer past a semi-infinite static and moving flat plate using nanofluids-by Haar wavelets. J. Nanofluids **10**(1), 106–117 (2021)
3. Bartle, R.G., Sherbert, D.R.: Introduction to Real Analysis. McGraw-Hill, Wiley-India edition (2012)
4. Daubechies, I.: Ten Lectures on Wavelets, Society for Industrial and Applied Mathematics (1992)
5. Haar, A.: Zur theorie der orthogonalen funktionsysteme. Math. Annal **69**, 331–371 (1910)
6. Hariharan, G., Kannan, K.: An overview of Haar wavelet method for solving differential and integral equations. World Appl. Sci. J. **23**(12), 1–14 (2013)
7. Lepik, U., Hein, H.: Haar Wavelets with Applications. Springer International Publishing (2014)
8. Majak, J., Shvartsman, B.S., Kirs, M., Pohlak, M., Herranen, H.: Convergence theorem for the Haar wavelet based disretization method. Compos. Struct. **126**, 227–232 (2015)
9. Malik, S.C., Arora, S.: Mathematical Analysis. New Age International Publishers, New Delhi (2012)
10. Mehra, M.: Wavelets and differential equations-a short review. AIP Conf. Proc. **1146**, 241–252 (2009)
11. Porwik, P., Lisowska, A.: The Haar wavelet transform in digital image processing: its status and achievements. Mach. Graph. Vis. **13**(1/2), 79–98 (2004)
12. Reddy, A.P., Bujurke, N.M.: Biorthogonal wavelets based preconditioners for sparse linear systems. The Math. Stud. **82**(1–4), 191–208 (2013)
13. Reddy, A.P.: Amendment to the result on the trace of Haar wavelets matrix. Wavel. Linear Algebra **7**(1), 13–22 (2020)
14. Ruch, D.K., Van Fleet, P.J.: Wavelet Theory: an Elementary Approach with Applications. Wiley (2009)
15. Shiralashetti, S.C., Kumbinarasaiah, S.: Some results on Haar wavelets matrix through linear algebra. Wavel. Linear Algebra **4**(2), 49–59 (2017)

Chapter 8
Gradual Normed Linear Spaces

Shyamal Debnath and Chiranjib Choudhury

Abstract In this chapter we will discuss gradual normed linear spaces and some related algebraic and topological properties of the space. We also discuss different convergence concepts of the sequences in the said space and the implication relations between them.

Keywords Gradual normed linear space · Convergence

8.1 Introduction and Background

In 1965, Zadeh [46] introduced the concept of fuzzy sets as an extension of classical set theory. Since then, fuzzy sets have found numerous applications in various fields of engineering and science. Within the study of fuzzy set theory, the term "fuzzy number" holds significance. Fuzzy numbers are an extension of intervals rather than conventional numbers and do not adhere to several algebraic properties of classical numbers. Consequently, there has been ongoing debate among researchers regarding the term "fuzzy number". Many researchers have opted to use the term "fuzzy intervals" instead. To address this confusion, Fortin et al. [14] proposed the concept of gradual real numbers as elements of fuzzy intervals in 2008. Gradual real numbers are defined by their assignment function, which maps the interval $(0, 1]$ to the set of all real numbers $\mathbb{R}$. Consequently, every real number can be regarded as a gradual number with a constant assignment function. Gradual real numbers satisfy all the algebraic properties of classical real numbers and have been effectively utilized in computation, fuzzy optimization, and various other problem domains.

S. Debnath (✉) · C. Choudhury
Department of Mathematics, Tripura University (A Central University), Suryamaninagar, Agartala 799022, India
e-mail: debnathshyamal@tripurauniv.ac.in

C. Choudhury
e-mail: chiranjib.mathematics@tripurauniv.in

B. Hazarika et al. (eds.), *Advances in Functional Analysis and Fixed-Point Theory*, Industrial and Applied Mathematics,
https://doi.org/10.1007/978-981-99-9207-2_8

Sadeqi and Azari [36] were the first to introduce the notion of gradual normed linear space in the year 2011. They investigated various properties of the space from both topological and algebraic points of view. Very recently, in [11], Ettefagh et al. presented some more interesting new results in this direction. For an extensive study on gradual real numbers [1, 10, 21, 30, 43] can be addressed where many more references can be found.

For the last several decades study of summability theory and investigation of various convergence concepts have been one of the most active areas of research in several branches of mathematics. The usual notion of convergence of sequences does not always capture the details of the sequences which are not convergent but in some sense behaves more or less like a convergent sequence. To include those sequences for investigation, in 1951 Fast [13] and Steinhaus [42] independently introduced the notion of statistical convergence using "natural density". Now the natural density of a set $P \subseteq \mathbb{N}$ is defined by $d(P) = \lim_{n\to\infty} \frac{|K_n|}{n}$, where $|K_n| = card\{k \in K : k \leq n\}$ provided that the limit exists. A sequence (x_k) is said to be statistically convergent [16] to l if for each $\kappa > 0$,

$$\lim_n \frac{1}{n} |\{k \leq n :| x_k - l |\geq \kappa\}| = 0$$

i.e., if the set $\{k \in \mathbb{N} :| x_k - l |\geq \kappa\}$ has natural density zero. Now since in case of usual convergence all but a finite number of elements should lie outside the interval $(l - \kappa, l + \kappa)$ and the natural density of a finite subset of $\mathbb{N}$ is zero, so statistical convergence is a generalization of usual convergence.

Statistical convergence turned out to be one of the most active areas of research in summability theory after the works of Fridy [15, 16], Salat [37], and many others where many more references can be found.

In another direction, the notion of statistical convergence was further extended to ideal convergence in 2001 by Kostyrko et al. [26]. They also showed that ideal convergence was not only an extension of statistical convergence but also some known methods of convergences can be obtained by considering particular ideals. Later on, several investigations in this direction have been occurred due to Lahiri and Das [22–24], Mursaleen and Mohiuddine [33], Savas and Gurdal [40], Tripathy and Hazarika [44], and many others. For an extensive view of several recent advancements in ideal convergence, we refer to [3–9, 17–20, 25, 27–29, 31, 32, 34, 35, 38, 39, 41, 45].

The concept of convergence of sequences in the gradual normed linear space can be found in the works of Sadeqi and Azari [36]. Following their work, Ettefagh et al. [12] investigated the properties of convergence of sequences in more detail. Very recently, Choudhury and Debnath [2] introduced the notion of ideal convergence in the gradual normed linear spaces. Therefore, the study of various methods of convergence of sequences in gradual normed linear spaces is very natural.

8.2 Gradual Numbers, Gradual Normed Linear Space, and Some Related Definitions

In this section, we will review some fundamental definitions and concepts related to gradual numbers and gradual normed linear spaces.

Definition 8.1 [14] The concept of a gradual real number, denoted as $\tilde{r}$, is characterized by an assignment function $\Psi_{\tilde{r}}$, which maps values from the interval (0, 1] to real numbers in $\mathbb{R}$. The collection of all gradual real numbers is represented as $G(\mathbb{R})$. A gradual number $\tilde{r}$ is considered non-negative if, for any σ belonging to the interval (0, 1], the corresponding assigned value $\Psi_{\tilde{r}}(\sigma)$ is greater than or equal to zero. The set comprising all non-negative gradual real numbers is denoted as $G^*(\mathbb{R})$.

The paper referenced as [14] introduces a definition for gradual operations performed on the elements of $G(\mathbb{R})$ in the following manner:

Definition 8.2 Consider any operation $*$ defined on the set of real numbers $\mathbb{R}$. Let $\tilde{r}_1$ and $\tilde{r}_2$ be elements of $G(\mathbb{R})$, each characterized by assignment functions $\Psi_{\tilde{r}_1}$ and $\Psi_{\tilde{r}_2}$ respectively. The gradual operation $\tilde{r}_1 * \tilde{r}_2$ in $G(\mathbb{R})$ is defined using the assignment function $\Psi_{\tilde{r}_1 * \tilde{r}_2}$, which is given by $\Psi_{\tilde{r}_1 * \tilde{r}_2}(\sigma) = \Psi_{\tilde{r}_1}(\sigma) * \Psi_{\tilde{r}_2}(\sigma)$ for all $\sigma \in (0, 1]$. Consequently, the gradual addition $\tilde{r}_1 + \tilde{r}_2$ and the gradual scalar multiplication $c\tilde{r}$ (where $c \in \mathbb{R}$) are defined as follows:

$$\Psi_{\tilde{r}_1+\tilde{r}_2}(\sigma) = \Psi_{\tilde{r}_1}(\sigma) + \Psi_{\tilde{r}_2}(\sigma) \quad and \quad \Psi_{c\tilde{r}}(\sigma) = c\Psi_{\tilde{r}}(\sigma), \quad \forall \sigma \in (0, 1].$$

The constant gradual real number $\tilde{p}$ is defined for any real number $p \in \mathbb{R}$ using the constant assignment function $\Psi_{\tilde{p}}(\sigma) = p$ for all $\sigma \in (0, 1]$. Specifically, the constant gradual numbers $\tilde{0}$ and $\tilde{1}$ are defined with assignment functions $\Psi_{\tilde{0}}(\sigma) = 0$ and $\Psi_{\tilde{1}}(\sigma) = 1$ respectively. It can be easily verified that $G(\mathbb{R})$, equipped with gradual addition and multiplication operations, forms a real vector space [14].

Definition 8.3 [36] Let X be a real vector space. The function $|| \cdot ||_G : X \to G^*(\mathbb{R})$ is said to be a gradual norm on X, if for every $\sigma \in (0, 1]$, the following three conditions are true for any $x, y \in X$

(G_1) $\Psi_{||x||_G}(\sigma) = \Psi_{\tilde{0}}(\sigma)$ iff $x = 0$;

(G_2) $\Psi_{||\lambda x||_G}(\sigma) = |\lambda| \Psi_{||x||_G}(\sigma)$ for any $\lambda \in \mathbb{R}$;

(G_3) $\Psi_{||x||_G}(\sigma) + \Psi_{||y||_G}(\sigma) \geq \Psi_{||x+y||_G}(\sigma)$.

The pair $(X, || \cdot ||_G)$ is referred to as gradual normed linear space (GNLS).

Example 8.1 [36] Suppose $X = \mathbb{R}^n$ and define $|| \cdot ||_G$ by

$$\Psi_{||x||_G}(\sigma) = e^{\sigma} \sum_{i=1}^{n} |x_i| \tag{8.1}$$

for $x = (x_1, x_2, \cdots, x_n) \in \mathbb{R}^n$, $\sigma \in (0, 1]$. Then, (8.1) satisfies (G_1), (G_2), and (G_3) and so $(\mathbb{R}^n, || \cdot ||_G)$ is a GNLS.

Example 8.2 [36] Let $C[0, 1]$ denote the space of all continuous function on $[0, 1]$ associated with usual linear operation. Then, the function, $||\cdot||_G : C[0, 1] \to G^*(\mathbb{R})$ defined by

$$\Psi_{||f||_G}(\sigma) = \begin{cases} \max\limits_{0\le x\le 1}\{|f(x)|\}, & \frac{1}{2} < \sigma \le 1 \\ \sqrt{\int_0^1 |f(x)|^2 dx}, & 0 < \sigma \le \frac{1}{2} \end{cases}$$

forms a gradual norm on $C[0, 1]$ and $(C[0, 1], ||\cdot||_G)$ is a GNLS.

Definition 8.4 [36] Let $(X, ||\cdot||_G)$ be a GNLS and $x_0 \in X$. Then for $\sigma \in (0, 1]$, the gradual neighborhood of x_0 having radius $\kappa > 0$ is defined as follows:

$$x_0 + N(\kappa, \sigma) = \{x \in X : \Psi_{||x-x_0||_G}(\sigma) < \kappa\}$$

Specifically, if we take $x_0 = 0$, the neighborhood $N(\kappa, \sigma) = \{x \in X : \Psi_{||x||_G}(\sigma) < \kappa\}$ corresponds to the neighborhood of the origin with a radius of κ.

Lemma 8.1 *[36] Let $(X, ||\cdot||_G)$ be a GNLS. Then, for any $\kappa, \kappa_1, \kappa_2 > 0$ and $\sigma, \sigma_1, \sigma_2 \in [0, 1)$,*
(i) $N(\kappa, \sigma) = \kappa N(1, \sigma)$;
(ii) $N(\kappa_1, \sigma) \subseteq N(\kappa_2, \sigma)$ whenever $\kappa_1 \le \kappa_2$;
(iii) For any $x \in X$, $\sigma_1 \le \sigma_2$ implies $N(\kappa, \sigma_1) \subseteq N(\kappa, \sigma_2)$, provided that the function $\Psi_{||x||_G}$ is decreasing.

Proof (i) For any $\kappa > 0$ and $\sigma \in [0, 1)$,

$$\begin{aligned} N(\kappa, \sigma) =& \{x \in X : \Psi_{||x||_G}(\sigma) < \kappa\} \\ =& \{x \in X : \frac{1}{\kappa}\Psi_{||x||_G}(\sigma) < 1\} \\ =& \{x \in X : \Psi_{||\frac{x}{\kappa}||_G}(\sigma) < 1\} \\ =& \kappa\{z \in X : \Psi_{||z||_G}(\sigma) < 1\} \\ =& \kappa N(1, \sigma). \end{aligned}$$

(ii) Let $x \in N(\kappa_1, \sigma)$, then we have $\Psi_{||x||_G}(\sigma) \le \kappa_1$. Now since $\kappa_1 \le \kappa_2$, so $\Psi_{||x||_G}(\sigma) \le \kappa_2$ and eventually $x \in N(\kappa_2, \sigma)$ from which the result follows.
(iii) Let $x \in N(\kappa, \sigma_1)$, then we have $\Psi_{||x||_G}(\sigma_1) \le \kappa$. Now since $\Psi_{||x||_G}$ is a decreasing function, so $\sigma_1 \le \sigma_2$ implies $\Psi_{||x||_G}(\sigma_1) \ge \Psi_{||x||_G}(\sigma_2)$ and therefore $\Psi_{||x||_G}(\sigma_2) \le \kappa$ holds and as a consequence $x \in N(\kappa, \sigma_2)$. □

8.3 Some Topological Properties of GNLS

In this section, we present some very recent works related to the topological properties of GNLS that has been proved in [11]. But before that we first recall some basic definitions and notions from [36] that are related to those properties.

In a GNLS $(X, ||\cdot||_G)$, let $P \subseteq X$. An element $x_0 \in X$ is considered an interior point of P if there exist $\kappa > 0$ and $\sigma \in (0, 1]$ so that the gradual neighborhood $x_0 + N(\kappa, \sigma)$ is completely contained within P. $Int(P)$ indicates the collection of all interior points of set P. A set $P \subseteq X$ is said to be gradual open (G.O.) if $Int(P) = P$. Also, x_0 is said to be a closure point of the set $P \subseteq X$, if for any $\sigma \in (0, 1]$, $\underline{(x_0 + N(\sigma, \sigma))} \cap P \neq \emptyset$. The set of all closure points of the set P is denoted by $\overline{P}$. Further P is said to be gradual closed (G.C.) set if $\overline{P} = P$.

Theorem 8.1 *In the context of a GNLS $(X, ||\cdot||_G)$ for any given $\sigma \in (0, 1]$ and $\kappa > 0$, the following holds:*
(i) $N(\kappa, \sigma)$ is G.O. provided that $\Psi_{||x||_G}$ is a decreasing function for all $x \in X$;
(ii) $N(\kappa, \sigma)$ is a convex set.

Proof (i) Since $||\cdot||_G$ is a gradual norm, so for any $x_0, y \in X$ and $\sigma \in (0, 1]$, $\Psi_{||x_0+y||_G}(\sigma) \leq \Psi_{||x_0||_G}(\sigma) + \Psi_{||y||_G}(\sigma)$. Now as the function $\Psi_{||x||_G}$ is decreasing, so for $\sigma_0 \in (0, \sigma]$ the above inequation yields

$$\Psi_{||x_0+y||_G}(\sigma) \leq \Psi_{||x_0||_G}(\sigma) + \Psi_{||y||_G}(\sigma_0).$$

Let $\kappa > 0$ be given and x_0 is a member of $N(\kappa, \sigma)$. Then, for any $\sigma \in (0, 1]$, $\Psi_{||x_0||_G} < \kappa$. Suppose $\Psi_{||x_0||_G} = \kappa - \kappa_0$. Then, for any $y \in N(\kappa_0, \sigma_0)$, we have

$$\Psi_{||x_0+y||_G}(\sigma) \leq \Psi_{||x_0||_G}(\sigma) + \Psi_{||y||_G}(\sigma_0) \leq \kappa.$$

Therefore, we must have $x_0 + N(\kappa_0, \sigma_0) \subseteq N(\kappa, \sigma)$ i.e., $x_0 \in Int N(\kappa, \sigma)$, and the proof is complete.

(ii) Let $x_0, y_0 \in N(\kappa, \sigma)$. Then, $\Psi_{||x_0||_G}(\sigma) \leq \kappa$ and $\Psi_{||y_0||_G}(\sigma) \leq \kappa$. Now for any $0 \leq \lambda \leq 1$ we have,

$$\begin{aligned}\Psi_{||\lambda x_0+(1-\lambda)y_0||_G}(\sigma) &\leq \Psi_{||\lambda x_0||_G}(\sigma) + \Psi_{||(1-\lambda)y_0||_G}(\sigma)\\ &= \lambda\Psi_{||x_0||_G}(\sigma) + (1-\lambda)\Psi_{||y_0||_G}(\sigma)\\ &\leq \lambda\kappa + (1-\lambda)\kappa = \kappa.\end{aligned}$$

Hence, $\lambda x_0 + (1-\lambda)y_0 \in N(\kappa, \sigma)$ and the set $N(\kappa, \sigma)$ is convex. □

Theorem 8.2 *Let $(X, ||\cdot||_G)$ be a GNLS and $P, Q \subseteq X$. Suppose $\Psi_{||x||_G}$ is decreasing for any $x \in X$. Then,*
(i) $Int(P) = X \setminus \overline{(X \setminus P)}$;
(ii) P is gradually open set if and only if $X \setminus P$ is gradually closed set;
(iii) $\overline{P}$ is G.C. set;
(iv) $Int(P)$ is gradually open;
(v) $\overline{P \cap Q} \subseteq \overline{P} \cap \overline{Q}$ and $\overline{P \cup Q} = \overline{P} \cup \overline{Q}$;
(vi) $Int(P) \cup Int(Q) \subseteq Int(P \cup Q)$ and $Int(P \cap Q) = Int(P) \cap Int(Q)$.

Proof (i) Suppose $x_0 \in Int(P)$. Then by definition, there exists some $N(\kappa_0, \sigma_0)$ such that $(x_0 + N(\kappa_0, \sigma_0)) \cap (X \setminus P) = \emptyset$. Let $\sigma = \min\{\sigma_0, \kappa_0\}$. Then, we have $N(\sigma, \sigma) \subseteq N(\kappa_0, \sigma_0)$ and consequently $(x_0 + N(\sigma, \sigma)) \cap (X \setminus P) = \emptyset$ i.e., $x_0 \notin \overline{X \setminus P}$. Thus we have, $x_0 \in X \setminus (\overline{X \setminus P})$ and the inclusion $Int(P) \subseteq X \setminus (\overline{X \setminus P})$ holds. The proof of the reverse inclusion is straightforward and is omitted.

(ii) Assume that $X \setminus P$ be a G.C. set. If we select $x_0 \in P$, it implies that $x_0 \notin X \setminus P$, and additionally, x_0 is not a closure point of $X \setminus P$. So for any $\sigma \in (0, 1]$, $(x_0 + N(\sigma, \sigma)) \cap (X \setminus P) = \emptyset$. In other words, $(x_0 + N(\sigma, \sigma)) \subset P$. This indicates that x_0 is an interior point of the set P. Hence P is G.O. set. To prove the converse part, we assume P to be a G.O. set. Suppose x_0 be a closure point of $X \setminus P$. Then, $(x_0 + N(\sigma, \sigma)) \cap (X \setminus P) = \emptyset$ holds and eventually x_0 is an interior point of P. Now since P is an open set, so $x_0 \in X \setminus P$.

(iii) Clearly, for any $x_0, y \in X$ and $\sigma \in (0, 1]$, $\Psi_{||x_0+y||_G}(\sigma) \leq \Psi_{||x_0||_G}(\sigma) + \Psi_{||y||_G}(\sigma)$. Now as the function $\Psi_{||x||_G}$ is decreasing, so for $\sigma_0 \in (0, \sigma]$ the above inequation yields $\Psi_{||x_0+y||_G}(\sigma) \leq \Psi_{||x_0||_G}(\sigma) + \Psi_{||y||_G}(\sigma_0)$. Thus for $x_0 \in \overline{\overline{P}}$ and every gradual neighborhood $N(\kappa, \sigma)$, there exists $N(\frac{\kappa}{2}, \sigma_0)$ satisfying $N(\frac{\kappa}{2}, \sigma_0) + N(\frac{\kappa}{2}, \sigma_0) \subseteq N(\kappa, \sigma)$. Now since $(x_0 + N(\frac{\kappa}{2}, \sigma_0)) \cap \overline{P} \neq \emptyset$, there exists $y_0 \in \overline{P}$ such that $y_0 \in x_0 + N(\frac{\kappa}{2}, \sigma_0)$. Then, $y_0 + N(\frac{\kappa}{2}, \sigma_0) \subseteq x_0 + N(\frac{\kappa}{2}, \sigma_0) + N(\frac{\kappa}{2}, \sigma_0) \subseteq x_0 + N(\kappa, \sigma)$. Again, since $y_0 \in \overline{P}$ and the inclusion $(y_0 + N(\frac{\kappa}{2}, \sigma_0)) \cap P \subseteq (x_0 + N(\kappa, \sigma)) \cap P$ holds, so $x_0 + N(\kappa, \sigma) \neq \emptyset$. Hence, $x_0 \in \overline{P}$ and the proof is complete.

(iv) Proof is omitted because it can be easily obtained from (i) and (iii).

(v) Let $x_0 \in \overline{P \cap Q}$. Then, for any $\sigma \in (0, 1]$, we have $(x_0 + N(\sigma, \sigma)) \cap (P \cap Q) \neq \emptyset$. This implies that for any $\sigma \in (0, 1]$, $(x_0 + N(\sigma, \sigma)) \cap P \neq \emptyset$ and $(x_0 + N(\sigma, \sigma)) \cap Q \neq \emptyset$. Consequently, $x_0 \in \overline{P} \cap \overline{Q}$. Hence, $\overline{P \cap Q} \subseteq \overline{P} \cap \overline{Q}$.

To prove the second part assume $x_0 \in \overline{P \cup Q}$. Then, for every $\sigma \in (0, 1]$, $(x_0 + N(\sigma, \sigma)) \cap (P \cup Q) \neq \emptyset$. Eventually one can conclude that $(x_0 + N(\sigma, \sigma)) \cap P \neq \emptyset$ or $(x_0 + N(\sigma, \sigma)) \cap Q \neq \emptyset$. Thus, $x_0 \in \overline{P} \cup \overline{Q}$. Therefore, $\overline{P \cup Q} \subseteq \overline{P} \cup \overline{Q}$. The reverse inclusion can be proved by applying similar technique.

(vi) Suppose $x_0 \in Int(P) \cup Int(Q)$. Then, $x_0 \in Int(P)$ or $x_0 \in Int(Q)$. Thus by definition, there exists some $\sigma_1, \sigma_2 \in (0, 1]$ and $\kappa_1, \kappa_2 > 0$ satisfying $(x_0 + N(\kappa_1, \sigma_1)) \subset P$ and $(x_0 + N(\kappa_2, \sigma_2)) \subset Q$. Now choose $\sigma \in (0, 1]$ and $\kappa > 0$ in such a way so that $\sigma < \min\{\sigma_1, \sigma_2\}$ and $\kappa < \min\{\kappa_1, \kappa_2\}$. By our assumption since every assignment function is decreasing, so $(x_0 + N(\kappa, \sigma)) \subset P \cup Q$ which means that $x_0 \in Int(P \cup Q)$.

The proof of the second part can be obtained by applying a similar technique and so omitted. □

Theorem 8.3 *Let $(X, ||\cdot||_G)$ be a GNLS. Then,*
(i) If $\{O_\alpha : \alpha \in \Lambda\}$ is an arbitrary collection of G.O. sets, then $\bigcup_\alpha O_\alpha$ is a G.O. set,

(ii) If $\{O_i\}_{i=1}^n$ *is an finite collection of G.O. sets, then* $\bigcap_{i=1}^n O_i$ *is a G.O. set provided that every assignment function is decreasing.*

Proof (i) Let $x_0 \in \bigcup_\alpha O_\alpha$. Then, there exists some α_0 such that $x_0 \in O_{\alpha_0}$. Now since O_{α_0} is G.O. set, so for any $\kappa > 0$, there exists $\sigma \in (0, 1]$ such that

$$(x_0 + N(\kappa, \sigma)) \subseteq O_{\alpha_0} \subseteq \bigcup_\alpha O_\alpha.$$

Hence $\bigcup_\alpha O_\alpha$ is a G.O. set.

(ii) Let $x_0 \in \bigcap_{i=1}^n O_i$. Then, $x_0 \in O_i$ for any $i = 1, 2, \cdots, n$. Now since each O_i is G.O. set, so for any $\kappa_i > 0$ and $\sigma_i \in (0, 1]$, there exists a gradual neighborhood $x_0 + N(\kappa_i, \sigma_i) \subseteq O_i$. Choose $\sigma \in (0, 1]$ and $\kappa > 0$ such that $\sigma < \min\{\sigma_1, \sigma_2, \cdots, \sigma_n\}$ and $\kappa < \min\{\kappa_1, \kappa_2, \cdots, \kappa_n\}$. Then since every assignment function is decreasing, so for any $i = 1, 2, \cdots, n$, the inclusion $N(\kappa, \sigma) \subseteq N(\kappa_i, \sigma_i)$ holds and consequently we have $(x_0 + N(\kappa, \sigma)) \subseteq (x_0 + N(\kappa_i, \sigma_i)) \subseteq O_i$ for any $i = 1, 2, \cdots, n$. Hence, $(x_0 + N(\kappa, \sigma)) \subseteq \bigcap_{i=1}^n O_i$ holds and the set $\bigcap_{i=1}^n O_i$ is a G.O. set. □

Theorem 8.4 *Let* $(X, ||\cdot||_G)$ *be a GNLS. Then,*
(i) If $\{C_\alpha : \alpha \in \Lambda\}$ *is an arbitrary collection of G.C. sets, then* $\bigcap_\alpha C_\alpha$ *is a G.C. set.*

(ii) If $\{C_i\}_{i=1}^n$ *is an finite collection of G.C. sets, then* $\bigcup_{i=1}^n C_i$ *is a G.C. set.*

Proof (i) Let C denotes the set $C = \bigcap_\alpha C_\alpha$. Now to prove C is G.C., it is sufficient to prove that $\overline{C} \subseteq C$. Consider $x_0 \in \overline{C}$. Then, by definition for any $\sigma \in (0, 1]$, we have $(x_0 + N(\sigma, \sigma)) \cap C \neq \emptyset$. This means that for any α, $(x_0 + N(\sigma, \sigma)) \cap C_\alpha \neq \emptyset$ and eventually, $x_0 \in \overline{C_\alpha}$. Now since for any α, the set C_α is G.C., so we must have $x_0 \in C_\alpha$ for any α. Consequently, $x_0 \in \bigcap_\alpha C_\alpha = C$. This completes the proof.

(ii) Since for every $i = 1, 2, \cdots, n$, the set C_i is G.C., so $\overline{C_i} = C_i$. Now to prove $\bigcup_{i=1}^n C_i$ is a G.C. set, it is enough to show that $\overline{\bigcup_{i=1}^n C_i} \subseteq \bigcup_{i=1}^n C_i$. Let x_0 be an element of $\overline{\bigcup_{i=1}^n C_i}$. Then, by definition for any $\sigma \in (0, 1]$, $(x_0 + N(\sigma, \sigma)) \cap (\bigcup_{i=1}^n C_i) \neq \emptyset$. This implies that for any $\sigma \in (0, 1]$, there exists some $1 \leq j \leq n$ such that $(x_0 + N(\sigma, \sigma)) \cap C_j \neq \emptyset$. Therefore, $x_0 \in \overline{C_j} = C_j$ and eventually $x_0 \in \bigcup_{i=1}^n C_i$. This completes the proof.

Theorem 8.5 *Consider a GNLS* $(X, ||\cdot||_G)$ *and let* x_0 *and* y_0 *be two members of* X *such that* $x_0 \neq y_0$. *Then, there exists two gradual neighborhoods of* x_0 *and* y_0 *that are disjoint from each other.*

Proof Let $x_0, y_0 \in X$ such that $x_0 \neq y_0$. Then, for any $\sigma \in (0, 1]$, $\Psi_{||x_0-y_0||_G}(\sigma) \neq A_{\bar{0}}(\sigma)$. Thus there exists a $\sigma_0 \in (0, 1]$ such that $\Psi_{||x_0-y_0||_G}(\sigma_0) > 0$. Choose κ_0 such that $\Psi_{||x_0-y_0||_G}(\sigma_0) > 2\kappa_0$. We claim that the neighborhoods $x_0 + N(\kappa_0, \sigma_0)$ and $y_0 + N(\kappa_0, \sigma_0)$ are disjoint. Otherwise if there exists some $z_0 \in X$ such that $z_0 \in (x_0 + N(\kappa_0, \sigma_0)) \cap (y_0 + N(\kappa_0, \sigma_0))$, then we have,

$$\Psi_{||x_0-y_0||_G}(\sigma_0) \leq \Psi_{||x_0-z_0||_G}(\sigma_0) + \Psi_{||y_0-z_0||_G}(\sigma_0) < 2\kappa_0 < \Psi_{||x_0-y_0||_G}(\sigma_0),$$

which is a contradiction. □

Theorem 8.6 *Let $(X, ||\cdot||_G)$ be a GNLS and suppose the assignment function $\Psi_{||x||_G}$ is decreasing for any $x \in X$. Then x_0 is a limit point of a set $Y \subseteq X$, if every gradual neighborhood of x_0 contains infinitely many elements of Y.*

Proof If possible suppose there exists some $\kappa > 0$ and $\sigma \in (0, 1]$ such that $(x_0 + N(\kappa, \sigma)) \cap Y = \{y_1, y_2, \cdots, y_n\}$. Choose $\kappa_0 < \kappa$ in such a way that $\kappa_0 < \min\limits_{1\leq i\leq n} \Psi_{||x_0-y_i||_G}(\sigma)$. Then, $(x_0 + N(\kappa, \sigma)) \supset (x_0 + N(\kappa_0, \sigma))$. Now we claim that $(x_0 + N(\kappa_0, \sigma)) \cap Y = \emptyset$. Otherwise if there exists some $p \in \{1, 2, \cdots, n\}$ such that $y_p \in (x_0 + N(\kappa_0, \sigma))$, then we have $\Psi_{||x_0-y_p||_G}(\sigma) < \kappa_0$, which is a contradiction to the fact that $\kappa_0 < \min\limits_{1\leq i\leq n} \Psi_{||x_0-y_i||_G}(\sigma)$. Therefore we must have $(x_0 + N(\kappa_0, \sigma)) \cap Y = \emptyset$. Now put $\sigma_0 < \min\{\kappa_0, \sigma\}$. Then since every assignment function is decreasing, so the inclusion $(x_0 + N(\kappa_0, \sigma)) \supset (x_0 + N(\sigma_0, \sigma_0))$ is true and consequently $(x_0 + N(\sigma_0, \sigma_0)) \cap B = \emptyset$. But this is a contradiction to our assumption that x_0 is a limit point of Y. Hence, the set $(x_0 + N(\kappa, \sigma)) \cap Y$ must be an infinite set. This completes the proof. □

Definition 8.5 [11] Let $(X, ||\cdot||)_G$ be a GNLS and let $C(\neq \emptyset)$ be a subset of X. Then, C is said to be gradual compact set if each cover $\{O_i\}_{i\in\Lambda}$ of G.O. sets for C contains a finite subcover, i.e., there exists finite number of $O_i (i = 1, 2, \cdots, n)$ such that $C \subseteq \bigcup\limits_{i=1}^{n} O_i$.

Theorem 8.7 *Let $(X, ||\cdot||)_G$ be a GNLS and suppose the assignment function $\Psi_{||x||_G}$ be decreasing for all $x \in X$. Then every gradual compact set is G.C.*

Proof Let $C \subseteq X$ be gradual compact. To prove C is G.C. it is enough to show that $X \setminus C$ is G.O. Let $p \in X \setminus C$ and for $q \in C$ and a fixed $\sigma_0 \in (0, 1]$, consider the neighborhoods $V_q = p + N(\kappa, \sigma_0)$ and $W_q = q + N(\kappa, \sigma_0)$ for p and q respectively where $\Psi_{||p-q||_G}(\sigma_0) > 2\kappa$. We claim that for every $q \in C$, the set $V_q \cap W_q$ is empty. Otherwise if $z \in V_q \cap W_q$, then we have $\Psi_{||z-p||_G}(\sigma_0) < \kappa$ and $\Psi_{||z-q||_G}(\sigma_0) < \kappa$ which altogether leads to the contradiction $\Psi_{||p-q||_G}(\sigma_0) < 2\kappa$. Now, since C is gradual compact, so the cover $\bigcup\limits_{q\in C} W_q$ of G.O. sets contains a finite subcover such that $C \subseteq \bigcup\limits_{i=1}^{n} W_{q_i}$. Let $V = \bigcap\limits_{i=1}^{n} V_{q_i}$. Then, by Theorem 8.3(ii), we can conclude that V is a G.O. set containing p. Therefore $V \cap C = \emptyset$ and eventually, $V \subseteq X \setminus C$. Hence, $X \setminus C$ is G.O. set. □

Theorem 8.8 *Let $(X, ||\cdot||)_G$ be a GNLS and $C \subseteq X$ be gradual compact. Then every G.C. subset of C is gradual compact.*

Proof Let $F \subset C$ represent a G.C. subset. We have a cover denoted by $\{O_i\}_{i\in\Lambda}$, consisting of G.O. subsets for F. According to Theorem 8.2, it follows that $X \setminus F$ is also a G.O. subset. As a result, the union of $\{O_i\}_{i\in\Lambda}$ and $(X \setminus F)$ forms a cover for the entire set C. As C is gradual compact, it implies that there exists a finite subcover, denoted by $\bigcup_{i=1}^{n} O_i \cup (X \setminus F)$, which covers the entire set C. This, in turn, implies that $\{O_i\}_{i=1}^{n}$ forms a finite cover for F. Consequently, F is gradual compact. □

Theorem 8.9 *Consider a GNLS $(X, ||\cdot||)_G$ and an infinite subset B of the gradual compact set C. Then, B has at least one limit point within C.*

Proof Suppose, for the sake of contradiction, that for any $x_0 \in C$, x_0 is not a limit point of B. By definition, this means that there exists some $\sigma \in (0, 1]$ depending on x_0 such that $(x_0 + N(\sigma, \sigma)) \cap B = \emptyset$. Consequently, for any arbitrary $x_0 \in B$, we have $(x_0 + N(\sigma, \sigma)) \cap B = \{x_0\}$.

This implies that there is no finite collection of infinite cover $\{x_0 + N(\sigma, \sigma)\}_{x_0\in B}$ consisting of G.O. sets that covers B.

Now, consider an arbitrary cover $\{O_i\}_{i\in\Lambda}$ of G.O. sets for $C \setminus B$. Then, the cover $\{x_0 + N(\sigma, \sigma)\}_{x_0\in B} \cup \{O_i\}_{i\in\Lambda}$ of G.O. sets for C contains no finite subcover. However, this contradicts the fact that C is gradual compact, which means every open cover of C must have a finite subcover.

Hence, our initial assumption that B has no limit point in C leads to a contradiction. Therefore, we can conclude that B must have a limit point in C. □

Theorem 8.10 *Let $(X, ||\cdot||)_G$ be a GNLS and the assignment function $\Psi_{||x||_G}$ is decreasing for any $x \in X$. Then, $(X, ||\cdot||)_G$ is a Hausdorff topological vector space, whose neighborhood base of origin is given by $\{N(\kappa, \sigma) : \kappa > 0, \sigma \in (0, 1]\}$.*

Proof Consider the neighborhoods $N(\kappa_1, \sigma_1)$ and $N(\kappa_2, \sigma_2)$. Then since for $\kappa_0 = \min\{\kappa_1, \kappa_2\}$ and $\sigma_0 = \min\{\sigma_1, \sigma_2\}$, the inclusion $N(\kappa_0, \sigma_0) \subseteq N(\kappa_1, \sigma_1) \cap N(\kappa_2, \sigma_2)$ holds, so the family $\{N(\kappa, \sigma) : \kappa > 0, \sigma \in (0, 1]\}$ forms a base of the topology on X.

Now if $x_0 \in X$ such that $x_0 \neq 0$, then there exists $\sigma_0 \in (0, 1]$ such that $\Psi_{||x_0||_G}(\sigma_0) \neq 0$. Therefore, there exists some $\kappa_0 > 0$ such that $\Psi_{||x_0||_G}(\sigma_0) > \kappa_0$. In other words, $x_0 \notin N(\kappa_0, \sigma_0)$. Hence, X is a Hausdroff space. To prove the rest, it is enough if we show that the vector operations namely the addition and the scalar multiplication are continuous. □

8.4 Convergence of Sequences in GNLS

8.4.1 Basic Definitions and Properties

Convergence of sequence always plays an important role in several branches on mathematics particularly, in mathematical analysis. Several notions of calculus also can be described by considering a convergent sequence. So from that point of view, in this section, we will concentrate on the convergence of sequences in the GNLS $(X, ||\cdot||_G)$ and will present some very recently obtained results mainly by Ettefagh et al. in [12].

Definition 8.6 [36] Let (x_k) be a sequence in the GNLS $(X, ||\cdot||_G)$. If for every $\sigma \in (0, 1]$ and $\kappa > 0$, there exists $k_0(= k_0(\kappa, \sigma)) \in \mathbb{N}$ such that

$$\Psi_{||x_k - x_0||_G}(\sigma) < \kappa, \ \forall n \geq k_0,$$

then (x_k) is said to be gradual convergent to $x_0 \in X$. Symbolically, $x_k \xrightarrow{||\cdot||_G} x_0$.

Theorem 8.11 *If a sequence (x_k) in the GNLS $(X, ||\cdot||_G)$ is gradual convergent, then it cannot converge to more than one limit.*

Proof If possible suppose $x_k \xrightarrow{||\cdot||_G} x_0$ and $x_k \xrightarrow{||\cdot||_G} y_0$ for $x_0, y_0 \in X$ with $x_0 \neq y_0$. Then by definition for every $\kappa > 0$ and $\sigma \in (0, 1]$, there exists $k_0(= k_0(\kappa, \sigma)) \in \mathbb{N}$ and $k_0'(= k_0'(\kappa, \sigma)) \in \mathbb{N}$ such that $\Psi_{||x_k - x_0||_G}(\sigma) \leq \frac{\kappa}{2}$ for all $k \geq k_0$ and $\Psi_{||x_k - y_0||_G}(\sigma) \leq \frac{\kappa}{2}$ for all $k \geq k_0'$. Let $k'' = \max\{k_0, k_0'\}$. Then, for any $k \geq k''$,

$$\Psi_{||x_0 - y_0||_G}(\sigma) \leq \Psi_{||x_0 - x_k||_G}(\sigma) + \Psi_{||x_k - y_0||_G}(\sigma) < \frac{\kappa}{2} + \frac{\kappa}{2} = \kappa.$$

Since $\kappa > 0$ is arbitrary, so $\Psi_{||x_0 - y_0||_G}(\sigma) = A_{\tilde{0}}(\sigma)$ and thus $x_0 = y_0$. □

Theorem 8.12 *Let (x_k) be any arbitrary sequence in the GNLS $(X, ||\cdot||_G)$. Then, $x_k \xrightarrow{||\cdot||_G} x_0$ for some $x_0 \in X$ if and only if every gradual neighborhood of x_0 contains all but finitely many elements of the sequence (x_k).*

Proof Let $x_k \xrightarrow{||\cdot||_G} x_0$ and suppose $x_0 + N(\kappa, \sigma)$ be an arbitrary neighborhood of x_0 where $\kappa > 0$ and $\sigma \in (0, 1]$. Then by definition of gradual convergence, there exists $k_0(= k_0(\kappa, \sigma)) \in \mathbb{N}$ such that $\Psi_{||x_k - x_0||_G}(\sigma) < \kappa$ for all $k \geq k_0$. This shows that $x_k \in (x_0 + N(\kappa, \sigma))$ for any $k \geq k_0$. Hence, $x_0 + N(\kappa, \sigma)$ contains all but finitely many elements of (x_k).

Conversely, suppose an arbitrary neighborhood $x_0 + N(\kappa, \sigma)$ of x_0 contains x_k for all but finitely many k. Then, there exists some $k_0(= k_0(\kappa, \sigma)) \in \mathbb{N}$ such that $x_k \in (x_0 + N(\kappa, \sigma))$ for all $k \geq k_0$. In other words, $\Psi_{||x_k - x_0||_G}(\sigma) < \kappa$ for all $k \geq k_0$. Hence, $x_k \xrightarrow{||\cdot||_G} x_0$. □

Definition 8.7 [12] Let (x_k) be a sequence in the GNLS $(X, ||\cdot||_G)$. If for any $\sigma \in (0, 1]$, there exists some $B = B(\sigma) > 0$ such that $\Psi_{||x_k||_G}(\sigma) < B, \ \forall n \in \mathbb{N}$, then (x_k) is said to be gradual bounded.

Theorem 8.13 *Every gradual convergent sequence in a GNLS $(X, ||\cdot||_G)$ is gradual bounded.*

Proof Let (x_k) be a sequence in a GNLS $(X, ||\cdot||_G)$ such that $x_k \xrightarrow{||\cdot||_G} x_0$. Then by definition for every $\sigma \in (0, 1]$, there exists $k_0(= k_0(\kappa, \sigma)) \in \mathbb{N}$ such that $\Psi_{||x_k - x_0||_G}(\sigma) < 1$, $\forall n \geq k_0$ i.e., $x_k \in (x_0 + N(1, \sigma))$. Now put

$$B(= B(\kappa, \sigma)) = \max\{1, \Psi_{||x_1 - x_0||_G}(\sigma), \Psi_{||x_2 - x_0||_G}(\sigma), \cdots, \Psi_{||x_{k_0} - x_0||_G}(\sigma)\}.$$

Then, for every $\sigma \in (0, 1]$ and for any $k \in \mathbb{N}$, $\Psi_{||x_k - x_0||_G}(\sigma) < B$. Hence, (x_k) is gradually bounded. □

Theorem 8.14 *Let (x_k) and (y_k) be two sequences in the GNLS $(X, ||\cdot||_G)$ such that $x_k \xrightarrow{||\cdot||_G} x_0$ and $y_k \xrightarrow{||\cdot||_G} y_0$. Then,*
(i) $x_k + y_k \xrightarrow{||\cdot||_G} x_0 + y_0$ and (ii) $cx_k \xrightarrow{||\cdot||_G} cx_0$ for any $c \in \mathbb{R}$.

Proof From the assumption, for any $\kappa > 0$ and $\sigma \in (0, 1]$, there exists $k_0(= k_0(\kappa, \sigma))$ and $k_0'(= k_0'(\kappa, \sigma))$ in $\mathbb{N}$ such that $\Psi_{||x_k - x_0||} < \frac{\kappa}{2}$ for all $k \geq k_0$ and $\Psi_{||y_k - y_0||} < \frac{\kappa}{2}$ for all $k \geq k_0'$. Let $k_0'' = \max\{k_0, k_0'\}$. then for all $k \geq k_0''$ we have

$$\Psi_{||x_k + y_k - x_0 - y_0||_G}(\sigma) \leq \Psi_{||x_k - x_0||_G}(\sigma) + \Psi_{||y_k - y_0||_G}(\sigma) < \frac{\kappa}{2} + \frac{\kappa}{2} = \kappa.$$

Hence, $x_k + y_k \xrightarrow{||\cdot||_G} x_0 + y_0$.
(ii) If $c = 0$, then the statement being proved is trivially true. Therefore, let us assume that $c \neq 0$. Now since $x_k \xrightarrow{||\cdot||_G} x_0$, so for any $\kappa > 0$ and $\sigma \in (0, 1]$, there exists $k_0(= k_0(\kappa, \sigma))$ such that $\Psi_{||x_k - x_0||} < \frac{\kappa}{|c|}$ for all $k \geq k_0$. Then we have,

$$\Psi_{||cx_k - cx_0||_G}(\sigma) = |c|\Psi_{||x_k - x_0||_G}(\sigma) < |c| \cdot \frac{\kappa}{|c|} = \kappa.$$

Hence, $cx_k \xrightarrow{||\cdot||_G} cx_0$. □

8.4.2 Gradual Cauchy Condition

Definition 8.8 [36] Let (x_k) be a sequence in the GNLS $(X, ||\cdot||_G)$. Then (x_k) is said to be gradual Cauchy, if for every $\sigma \in (0, 1]$ and $\kappa > 0$, there exists $N(= N_\kappa(\sigma)) \in \mathbb{N}$ such that $\Psi_{||x_i - x_j||_G}(\sigma) < \kappa$, $\forall i, j \geq N$.

Theorem 8.15 *In a GNLS $(X, ||\cdot||_G)$, it holds that every gradual convergent sequence is also a gradual Cauchy sequence.*

Proof Let (x_k) be a sequence in the GNLS $(X, ||\cdot||_G)$ such that (x_k) gradual converges to x_0 with respect to the gradual norm $||\cdot||_G$. Then by definition for every $\sigma \in (0, 1]$, there exists $N(= N(\kappa, \sigma)) \in \mathbb{N}$ such that $\Psi_{||x_k - x_0||_G}(\sigma) < \frac{\kappa}{2}$, $\forall n \geq N$. Now for any $i, j \geq N$,

$$\begin{aligned}\Psi_{||x_i-x_j||_G}(\sigma) &= \Psi_{||x_i-x_0+x_0-x_j||_G}(\sigma)\\ &\le \Psi_{||x_i-x_0||_G}(\sigma)+\Psi_{||x_j-x_0||_G}(\sigma)\\ &\le \frac{\kappa}{2}+\frac{\kappa}{2}=\kappa.\end{aligned}$$

Hence, (x_k) is gradual Cauchy. □

Definition 8.9 [12] Let $(X, ||\cdot||_G)$ be a GNLS and $B \subseteq X$. Then, the gradual diameter of B is denoted and defined by

$$A_{G-diamB}(\sigma) = \sup\{\Psi_{||x-y||_G}(\sigma) : x, y \in B\},\ \sigma \in (0, 1].$$

Theorem 8.16 *A sequence (x_k) in the GNLS $(X, ||\cdot||_G)$ is gradual Cauchy iff for any $\sigma \in (0, 1]$, $\lim_{j\to\infty} A_{G-diamE_j}(\sigma) = A_{\tilde{0}}(\sigma)$ where $E_j = \{x_j, x_{j+1}, \cdots\}$.*

Proof Suppose (x_k) is a gradual Cauchy sequence in X. Then, by definition for every $\sigma \in (0, 1]$ and $\kappa > 0$, there exists $N(= N_\kappa(\sigma)) \in \mathbb{N}$ satisfying

$$\Psi_{||x_i-x_j||_G}(\sigma) < \kappa,\ \forall i, j \ge N.$$

Now for every $\sigma \in (0, 1]$, and $k \ge N$, we choose $x_i, x_j \in E_k$ arbitrarily. Then, we have $\Psi_{||x_i-x_j||_G}(\sigma) < \kappa$. Consequently, $A_{G-diamE_k}(\sigma) < \kappa$ and $\lim_{j\to\infty} A_{G-diamE_j}(\sigma) = A_{\tilde{0}}(\sigma)$ holds.

To prove the converse part, let $\lim_{j\to\infty} A_{G-diamE_j}(\sigma) = A_{\tilde{0}}(\sigma)$ holds for every $\sigma \in (0, 1]$. Then, for any $\sigma \in (0, 1]$ and $\kappa > 0$, there exists $N(\sigma) \in \mathbb{N}$ such that for any $k \ge N = N(\kappa, \sigma)$, $A_{G-diamE_k}(\sigma) < \kappa$. Now let $i, j \in \mathbb{N}$ be such that $i, j \ge N$. Then, $x_i, x_j \in E_N(\sigma)$ and as a consequence we have,

$$\Psi_{||x_i-x_j||_G}(\sigma) \le A_{G-diamE_k}(\sigma) < \kappa.$$

Hence, (x_k) is a gradual Cauchy sequence. □

8.5 $\mathscr{I}$ and $\mathscr{I}^*$−convergence of Sequences in GNLS

8.5.1 Basic Definitions and Examples

Definition 8.10 [26] Let X is a non-empty set. A family of subsets $\mathscr{I} \subset P(X)$ is called an ideal in X if and only if

(i) $\emptyset \in \mathscr{I}$;

(ii) for each $A, B \in \mathscr{I}$ implies $A \cup B \in \mathscr{I}$;

(iii) for each $A \in \mathscr{I}$ and $B \subset A$ implies $B \in \mathscr{I}$.

Definition 8.11 [26] Let X be a non-empty set. A family of subsets $\mathscr{F} \subset P(X)$ is called a filter on X if and only if
(i) $\emptyset \notin \mathscr{F}$;
(ii) for each $A, B \in \mathscr{F}$ implies $A \cap B \in \mathscr{F}$;
(iii) for each $A \in \mathscr{F}$ and $B \supset A$ implies $B \in \mathscr{F}$.

Some standard examples of ideal are given below.

Example 8.3 (i) The set $\mathscr{I}_f = \{A \subseteq \mathbb{N} : |A| < \infty\}$ is an admissible ideal in $\mathbb{N}$.
(ii) The set $\mathscr{I}_d = \{A \subseteq \mathbb{N} : d(A) = 0\}$ is an admissible ideal in $\mathbb{N}$.
(iii) The set $\mathscr{I}_c = \{A \subseteq \mathbb{N} : \sum_{a \in A} a^{-1} < \infty\}$ is an admissible ideal in $\mathbb{N}$.
(iv) Assume that $\mathbb{N} = \bigcup\limits_{p=1}^{\infty} D_p$ is a decomposition of the set of natural numbers, where for $i \neq j$, $D_i \cap D_j = \emptyset$. Then, the collection $\mathscr{I}$ of all subsets of $\mathbb{N}$ that intersect only a finite number of D_p's forms an ideal in $\mathbb{N}$.

Additional significant examples can be found in the references [17] and [25].

A non-trivial ideal $\mathscr{I}$ is defined as an ideal that is not empty ($\mathscr{I} \neq \emptyset$) and does not contain the entire set X ($X \notin \mathscr{I}$). The filter associated with this ideal, denoted as $\mathscr{F}(\mathscr{I})$, is defined as $\mathscr{F} = \mathscr{F}(\mathscr{I}) = \{X - A : A \in \mathscr{I}\}$. An admissible ideal in X is a non-trivial ideal $\mathscr{I}$ in $P(X)$ (the power set of X) that satisfies $\mathscr{I} \supset \{\{x\} : x \in X\}$. In Sect. 8.5.2, the notation $\mathscr{I}$ will specifically represent a non-trivial admissible ideal in $\mathbb{N}$.

Definition 8.12 [26] A sequence (x_k) of real numbers is said to be $\mathscr{I}$-convergent to l, if

$$\forall \kappa > 0, \{k \in \mathbb{N} :| x_k - l |\geq \kappa\} \in \mathscr{I}.$$

The l value is referred to as the $\mathscr{I}$-limit of the sequence (x_k) and is denoted as $\mathscr{I} - \lim x_k = l$.

Definition 8.13 [26] A sequence (x_k) of real numbers is said to be $\mathscr{I}^*-$convergent to l, if there exists a set $M = \{m_1 < m_2 < \cdots < m_k < \cdots\} \in \mathscr{F}(\mathscr{I})$ such that $\lim\limits_{k \in M} x_k = l$.

Definition 8.14 [34] Let (x_k) be a real-valued sequence. If for every $\kappa > 0$, there exists a $N \in \mathbb{N}$ such that $\{k \in \mathbb{N} : |x_k - x_N| \geq \kappa\} \in \mathscr{I}$, then (x_k) is considered to be $\mathscr{I}-$Cauchy.

Definition 8.15 [34] Let (x_k) be a real-valued sequence. If there exists a set $M = \{m_1 < m_2 < \cdots < m_i < \cdots\} \subset \mathbb{N}$, $M \in \mathscr{F}(\mathscr{I})$ so that the subsequence (x_{m_k}) is a Cauchy sequence (i.e., $\lim\limits_{i,j \to \infty} | x_{m_i} - x_{m_j} |= 0$), then (x_k) is said to be $\mathscr{I}^*-$Cauchy.

Definition 8.16 [26] Let $\mathscr{I}$ represents an ideal. If for every countable family of mutually disjoint sets $\{U_n\}_{n \in \mathbb{N}}$ from $\mathscr{I}$, there exists a countable family of sets $\{V_n\}_{n \in \mathbb{N}}$ such that the cardinality of $U_j \Delta V_j$ is finite for every $j \in \mathbb{N}$ and $\bigcup\limits_{j=1}^{\infty} V_j \in \mathscr{I}$, then $\mathscr{I}$ is said to satisfy the condition AP.

8.5.2 $\mathscr{I}$ and $\mathscr{I}^*$−convergence in GNLS, Examples, and Properties

We now move our attention to the ideal convergence in GNLS. Recently, in [2], Choudhury and Debnath have introduced the notion of $\mathscr{I}$ and $\mathscr{I}^*$ convergence of sequences. In this section, we mainly concentrate on the results obtained so far in this direction.

Definition 8.17 [2] Let (x_k) be a sequence in the GNLS $(X, ||\cdot||_G)$. If

$$\forall \kappa > 0, \sigma \in (0, 1], \ \{k \in \mathbb{N} : \Psi_{||x_k - x_0||_G}(\sigma) \geq \kappa\} \in \mathscr{I}$$

holds then (x_k) is said to be gradual $\mathscr{I}$−convergent to x_0. Symbolically, $x_k \xrightarrow{\mathscr{I}-||\cdot||_G} x_0$.

Definition 8.18 [2] Let (x_k) be a sequence in the GNLS $(X, ||\cdot||_G)$. If there exists a set $M = \{m_1 < m_2 < \cdots < m_k < \cdots\} \in \mathscr{F}(\mathscr{I})$ so that the subsequence (x_{m_k}) is gradual convergent to x_0, then (x_k) is said to be gradually $\mathscr{I}^*$−convergent to x_0. Symbolically, $x_k \xrightarrow{\mathscr{I}^*-||\cdot||_G} x_0$.

Example 8.4 [2] Consider the GNLS $(\mathbb{R}^n, ||\cdot||_G)$, where $||\cdot||_G$ is the gradual norm defined in Example 8.1. Suppose $\mathscr{I}$ denote the ideal considered in Example 8.3(iv). Then, the sequence (x_k) defined by

$$x_k = \left(0, 0, \cdots, 0, \frac{1}{p}\right), \ if \ k \in D_p$$

is gradual $\mathscr{I}$−convergent to $\mathbf{0}$ where $\mathbf{0} = (0, 0, \cdots, 0) \in \mathbb{R}^n$.

Justification. It is clear that $\Psi_{||x_k - \mathbf{0}||_G}(\sigma) = \frac{1}{p}e^{\sigma}$ for $k \in D_p$. Let $\kappa > 0$ be given. Choose $m \in \mathbb{N}$ such that $\frac{1}{m}e^{\sigma} < \kappa$. Then,

$$\{k \in \mathbb{N} : \Psi_{||x_k - \mathbf{0}||_G}(\sigma) \geq \frac{1}{m}e^{\sigma}\} \supseteq \{k \in \mathbb{N} : \Psi_{||x_k - \mathbf{0}||_G}(\sigma) \geq \kappa\}. \tag{8.2}$$

Again, since $\Psi_{||x_k - \mathbf{0}||_G}(\sigma) = \frac{1}{p}e^{\sigma}$ for $k \in D_p$, so we have

$$\{k \in \mathbb{N} : \Psi_{||x_k - \mathbf{0}||_G}(\sigma) \geq \frac{1}{m}e^{\sigma}\} = D_1 \cup D_2 \cup \cdots \cup D_m. \tag{8.3}$$

From (8.2) and (8.3), we have $\{k \in \mathbb{N} : \Psi_{||x_k - \mathbf{0}||_G}(\sigma) \geq \kappa\} \subseteq D_1 \cup D_2 \cup \cdots \cup D_m \in \mathscr{I}$.
Hence, $x_k \xrightarrow{\mathscr{I}-||\cdot||_G} \mathbf{0}$.

Theorem 8.17 *Let (x_k) be a sequence in the GNLS $(X, ||\cdot||_G)$. If $x_k \xrightarrow{||\cdot||_G} x_0$ then, $x_k \xrightarrow{\mathscr{I}-||\cdot||_G} x_0$.*

Proof The proof is straightforward, so omitted. □

The converse of Theorem 8.17 is not true, and this fact is illustrated by Example 8.5.

Example 8.5 [2] Consider the GNLS $(\mathbb{R}^n, ||\cdot||_G)$, where $||\cdot||_G$ is the gradual norm defined in Example 8.1. Consider the sequence (x_k) in $\mathbb{R}^n$ defined as

$$x_k = \begin{cases} (0, 0, \cdots, 0, n) & if\ k = s^3, s \in \mathbb{N} \\ \mathbf{0} & otherwise. \end{cases}$$

Then for any $\kappa > 0$ and $\sigma \in (0, 1]$,

$$\{k \in \mathbb{N} : \Psi_{||x_k - \mathbf{0}||_G}(\sigma) \geq \kappa\} \subseteq \{1^3, 2^3, 3^3, \cdots\} \in \mathscr{I}_d.$$

Hence $x_k \xrightarrow{\mathscr{I}_d - ||\cdot||_G} \mathbf{0}$ in $\mathbb{R}^n$.

Theorem 8.18 *Let (x_k) be any sequence in the GNLS $(X, ||\cdot||_G)$. Then, if (x_k) is gradual $\mathscr{I}-$convergent, then it cannot converge to more than one limit.*

Proof Suppose $x_k \xrightarrow{\mathscr{I} - ||\cdot||_G} x_0$ and $x_k \xrightarrow{\mathscr{I} - ||\cdot||_G} y_0$ for some $x_0 \neq y_0$ in X. Let $\kappa > 0$ be given. Then, for any $\sigma \in (0, 1]$ and $\kappa > 0$, we have, $P_1(\sigma, \kappa)$, $P_2(\sigma, \kappa) \in \mathscr{I}$ where $P_1(\sigma, \kappa) = \{k \in \mathbb{N} : \Psi_{||x_k - x_0||_G}(\sigma) \geq \kappa\}$ and $P_2(\sigma, \kappa) = \{k \in \mathbb{N} : \Psi_{||x_k - y_0||_G}(\sigma) \geq \kappa\}$. Then, $P_1(\sigma, \kappa) \cup P_2(\sigma, \kappa) \in \mathscr{I}$. Choose $m \in \mathbb{N} \setminus (P_1(\sigma, \kappa) \cup P_2(\sigma, \kappa))$, then $\Psi_{||x_m - x_0||_G}(\sigma) < \kappa$ and $\Psi_{||x_m - y_0||_G}(\sigma) < \kappa$. Hence $\Psi_{||x_0 - y_0||_G}(\sigma) \leq \Psi_{||x_m - x_0||_G}(\sigma) + \Psi_{||x_m - y_0||_G}(\sigma) < \kappa + \kappa = 2\kappa$. Since κ is arbitrary, so we must have $\Psi_{||x_0 - y_0||_G}(\sigma) = A_{\tilde{0}}(\sigma)$, which gives $x_0 = y_0$. □

Theorem 8.19 *Let (x_k) and (y_k) be two sequences in the GNLS $(X, ||\cdot||_G)$ such that $x_k \xrightarrow{\mathscr{I} - ||\cdot||_G} x_0$ and $y_k \xrightarrow{\mathscr{I} - ||\cdot||_G} y_0$. Then,*
(i) $x_k + y_k \xrightarrow{\mathscr{I} - ||\cdot||_G} x_0 + y_0$ and (ii) $cx_k \xrightarrow{\mathscr{I} - ||\cdot||_G} cx_0$.

Proof (i) From the assumption we have for given $\kappa > 0$, $P_1(\sigma, \kappa)$, $P_2(\sigma, \kappa) \in \mathscr{I}$, where $P_1(\sigma, \kappa) = \{k \in \mathbb{N} : \Psi_{||x_k - x_0||_G}(\sigma) \geq \frac{\kappa}{2}\}$ and $P_2(\sigma, \kappa) = \{k \in \mathbb{N} : \Psi_{||y_k - y_0||_G}(\sigma) \geq \frac{\kappa}{2}\}$. Now as the inclusion $(\mathbb{N} \setminus P_1(\sigma, \kappa)) \cap (\mathbb{N} \setminus P_2(\sigma, \kappa)) \subseteq \{k \in \mathbb{N} : \Psi_{||x_k + y_k - x_0 - y_0||_G}(\sigma) < \kappa\}$ holds, so we must have

$$\{k \in \mathbb{N} : \Psi_{||x_k + y_k - x_0 - y_0||_G}(\sigma) \geq \kappa\} \subseteq P_1(\sigma, \kappa) \cup P_2(\sigma, \kappa) \in \mathscr{I};$$

and consequently, $x_k + y_k \xrightarrow{\mathscr{I} - ||\cdot||_G} x_0 + y_0$.
(ii) If $c = 0$, then the proof becomes trivial. Hence, let us assume that $c \neq 0$ in order to proceed with the proof. Then since $x_k \xrightarrow{\mathscr{I} - ||\cdot||_G} x_0$, we have for given $\kappa > 0$, $P_1(\sigma, \kappa) \in \mathscr{I}$ where $P_1(\sigma, \kappa) = \{k \in \mathbb{N} : \Psi_{||x_k - x_0||_G}(\sigma) \geq \frac{\kappa}{|c|}\}$. Now since $\Psi_{||cx_k - cx_0||_G}(\sigma) = |c|\Psi_{||x_k - x_0||_G}(\sigma)$ holds for any $c \in \mathbb{R}$, we must have $P_2(\sigma, \kappa) \subseteq P_1(\sigma, \kappa)$ where $P_2(\sigma, \kappa) = \{k \in \mathbb{N} : \Psi_{||cx_k - cx||_G}(\sigma) \geq \kappa\}$, which subsequently implies $P_2(\sigma, \kappa) \in \mathscr{I}$. □

Theorem 8.20 *A sequence (x_k) in the GNLS $(X, ||\cdot||_G)$ is gradual $\mathscr{I}-$convergent to x_0 provided that every subsequence of (x_k) is gradually $\mathscr{I}-$convergent to x_0.*

Proof Let us assume the contrary. Then there exists some $\kappa > 0$ and $\sigma \in (0, 1]$ such that $P(\sigma, \kappa) \notin \mathscr{I}$, where $P(\sigma, \kappa) = \{k \in \mathbb{N} : \Psi_{||x_k - x_0||_G}(\sigma) \geq \kappa\}$. So $C(\sigma, \kappa)$ must contain infinite number of elements. Let $P(\sigma, \kappa) = \{p_1 < p_2 < \cdots < p_j < \cdots\}$. Define a sequence (y_j) as $y_j = x_{p_j}$ for $j \in \mathbb{N}$. Then, (y_j) is not gradually $\mathscr{I}-$convergent to x_0, which is a contradiction. □

Remark 8.1 The converse of the above theorem is not true. Example 8.5 works as a counter example.

Theorem 8.21 *Let (x_k) and (y_k) be two sequences in the GNLS $(X, ||\cdot||_G)$ such that $y_k \xrightarrow{||\cdot||_G} y_0$ and $\{k \in \mathbb{N} : x_k \neq y_k\} \in \mathscr{I}$. Then, $x_k \xrightarrow{\mathscr{I}-||\cdot||_G} y_0$.*

Proof Let $\{k \in \mathbb{N} : x_k \neq y_k\} \in \mathscr{I}$ holds and $y_k \xrightarrow{||\cdot||_G} y_0$. Then by definition for any $\kappa > 0$ and $\sigma \in (0, 1]$, the set $\{k \in \mathbb{N} : \Psi_{||y_k - y_0||_G}(\sigma) \geq \kappa\}$ contains finite number of elements and eventually

$$\{k \in \mathbb{N} : \Psi_{||y_k - y_0||_G}(\sigma) \geq \kappa\} \in \mathscr{I} \tag{8.4}$$

holds as $\mathscr{I}$ is admissible. Now since the inclusion

$$\{k \in \mathbb{N} : \Psi_{||y_k - y_0||_G}(\sigma) \geq \kappa\} \cap \{k \in \mathbb{N} : x_k \neq y_k\} \supseteq \{k \in \mathbb{N} : \Psi_{||x_k - y_0||_G}(\sigma) \geq \kappa\}$$

holds, so using Eq. (8.4) we obtain,

$$\{k \in \mathbb{N} : \Psi_{||x_k - y_0||_G}(\sigma) \geq \kappa\} \in \mathscr{I}.$$

Hence $x_k \xrightarrow{\mathscr{I}-||\cdot||_G} y_0$ and the proof is complete. □

Theorem 8.22 *Let (x_k) be a sequence in the GNLS $(X, ||\cdot||_G)$ such that $x_k \xrightarrow{\mathscr{I}^*-||\cdot||_G} x_0$. Then $x_k \xrightarrow{\mathscr{I}-||\cdot||_G} x_0$.*

Proof Let us assume that $x_k \xrightarrow{\mathscr{I}^*-||\cdot||_G} x_0$. Then, there exists $M = \{m_1 < m_2 < \cdots < m_k < \cdots\} \in \mathscr{F}(\mathscr{I})$ such that for every $\kappa > 0$ and $\sigma \in (0, 1]$, there exists $N(= N_\kappa(\sigma)) \in \mathbb{N}$ such that $\Psi_{||x_{m_k} - x_0||_G}(\sigma) < \kappa \; \forall k > N$. Since $\mathscr{I}$ is admissible, we must have $\{k \in \mathbb{N} : \Psi_{||x_k - x_0||_G}(\sigma) \geq \kappa\} \subseteq (\mathbb{N} \setminus M) \cup \{m_1, m_2, \cdots, m_N\} \in \mathscr{I}$. Hence $x_k \xrightarrow{\mathscr{I}-||\cdot||_G} x_0$. □

Remark 8.2 The converse of the above theorem is not true in general. Consider Example 8.4. It was shown that $x_k \xrightarrow{\mathscr{I}-||\cdot||_G} \mathbf{0}$. But the same sequence is not gradually $\mathscr{I}^*-$convergent to $\mathbf{0}$. Because for any $P \in \mathscr{I}$ there exists $p \in \mathbb{N}$ such that $P \subseteq \bigcup_{j=1}^{p} D_j$ and as a consequence $D_{p+1} \subseteq \mathbb{N} \setminus P$. Let M denote the set $\mathbb{N} \setminus P$, then $M \in \mathscr{F}(\mathscr{I})$ and (x_{m_k}) is gradual convergent to $\left(0, 0, \cdots, 0, \frac{1}{p+1}\right)$, not to $\mathbf{0}$. Hence, (x_k) is not gradually $\mathscr{I}^*-$convergent to $\mathbf{0}$.

The next theorem gives a condition under which gradual $\mathscr{I}-$convergence of a sequence implies gradual $\mathscr{I}^*-$convergence.

Theorem 8.23 *Let $\mathscr{I}$ be an ideal in $\mathbb{N}$ which satisfies the condition AP and (x_k) be a sequence in the GNLS $(X, ||\cdot||_G)$ such that $x_k \xrightarrow{\mathscr{I}-||\cdot||_G} x_0$. Then $x_k \xrightarrow{\mathscr{I}^*-||\cdot||_G} x_0$.*

Proof Let us assume that $x_k \xrightarrow{\mathscr{I}-||\cdot||_G} x_0$. Then, for every $\sigma \in (0, 1]$ and $\eta > 0$, the set $P(\sigma, \eta) = \{k \in \mathbb{N} : \Psi_{||x_k - x_0||_G}(\sigma) \geq \eta\} \in \mathscr{I}$. This enables us to construct a countable family of mutually disjoint sets $\{P_m(\sigma)\}_{m\in\mathbb{N}}$ in $\mathscr{I}$ by considering

$$P_1(\sigma) = \left\{k \in \mathbb{N} : \Psi_{||x_k - x_0||_G}(\sigma) \geq 1\right\}$$

and

$$P_m(\sigma) = \left\{k \in \mathbb{N} : \frac{1}{m} \leq \Psi_{||x_k - x_0||_G}(\sigma) < \frac{1}{m-1}\right\} = P\left(\sigma, \frac{1}{m}\right) \setminus P\left(\sigma, \frac{1}{m-1}\right), \text{ for } m \geq 2.$$

Now since $\mathscr{I}$ satisfies the condition AP, so for the above countable collection $\{P_m(\sigma)\}_{m\in\mathbb{N}}$, there exists another countable family of subsets $\{Q_m(\sigma)\}_{m\in\mathbb{N}}$ of $\mathbb{N}$ satisfying

$$P_j(\sigma) \Delta Q_j(\sigma) \text{ is finite } \forall j \in \mathbb{N} \text{ and } Q(\sigma) = \bigcup_{j=1}^{\infty} Q_j(\sigma) \in \mathscr{I}. \tag{8.5}$$

Let $\kappa > 0$ be arbitrary. Then, we can choose a positive integer m such that $\frac{1}{m+1} < \kappa$ and eventually the following inclusion is satisfied:

$$\{k \in \mathbb{N} : \Psi_{||x_k - x_0||_G}(\sigma) \geq \kappa\} \subseteq \{k \in \mathbb{N} : \Psi_{||x_k - x_0||_G}(\sigma) \geq \frac{1}{m+1}\} = \bigcup_{j=1}^{m+1} P_j(\sigma) \in \mathscr{I}.$$

Then from (8.5), we can choose an integer $k_0 \in \mathbb{N}$ satisfying the following inclusion:

$$\left(\bigcup_{j=1}^{m+1} Q_j(\sigma)\right) \cap \{k \in \mathbb{N} : k > k_0\} = \left(\bigcup_{j=1}^{m+1} P_j(\sigma)\right) \cap \{k \in \mathbb{N} : k > k_0\}.$$

Choose $k \in \mathbb{N} \setminus Q(\sigma) \in \mathscr{F}(\mathscr{I})$ such that $k > k_0$. Then we must have $k \notin \bigcup_{j=1}^{m+1} Q_j(\sigma)$ and hence $k \notin \bigcup_{j=1}^{m+1} P_j(\sigma)$. Thus we have, $\Psi_{||x_k - x_0||_G}(\sigma) < \frac{1}{m+1} < \kappa$. Hence we have $x_k \xrightarrow{\mathscr{I}^*-||\cdot||_G} x_0$. □

8.5.3 Gradual $\mathscr{I}$ and $\mathscr{I}^*$–Cauchy Condition

In this portion, our attention will be directed toward the gradual $\mathscr{I}$ and $\mathscr{I}^*$-Cauchy sequences and their correlation.

Definition 8.19 [2] Let (x_k) be a sequence in the GNLS $(X, ||\cdot||_G)$. If for any $\kappa > 0$ and $\sigma \in (0, 1]$, there exists a natural number $N(= N_\kappa(\sigma))$ such that

$$C(\sigma, \kappa) = \{k \in \mathbb{N} : \Psi_{||x_k - x_N||_G}(\sigma) \geq \kappa\} \in \mathscr{I}$$

holds, then (x_k) is said to be gradually $\mathscr{I}$–Cauchy.

Definition 8.20 [2] Let (x_k) be a sequence in the GNLS $(X, ||\cdot||_G)$. If there exists a set $M = \{m_1 < m_2 < \cdots < m_k < \cdots\} \in \mathscr{F}(\mathscr{I})$ such that the subsequence (x_{m_k}) is gradual Cauchy sequence, then (x_k) is said to be gradually $\mathscr{I}^*$–Cauchy.

Theorem 8.24 *Suppose we have a GNLS $(X, ||\cdot||_G)$. In this case, if a sequence in X is gradually $\mathscr{I}$-convergent, then it is also gradually $\mathscr{I}$-Cauchy.*

Proof Consider a sequence (x_k) in X satisfying $x_k \xrightarrow{\mathscr{I}-||\cdot||_G} x_0$. Then, for any $\sigma \in (0, 1]$ and $\kappa > 0$,

$$P(\sigma, \kappa) = \{k \in \mathbb{N} : \Psi_{||x_k - x_0||_G}(\sigma) \geq \kappa\} \in \mathscr{I}.$$

It is evident that, $\mathbb{N} \setminus P(\sigma, \kappa) \in \mathscr{F}(\mathscr{I})$ and, as a result, it is non-empty. Let's choose $N(= N_\kappa(\sigma)) \in \mathbb{N} \setminus P(\sigma, \kappa)$. By doing so, we can assert that $\Psi_{||x_k - x_N||_G}(\sigma) < \kappa$.

Assume $Q(\sigma, \kappa) = \{k \in \mathbb{N} : \Psi_{||x_k - x_N||_G}(\sigma) \geq 2\kappa\}$. Now we will demonstrate the veracity of the following inclusion:

$$Q(\sigma, \kappa) \subseteq P(\sigma, \kappa).$$

Let $p \in Q(\sigma, \kappa)$. Then,

$$2\kappa \leq \Psi_{||x_p - x_N||_G}(\sigma) \leq \Psi_{||x_p - x_0||_G}(\sigma) + \Psi_{||x_0 - x_N||_G}(\sigma) < \Psi_{||x_p - x_0||_G}(\sigma) + \kappa,$$

which implies $p \in P(\sigma, \kappa)$. Hence, we can conclude that $Q(\sigma, \kappa) \in \mathscr{I}$, which implies that (x_k) is a gradually $\mathscr{I}$-Cauchy sequence. □

Theorem 8.25 *Let (x_k) be a sequence in the gradual normed linear space (GNLS) $(X, ||\cdot||_G)$. If (x_k) is gradually $\mathscr{I}^*$-Cauchy, then it is also gradually $\mathscr{I}$-Cauchy.*

Proof Assume that (x_k) is gradually $\mathscr{I}^*$-Cauchy. By definition, there exists a set $M = \{m_1 < m_2 < \cdots < m_k < \cdots\} \in \mathscr{F}(\mathscr{I})$ so that for any $\kappa > 0$, there exists $i_0 = i_0(\sigma, \kappa) \in \mathbb{N}$ satisfying $\Psi_{||x_{m_i} - x_{m_j}||_G}(\sigma) < \kappa$ for all $i, j > i_0$.

Let $N(= N_\kappa(\sigma)) = m_{i_0+1}$, and consider any $\kappa > 0$.
We have:

$$\{k \in \mathbb{N} : \Psi_{||x_k - x_N||_G}(\sigma) \geq \kappa\} \subseteq (\mathbb{N} \setminus M) \cup \{m_1, m_2, \cdots, m_{i_0}\} \in \mathscr{I}.$$

Thus, (x_k) is gradually $\mathscr{I}$-Cauchy. □

8.6 Conclusion

This chapter begins by introducing the concept of gradual normed linear space and exploring some of its algebraic and topological properties. The focus then shifts to the conventional convergence of sequences in gradual normed linear spaces. Finally, one of the extensions of conventional convergence, known as ideal convergence of sequences in gradual normed linear spaces, is discussed. It is important to note that research on the various notions of sequence convergence in gradual normed linear spaces is still in its early stages and has not yet made significant progress. However, the existing research indicates a notable similarity in the behavior of sequence convergence in gradual normed linear spaces.

References

1. Aiche, F., Dubois, D.: Possibility and gradual number approaches to ranking methods for random fuzzy intervals. Commun. Comput. Inf. Sci. **299**, 9–18 (2012)
2. Choudhury, C., Debnath, S.: On $\mathscr{I}$-convergence of sequences in gradual normed linear spaces. Facta Univ. Ser. Math. Inform. **36**(3), 595–604 (2021)
3. Das, P., Dutta, S., Pal, S.K.: On $\mathscr{I}$ and $\mathscr{I}^*-$equal convergence and an Egoroff like theorem. Mat. Vesnik **66**(2), 165–173 (2014)
4. Das, P., Ghosal, S.K.: Some further results on $I-$Cauchy sequences and condition (AP). Comput. Math. Appl. **59**(8), 2597–2600 (2010)
5. Das, P., Malik, P.: On extremal $\mathscr{I}-$limit points of double sequences. Tatra Mt. Math. Publ. **40**, 91–102 (2008)
6. Das, P., Savas, E.: On $\mathscr{I}-$convergence of nets in locally solid Riesz spaces. Filomat **27**(1), 84–89 (2014)
7. Debnath, S., Choudhury, C.: On $\mathscr{I}$-statistically ϕ-convergence. Proyecciones **40**(3), 593–604 (2021)
8. Debnath, S., Rakshit, D.: On I-statistical convergence. Iran. J. Math. Sci. Inform. **13**(2), 101–109 (2018)
9. Dems, K.: On $\mathscr{I}-$Cauchy sequences. Real Anal. Exchange **30**(1), 123–128 (2004–2005)
10. Dubois, D., Prade, H.: Gradual elements in a fuzzy set. Soft Comput. **12**(2), 165–175 (2007)
11. Ettefagh, M., Azari, F.Y., Etemad, S.: On some topological properties in gradual normed spaces. Facta Univ. Ser. Math. **35**(3), 549–559 (2020)
12. Ettefagh, M., Etemad, S., Azari, F.Y.: Some properties of sequences in gradual normed spaces. Asian-Eur. J. Math. **13**(4), 2050085 (2020)
13. Fast, H.: Sur la convergence statistique. Colloq. Math. **2**, 241–244 (1951)

14. Fortin, J., Dubois, D., Fargier, H.: Gradual numbers and their application to fuzzy interval analysis. IEEE Trans. Fuzzy Syst. **16**(2), 388–402 (2008)
15. Fridy, J.A.: Statistical limit points. Proc. Amer. Math. Soc. **118**(4), 1187–1192 (1993)
16. Fridy, J.A.: On statistical convergence. Analysis **5**, 301–313 (1985)
17. Gogola, J., Macaj, M., Visnyai, T.: On $I_c^{(q)}-$convergence. Ann. Math. Inform. **38**, 27–36 (2011)
18. Hazarika, B.: On ideal convergent sequences in fuzzy normed linear spaces. Afr. Mat. **25**(4), 987–999 (2014)
19. Hazarika, B., Esi, A.: On asymptotically Wijsman lacunary statistical convergence of set sequences in ideal context. Filomat **31**(9), 2691–2703 (2017)
20. Hazarika, B., Kumar, V., Lafuerza-Guillen, B.: Generalized ideal convergence in intuitionistic fuzzy normed linear spaces. Filomat **27**(5), 811–820 (2013)
21. Indumathi, A., Esi, A., Subramanian, N.: Rough triple sequences in gradual normed spaces. Honam Math. J. **41**(4), 757–768 (2019)
22. Lahiri, B.K., Das, P.: Further results on $\mathscr{I}-$limit superior and limit inferior. Math. Commun. **8**(2), 151–156 (2003)
23. Lahiri, B.K., Das, P.: I and I^*-convergence in topological spaces. Math. Bohem. **130**(2), 153–160 (2005)
24. Lahiri, B.K., Das, P.: On I and I^*-convergence of nets. Real Anal. Exch. **33**(2), 431–442 (2007–2008)
25. Kostyrko, P., Macaj, M., Salat, T., Sleziak, M.: I-convergence and extremal I-limit points. Math. Slovaca **55**(4), 443–464 (2005)
26. Kostyrko, P., Salat, T., Wilczynski, W.: I-convergence. Real Anal. Exch. **26**(2), 669–686 (2000–2001)
27. Kucukaslan, M., Altinok, M.: Ideal limit superior-inferior. Gazi Univ. J. Sci. **30**(1), 401–411 (2017)
28. Kumar, V.: On $\mathscr{I}$ and $\mathscr{I}^*-$convergence of double sequences. Math. Commun. **12**(2), 171–181 (2007)
29. Kumar, V., Kumar, K.: On the ideal convergence of sequence of fuzzy numbers. Inform. Sci. **178**(24), 4670–4678 (2008)
30. Lietard, L., Rocacher, D.: Conditions with aggregates evaluated using gradual numbers. Control Cybernet. **38**(2), 395–417 (2009)
31. Mursaleen, M., Alotaibi, A.: On $\mathscr{I}-$convergence in random 2-normed spaces. Math. Slovaca **61**(6), 933–940 (2011)
32. Mursaleen, M., Debnath, S., Rakshit, D.: $I-$statistical limit superior and $I-$statistical limit inferior. Filomat **31**(7), 2103–2108 (2017)
33. Mursaleen, M., Mohiuddine, S.A.: On ideal convergence in probabilistic normed spaces. Math. Slovaca **62**(1), 49–62 (2012)
34. Nabiev, A., Pehlivan, S., Gurdal, M.: On $I-$Cauchy sequences. Taiwanese J. Math. **11**(2), 569–576 (2007)
35. Nuray, F., Gok, H., Ulusu, U.: $I_\sigma-$convergence. Math. Commun. **16**(2), 531–538 (2011)
36. Sadeqi, I., Azari, F.Y.: Gradual normed linear space. Iran. J. Fuzzy Syst. **8**(5), 131–139 (2011)
37. Salat, T.: On statistically convergent sequences of real numbers. Math. Slovaca. **30**, 139–150 (1980)
38. Salat, T., Tripathy, B.C., Ziman, M.: On some properties of $I-$convergence. Tatra Mt. Math. Publ. **28**(2), 274–286 (2004)
39. Savas, E.: Generalized asymptotically $\mathscr{I}-$lacunary statistical equivalent of order α for sequences of sets. Filomat **31**(6), 1507–1514 (2017)
40. Savas, E., Gurdal, M.: Ideal convergent function sequences in random 2-normed spaces. Filomat **30**(3), 557–567 (2016)
41. Sengul, H., Et, M.: On $I-$lacunary statistical convergence of order α of sequences of sets. Filomat **31**(8), 2403–2412 (2017)
42. Steinhaus, H.: Sur la convergence ordinaire et la convergence asymptotique. Colloq. Math. **2**, 73–74 (1951)

43. Stock, E.A.: Gradual numbers and fuzzy optimization, Ph. D. Thesis, University of Colorado Denver, Denver, America (2010)
44. Tripathy, B.C., Hazarika, B.: Paranorm I-convergent sequence spaces. Math. Slovaca **59**(4), 485–494 (2009)
45. Tripathy, B., Tripathy, B.C.: On $I-$convergent double sequences. Soochow J. Math. **31**(4), 549–560 (2005)
46. Zadeh, L.A.: Fuzzy sets. Inf. Control **8**(3), 338–353 (1965)

Chapter 9
Spectral Properties of Class of Sequences with Respect to Matrix Operators

Amar Jyoti Dutta and Binod Chandra Tripathy

Abstract Let us consider X to be a linear space, the set of all bounded linear operators defined on X into itself. We denote it by $B(X)$. Let $T \in B(X)$ and X is a Banach space then the adjoint operator T^* of T is defined as a bounded linear operator on the dual of X which is denoted by X^* and defined by $(T^*\Phi)(x) = \Phi(Tx)$ for all $\Phi \in X^*$ and $x \in X$. Let $T \in B(X)$ and $T : D(T) \to X$ generate a complex number α of the operator $(T - \alpha I)$ defined on the domain $D(T)$, which is denoted by T_α. Then $(T - \alpha I)^{-1} = T_\alpha^{-1}$ is called the resolvent operator of T_α. By default α is a complex number such that

(R1) the inverse T_α^{-1} exists,
(R2) the inverse T_α^{-1} is bounded and,
(R3) the inverse T_α^{-1} is defined on a set which is dense in X.

The resolvent set denoted by $\rho(T)$ is the set of all such values α of T and its complement, written as $\mathbb{C} \setminus \rho(T)$ is called the spectrum of the operator T denoted by $\sigma(T)$. The spectrum of an operator is partitioned into the following three disjoint sets:

1. The point spectrum is the set such that T_α^{-1} does not exist. It is denoted by $\sigma_p(T)$ and called the eigen value of T.
2. The continuous spectrum is the set such that T_α^{-1} exists and satisfies the above property (R3) but not (R2). It is denoted by $\sigma_c(T)$.
3. The residual spectrum is the set such that $(T_\alpha I)^{-1}$ exists but does not satisfy the property (R3). It is denoted by $\sigma_r(T)$.

Keywords Point spectrum · Continuous spectrum · Residual spectrum

A. J. Dutta (✉)
Department of Mathematics, Pragjyotish College, Guwahati, Assam, India
e-mail: amarjyotics@gmail.com

B. C. Tripathy
Department of Mathematics, Tripura University, Agartala, Tripura, India

B. Hazarika et al. (eds.), *Advances in Functional Analysis and Fixed-Point Theory*, Industrial and Applied Mathematics,
https://doi.org/10.1007/978-981-99-9207-2_9

9.1 Introduction

In summability theory, different classes of matrices have been investigated. Characterization of matrix classes are found in Tripathy and Sen [49], Tripathy [39], Rath and Tripathy [32] and many others. There are particular types of summability methods like Nörlund, Riesz, Euler, Abel, etc. Matrix methods have been studied from different aspects recently by Tripathy and Baruah [40] and others. Spectra of some particular type of matrix operator have been investigated from different aspects by Rath and Tripathy [32], Okutoyi [30], Rhoades [34] are a few to be named. In functional analysis, the spectrum of an operator generalizes the notion of eigenvalues for matrices. The procedure of calculation to determine the fine spectrum of an operator involves different steps of calculation of the spectrum of that operator. Many renowned authors have studied the spectrum and fine spectrum of different linear operators on some particular limitation matrices over some sequence spaces. Altay and Başar [4–6] works on the spectra of difference operator and generalized difference operator $B(r, s)$ on the classes of sequence c_0, c and ℓ_p. Okutoyi [30, 31] determined the spectrum of the Cesàro operator C_1 on the sequence space bv and bv_0. Akhmedov and Başar [3, 4] determined the fine spectra of the difference operator Δ over the class of sequence $bv_p (1 \leq p < \infty)$. Yildirim [52] worked on the spectrum of the Rhally operator over the class of sequences c_0 and c. Bilgiç and Furkan [12] works on the operators $B(r, s, t)$ and $B(r, s)$ over the class of sequences ℓ_1, bv, ℓ_p and bv_p. Altay and Karakus [8] determined the fine spectrum of the operator, Zweier Matrix. Furkan et al. [19] determined the fine spectrum of the operator $B(r, s)$ over the classes of sequence ℓ_1 and bv.

Recently Furkan et al. [18] determined the spectrum of the generalized forward difference operator Δ^r over the class of sequence ℓ_1. Altun [9] worked on the fine spectra of the tridiagonal symmetric matrices over the classes of sequence ℓ_1, c_0, c and ℓ_∞. Karakaya and Altun [25] determined the fine spectrum of upper triangular double-band matrix over the classes of sequence c_0 and c. Spectrum and fine spectra of some other matrix operators have been investigated recently by Tripathy and Pal [43–47], Tripathy and Saikia [48] Akhmedov and Başar [2], Altun and Karakaya [10], Başar and Altay [11], Das and Tripathy [16], Kayaduman and Furkan [23], Rhoades [35–37], Rhoades and Yildirim [38], Tripathy and Das [41, 42] and many others.

Different matrix methods have been studied as operators in detail by Grosse-Erdmann [22], Cartlidge [13] Malkowsky [28], Malkowsky and Parashar [29] and others. We shall follow the monographs Wilansky [51] and Kreyszig [26].

9.2 Preliminaries and Basic Results

By a sequence space, we mean a linear subspace of space of w, the class of all real or complex sequences.

Throughout this chapter we write $\ell_\infty, c, c_0, \ell_p\,(1 \le p < \infty), bv, cs$ (or γ) and bv for the spaces of all bounded, convergent, null, p-absolutely summable, convergent series and bounded variation sequences of real or complex terms, respectively.

It is observed that the continuous dual, bv^*, of bv, is equal to $c \bigoplus bs$ where $\bigoplus$ denotes the direct sum and bs denotes the space of sequences with bounded partial sums.

Let us consider $A = (a_{nk})$ to be an infinite matrix of complex numbers a_{nk}, where $n, k \in N$. We define the following transformation.

$$(Ax)_n = \sum_{k=0}^{\infty} a_{nk}x_k,\ (n \in N, x \in D_{00}(A)),$$

here, by $D_{00}(A)$ we denote the subspace of w consisting of (x_k) such that the sum exist for each $k \in \mathbb{N}$. That is, if μ is a normed sequence space, it can be written as $D_\mu(A)$ for $x \in w$, which the sum $(Ax)_n$ defined above converges in the norm of μ. We write $(\lambda : \mu) = \{A : \lambda \subseteq D_\mu(A)\}$

Let μ and ν be two sequence spaces and $A = (a_{nk})$ be an infinite matrix of real or complex numbers a_{nk}, where $n, k \in N_0 = 0, 1, 2, ...$ Then, we say that the matrix A defines a mapping from μ into ν, if for every sequence $x = (x_n) \in \mu$ the sequence $Ax = (Ax)_n$ is in where $(Ax)_n = \sum_{k=0}^{\infty} a_{nk}x_k$ $(n \in N_0)$ and $x \in \mu$), provided the right-hand side converges for every $n \in N$ and $x \in \mu$, It is denoted by $A : \mu \to \nu$.

Let X be a linear space and $B(X)$ denote the set of all bounded linear operators defined on X into itself. If $T \in B(X)$, then the adjoint operator T^* of T is a bounded linear operator on the dual X^* of X defined by $(T^*\phi)(x) = \phi(Tx)$ for all $\phi \in X^*$ and $x \in X$. Where X is a Banach space and by $D(T)$ and $R(T)$ we represent the domain and range of T respectively.

Let $X \ne \{\theta\}$ be a complex normed linear space and $T : D(T) \to X$ is a linear operator defined on $D(T) \subset X$, the domain of X. For the operator $T \in B(X)$ we introduce a complex number α with $(T - \alpha I)$ defined on the same domain $D(T)$ as T. We denote it by T_α and I is the identity operator. The inverse operator $(T - \alpha I)^{-1}$, is known as the resolvent operator of T_α and we denote it by T_α^{-1}. It is contextual to state that by closed graph theorem, the operator defined by $(T - \alpha I)^{-1}$ is always bounded.

Now we can define any regular value of T_α as a complex number α of T, where

(R_1) the inverse $(T_\alpha)^{-1}$ exists,
(R_2) the inverse $(T_\alpha)^{-1}$ is bounded
(R_3) and the inverse $(T_\alpha)^{-1}$ is defined on a dense set in X.

The set of all such regular values α of T is called the resolvent set $\rho(T, X)$ of T. The complement $C \setminus \rho(T, X)$ of the resolvent set is called the spectrum of T. It is denoted by $\sigma(T, X)$. Therefore we can say that the spectrum $\sigma(T)$ consists of those values $\alpha \in T$ for which T_α is not invertible.

9.3 Classification of Spectrum

We can classify the spectrum $\sigma(T)$, into the following three categories:

(*i*) Point(discrete) spectrum: The point spectrum is the set, where $(T_\alpha I)^{-1}$ does not exist. Here T_α is not one to one and any $\alpha \in \sigma_p(T)$ is called the eigen value of T. We denote it by $\sigma_p(T)$.

(*ii*) Continuous spectrum: The continuous spectrum is the set, where $(T_\alpha I)^{-1}$ exists and satisfies the property (R_3) but not (R_2), that is T_α is one to one but not onto. We denote it by $\sigma_c(T)$.

(*iii*) Residual spectrum: The residual spectrum is the set, where $(T_\alpha I)^{-1}$ exists (it may be bounded) but does not satisfy the property (R_3). In this case either the domain of $(T_\alpha I)^{-1}$ is not dense in X or the range of T_α is not dense in X. We denote it by $\sigma_r(T)$.

In finite-dimensional case, the continuous spectrum coincides with the residual spectrum and equal to the empty set and in this case the spectrum consists only the point spectrum only.

According to Goldberg [21], if $T_\alpha = (T_\alpha I) \in B(X)$, X a Banach space, then there are three possibilities for $R(T)$ and T^{-1}:

(I) $R(T_\alpha) = X$, $\quad$ (1) $(T_\alpha I)^{-1}$ exists and continuous,

(II) $\overline{R(T_\alpha)} = X$, but $R(T_\alpha)) \neq X$, $\quad$ (2) $(T_\alpha I)^{-1}$ exists, but discontinuous,

(III) $\overline{R(T_\alpha)} \neq X$ $\quad$ (3) $\overline{R(T_\alpha)}$ does not exist.

If the above consequences are combined in all possible ways, there are nine different states that are created which are given in the following table.

		I	II	III
		$R(T) = X$	$\overline{R(T)} = X$, and $R(T) \neq X$	$\overline{R(T)} \neq X$
1	T^{-1} exists and bounded	$\rho(T, X)$	....	$\sigma_r(T, X)$
2	T^{-1} exists and unbounded	$\sigma_c(T, X)$	$\sigma_c(T, X)$	$\sigma_r(T, X)$
3	T^{-1} does not exist	$\sigma_p(T, X)$	$\sigma_p(T, X)$	$\sigma_p(T, X)$

These are labeled as I_1, I_2, I_3, II_1, II_2, II_3, III_1, III_2 and III_3. If α is a complex number such that $T_\alpha \in I_1$ or $T_\alpha \in II_1$, then α is in the resolvent set $\rho(T, X)$. All scalar values α not in $\rho(T, X)$ comprises the spectrum of T and it is denoted by $\sigma(T, X)$. The classification of $\sigma(T, X)$ gives rise to the fine spectrum of T, represented by $I_2\sigma(T, X)$, $I_3\sigma(T, X)$, $II_2\sigma(T, X)$, $II_3\sigma(T, X)$, $III_1\sigma(T, X)$, $III_2\sigma(T, X)$, $III_3\sigma(T, X)$. For example if the operator is in the stage II_2 then it implies that $R(T) \neq \overline{R(T)} = X$ and T^{-1} exists but is discontinuous and we write it as $\alpha \in II_1\sigma(T, X)$.

9.4 Some Basic Results on Spectra of Matrix Operators

Lemma 9.1 (One may refer to Goldberg [21]) *The matrix* $A = (a_{nk})$ *gives rise to a bounded linear operator* $T \in B(\ell_1)$ *from* ℓ_1 *to itself if and only if the supremum of* ℓ_1 *norms of the columns of A is bounded.*

Corollary 9.1 $U : \ell_1 \to \ell_1$ *is a bounded linear operator with* $\| U(r, s) \|_{(\ell_1,\ell_1)} = |r| + |s|$.

Lemma 9.2 (Goldberg [21]) *The adjoint operator* T^* *of* T *is onto if and only if* T *has a bounded inverse.*

We have the following well known result.

Lemma 9.3 *Let* X *is any Banach space and* $T \in B(X)$. *Then the spectrum of* T^* *is identical with the spectrum of* T. *Furthermore* $R_\lambda(T) = (T - \lambda I)^{-1}$ *and* $\rho(T) = \{\lambda \in C : (T - \lambda I)^{-1} \text{exist}\}$.

Lemma 9.4 *The matrix* $A = (a_{nk})$, *derives a bounded linear operator* $T \in B(bv)$, *which is defined from bv to itself, if and only if* $\sup_i \sum_{n=0}^{\infty} | \sum_{k=i}^{\infty} (a_{nk} - a_{n-1,k})| < \infty$.

Bilgiç and Furkan [12] determined the point spectrum, continuous spectrum, and residual spectrum of the operator $B(r, s, t)$ over the classes of sequence ℓ_1 and bv and derived the following results.

Theorem 9.1 *Let s be a complex number such that* $\sqrt{s^2} = -s$ *and define the set s by*

$$s = \left\{ \alpha \in C : \left| \frac{2(r - \alpha)}{-s + \sqrt{s^2 - 4t(r - \alpha)}} \right| \leq 1 \right\}.$$

Then, $\sigma(B(r, s, t), \ell_1) = S$.

Theorem 9.2 $\sigma_p(B(r, s, t), \ell_1) = \emptyset$.

Theorem 9.3 *(i)* $\sigma_p(B(r, s, t)^*, \ell_1^*) = S$,
(ii) $\sigma_r B(r, s, t), \ell_1) = S$,
(iii) $\sigma_c(B(r, s, t), \ell_1) = \phi$.

Theorem 9.4 *If* $|t| < |s|$, *then* $r \in III_1 \ \sigma(B(r, s, t), \ell_1)$. *If* $|t| \geq |s|$, *then* $r \in II I_2 \sigma(B(r, s, t), \ell_1)$.

Theorem 9.5 *If* $r \neq \alpha$ *and* $\alpha \in \sigma_r(B(r, s, t), \ell_1)$ *then* $\alpha \in II I_2 \sigma(B(r, s, t), \ell_1)$.

Theorem 9.6 *The spectrum of bv is given by* $\sigma(B(r, s, t), bv) = S$, *where S is the set mentioned in Theorem 9.1.*

Theorem 9.7 *The point spectrum of bv is given by* $\sigma_p(B(r, s, t), bv) = \emptyset$.

Wenger [50] works on the fine spectrum of the integer power of the Cesàro operator in c and Rhoades [34] generalized this result to the weighted mean methods. The fine spectrum of the Cesàro operator on the sequence space ℓ_p has been studied by Gonzàlez [20], where $1 < p < \infty$. The spectrum of the Cesàro operator on the classes of sequence c_0, bv and $bv_0 = bv \backslash c_0$ was investigated by Reade [33], Akhmedov and Başar [1], and Okutoyi [31], respectively. The fine spectrum of the Rhally operators over the classes of sequence c_0 and c, under assumption that $\lim_{n\to\infty}(n+1)a_n = L \neq 0$, was examined by Yildirim [52]. Further, Coşkun [14] have studied on the spectrum of the operator p-Cesàro over the class of sequence c_0. Malafosse [27] and Altay and Başar [2] have works on the spectrum of the difference operator on the sequence spaces s_r and c_0, c, where s_r denotes the Banach space of all sequences $x = (x_k)$ normed by $\|x\|_{s_r} = \sup\limits_{k\in N} \frac{|x_k|}{r^k}$, $(r > 0)$.

9.5 Spectra of Zweier Matrices

Altay and Karakus [8] works on the spectrum of the Zweier matrix the class of sequence ℓ_1 and bv.

The Zweier matrix Z^s is a band matrix define by the following matrix representation,

$$Z^s = \begin{bmatrix} s & 0 & 0\ 0 \ldots \\ 1-s & s & 0\ 0 \ldots \\ 0 & 1-s & s\ 0 \ldots \\ . & . & .\ . \ldots \end{bmatrix},$$

where s is a real number such that $s \neq 0, 1$.

Let us define the set D_s by $D_s = \{\alpha \in C : |\alpha - s| \leq |1 - s|\}$.

Lemma 9.5 *The matrix $A = (a_{nk})$ derive a bounded linear operator $T \in B(\ell_1)$, define from ℓ_1 to itself if and only if the supremum of ℓ_1 norms of the columns of A is bounded.*

Corollary 9.2 *$Z^s : \ell_1 \to \ell_1$ is a bounded linear operator with the norm defined by $\|Z^s\|_{(\ell_1,\ell 1)} = |s| + |1 - s|$.*

Lemma 9.6 *The matrix $A = (a_{nk})$ derive a bounded linear operator $T \in B(bv)$, defined from bv to itself if and only if* $\sup\limits_{\ell} \sum\limits_{n=0}^{\infty} \sum\limits_{k=l}^{\infty} |(a_{nk} - a_{n-1,k})| < 1$.

Corollary 9.3 *$Z^s : bv \to bv$ is a bounded linear operator with the norm defined by $\|Z^s\|_{(\ell_1,\ell_1)} = \|Z^s\|_{(bv,bv)}$.*

If $T : \ell_1 \to \ell_1$ is a bounded linear operator with matrix A, then we know that the adjoint operator $T^* : \ell_1^* \to \ell_1^*$ is defined by the transpose of the matrix A. The dual space of ℓ_1 is isomorphic to ℓ_∞, the space of bounded sequences with the norm $\|x\| = \sup_k |x_k|$.

The following results are due to Altay and Karaku [8].

Theorem 9.8 $\sigma(Z^s, \ell_1) = D_s$.

Theorem 9.9 $\sigma_p(Z^s, \ell_1) = \emptyset$.

Theorem 9.10 $\sigma_p((Z^s)^*, \ell_1^*) = D_s$.

Lemma 9.7 *The operator T has a dense range if and only if T^* is one to one.*

Lemma 9.8 *The adjoint operator of T is onto if and only if T has a bounded inverse, denoted by T^*.*

Theorem 9.11 $\sigma_r(Z^s, \ell_1) = D_s$.

Theorem 9.12 $s \in III_1\sigma(Z^s, \ell_1)$.

Theorem 9.13 *If $\alpha \neq s$ and $\alpha \in D_s$, then $\alpha \in III_2\sigma(Z^s, \ell_1)$.*

Theorem 9.14 $\sigma_c(Z^s, \ell_1) = \emptyset$.

Theorem 9.15 $\sigma_p((Z^s)^*, bv^*) = D_s$.

Theorem 9.16 *(a)* $\sigma(Z^s, bv) = D_s$.
(b) $\sigma_p(Z^s, bv) = \emptyset$.
(c) $\sigma_r(Z^s, bv) = D_s$.
(d) $\sigma_c(Z^s, bv) = \emptyset$.
(e) $s \in III_1\sigma(Z^s, bv)$.
(f) If $\alpha \neq s$ and $\alpha \in \sigma_r(Z^s, bv)$, then $\alpha \in III_2\sigma(Z^s, bv)$.

9.6 Spectra of Matrix Operators Generated by Difference Operator on Sequences Spaces

Altay and Başar [7] works on the spectrum of the difference operator Δ over the class of sequence ℓ_p which was based on the works of Akhmedov and Başar [3] and Altay and Başar [7].

Lemma 9.9 *If X is a normed linear space and $T \in B(X)$, then $\sigma(T, X) = \sigma(T^*, X^*)$.*

If $T : \ell_p \to \ell_p$ define a bounded linear operator with the matrix representation A, then the adjoint operator $T^* : \ell_p^* \to \ell_p^*$ is the transpose of the matrix A.

Theorem 9.17 *The spectrum of ℓ_∞ is given by $\sigma(\Delta^*, \ell_\infty) = \{\alpha \in C : |\alpha - 1| \leq 1\}$.*

Theorem 9.18 *The spectrum of ℓ_p is given by $\sigma(\Delta, \ell_p) = \{\alpha \in C : |\alpha - 1| \leq 1\}$.*

Theorem 9.19 *The point spectrum of ℓ_p is given by $\sigma_p(\Delta, \ell_p) = \emptyset$.*

Theorem 9.20 *The point spectrum of the adjoint operator of the class of sequence ℓ_p is given by $\sigma_p(\Delta^*, \ell_p^*) = \{\alpha \in C : |\alpha - 1| \leq 1\}$.*

Lemma 9.10 *The operator T has a dense range if and only if T^* is one to one.*

Theorem 9.21 *The spectrum and the residual spectrum of ℓ_p is given by $\sigma_r(\Delta, \ell_p) = \sigma(\Delta, \ell_p) = \{\alpha \in C : |\alpha - 1| \leq 1\}$.*

The class of difference sequences $\ell_\infty(\Delta)$, $c(\Delta)$ and $c_0(\Delta)$, which consisting of all sequences $x = (x_k)$ such that $\Delta x = (x_k - x_{k+1})$ is in the sequence spaces ℓ_∞, c and c_0, were defined by Kizmaz [24]. Let λ denotes the one of the sequence spaces ℓ_∞, c and c_0. The concept was generalized by Colak and Et [15]. They defined the sequence spaces $\Delta_m \lambda$ of mth order difference sequences by

$$\Delta^m \lambda = \{x = (x_k) \in w : \Delta^m x \in \lambda\},$$

where $m \in N$, $\Delta^1 x = (x_k - x_{k+1})$ and $\Delta^m x = D^1(D^{m-1}x)$. Malkowsky and Parashar [29] also defined the spaces $\Delta^{(m)}\lambda$ of mth order difference sequences by

$$\Delta^{(m)}\lambda = \{x = (x_k) \in w : \Delta^{(m)}x \in \lambda,$$

where $m \in N$, $\Delta x = x_k - x_{k-1}$ and $D^{(m)}x = \Delta(\Delta^{(m-1)}x)$.

9.7 Spectra of Matrix Operator on bv

Recently Başar and Altay [7] defined the class of sequence p-bounded variation for $1 \leq p < \infty$, which is the difference space of p-summable sequences, given by

$$bv_p = \left\{x = (x_k) \in w : \sum_k |x_k - x_{k-1}|^p < \infty\right\}$$

Theorem 9.22 *The class of sequence bv_p is a linear space with co-ordinate wise addition and scalar multiplication which is the p-normed space with the p-norm $\|x\|_{bv_p} = \sum_k |x_k - x_{k-1}|^p$.*

Theorem 9.23 *The class of sequence bv_p is linearly isomorphic to ℓ_p i.e., $bv_p \cong \ell_p$.*

Theorem 9.24 *The inclusion $\ell_p \subset bv_p$ strictly holds.*

Lemma 9.11 *$A \in (\ell_p : c)$ if and only if*

$$\sup_{k,n \in N} |a_{nk}|^p < \infty. \tag{9.1}$$

$$\lim_{n \to \infty} a_{nk} = \alpha_k, \text{ for all } k \in N.$$

$$\sup_{k,n \in N} |a_{nk} - a_k|^p < \infty.$$

Lemma 9.12 *The matrix operator $A \in (\ell_p : \ell_\infty)$ if and only if (9.1) holds.*

Theorem 9.25 *For the class of sequence $(bv_p)^\beta = cs$.*

Theorem 9.26 *The γ-dual of the class of sequence bv_p is the set*

$$D = \left\{ a = (a_k) \in w : \sup_{k,n \in N} |\sum_{j=k}^{n} a_j| < \infty \right\}.$$

Theorem 9.27 *Let μ denote any class of sequence and A_n is the n_{th} row of an infinite matrix $A = (a_{nk})$. If the elements of the matrices $B = (b_{nk})$ and $A = (a_{nk})$ are connected by the relation*

$$b_{nk} = \sum_{j=k}^{\infty} a_{nj};\ (k, n \in N).$$

Then, $A \in (bv_p : \mu)$ if and only if $A_n \in cs$ for all $n \in N$. and $B \in (\ell_p : \mu)$.

Theorem 9.28 *Let λ, μ be any two classes of sequence and A be an infinite matrix and B a triangle matrix. Then, $A \in (\mu : \lambda_B)$ if and only if $BA \in (\mu : \lambda)$.*

By suitably selecting $\mu = \ell_p$ and $B = \Delta$ one can change the roles of the spaces bv_p and ℓ_p with μ and derive the following easy consequence:

Theorem 9.29 *Let the elements of the matrices $A = (a_{nk})$ and $C = (c_{nk})$ are connected by the relation $c_{nk} = a_{nk} - a_{n-1,k}$; $(k, n \in N)$ and μ class of sequence. Then $A \in (\mu : bv_p)$ if and only if $C \in (\mu : \ell_p)$.*

9.8 Spectra of Multidiagonal Matrix Operator

Altun [9] determine the fine spectra of the operator,with tridiagonal symmetric matrix, over the class of sequences c_0, c, ℓ_1, and ℓ_∞.

A tridiagonal symmetric infinite matrix is of the form

$$S = S(q,r) = \begin{bmatrix} q & r & 0 & 0 & 0 & 0 & \dots \\ r & q & r & 0 & 0 & 0 & \dots \\ 0 & r & q & r & 0 & 0 & \dots \\ 0 & 0 & r & q & r & 0 & \dots \\ . & . & . & . & . & . & \dots \\ . & . & . & . & . & . & \dots \end{bmatrix},$$

where $q, r \in C$.

Theorem 9.30 *Let T be an operator with the associated matrix $A = (a_{nk})$.*

(i) The operator $T \in B(c)$ if and only if

$$\|A\| = \sup_n \sum_{k=1}^{\infty} |a_{nk}| < \infty, \tag{9.2}$$

$$a_k = \lim n \to \infty a_{nk} \ \textit{exist for each} \ k \in N, \tag{9.3}$$

$$a = \lim n \to \infty \sum_{k=1}^{\infty} a_{nk} \ \textit{exists.}$$

(ii) The operator $T \in B(c_0)$ if and only if (9.2) and (9.3) with $a_k = 0$ for each k.
(iii) The operator $T \in B(\ell_\infty)$ if and only if (9.2). The operator norm of T is

$$\|T\|_{(\ell_\infty:\ell_\infty)} = \|T\|_{(c:c)} = \|T\|_{(c_0:c_0)} = \|A\|$$

(iv) The operator $T \in B(\ell_1)$ if and only if

$$\|A^t\| = \sup_k \sum_{n=1}^{\infty} |a_{nk}| < \infty.$$

Corollary 9.4 *Let $\mu \in (c_0, c, \ell_1, \ell_\infty)$. $\|S(q,r)\| : \mu \to \mu$ is a bounded linear operator and $\|S(q,r)\|_{(\mu:\mu)} = |q| + 2|r|$.*

The following results are the consequence of the above results.

Theorem 9.31 *The point spectrum, $\sigma_p(S, \mu) = \phi$ for $\mu \in \{\ell_1, c_0, c\}$.*

Theorem 9.32 *The point spectrum, $\sigma_p(S, \ell_\infty) = (q - 2r, q + 2r)$.*

Corollary 9.5 *The spectrum, $\sigma(S, \mu) \subset [q - 2r, q + 2r]$ for $\mu \in \{\ell_1, c_0, c, \ell_\infty\}$.*

Theorem 9.33 *The spectrum, $\sigma(S, \mu) = [q - 2r, q + 2r]$ for $\mu \in \{\ell_1, c_0, c, \ell_\infty\}$.*

Lemma 9.13 *The operator T has a dense range if and only if T is one to one.*

Corollary 9.6 *If $T \in (\mu : \mu)$ then $\sigma_r(T, \mu) = \sigma_p(T^*, \mu^*) \backslash \sigma_p(T, \mu)$.*

Theorem 9.34 *The residual spectrum $\sigma_r(S, c_0) = \emptyset$.*

Theorem 9.35 *The residual spectrum $\sigma_r(S, \ell_1) = (q - 2r, q + 2r)$.*

Theorem 9.36 *The residual spectrum $\sigma_r(S, c) = \{q + 2r\}$.*

Theorem 9.37 *For an operator S, the following are true,*

$$\sigma_c(S, \ell_1) = \{q - 2r, q + 2r\}.$$

$$\sigma_c(S, c_0) = [q - 2r, q + 2r].$$

$$\sigma_c(S, c) = [q - 2r, q + 2r).$$

Theorem 9.38 *Let us consider $\mu \in \{c_0, c, \ell_1, \ell_\infty\}$ such that the resolvent operator S^{-1} over μ exists and continuous. Then the domain of S^{-1} will be the whole space μ if and only if $0 \notin [q - 2r, q + 2r]$. The matrix representation (s_{nk}) of S^{-1} is defined by*

$$s_{nk} = \frac{1}{r(\alpha_1^2 - 1)} \begin{cases} \alpha_1^{n-k+1} - \alpha_1^{n+k+3}, & if\ n \geq k, \\ \alpha_1^{-n+k+1} - \alpha_1^{n+k+3}, & if\ n < k, \end{cases}$$

where α_1 is the root of the polynomial $P(x) = rx^2 + qx + r$ with $|\alpha_1| < 1$.

Theorem 9.39 *Let α_1 be the root of $P(x) = rx^2 + qx + r$ with $|\alpha_1| < 1$ with $0 \notin [q - 2r, q + 2r]$. Then for $\mu \in \{c_0, c, \ell_1, \ell_\infty\}$,*

$$\|S^{-1}\|_{\mu:\mu} = \frac{|\alpha_1| + |\alpha_1|^2}{|r||1 - \alpha_1^2(1 - |\alpha_1|)}.$$

9.9 Spectra of Generalized Difference Operator on γ

Dutta and Tripathy [17] determined the spectra of the generalized difference operator $B(r, s)$ over the sequence space γ which is the class of all convergent series. It is well-established that the class γ of all convergent series is defined by

$$\gamma = \{(x_k) \in w : (\sum_{k=0}^{n} x_k) \in c\}.$$

The generalized difference operator $B(r, s)$ is represented by

$$B(r, S) = \begin{bmatrix} r & 0 & 0 & 0 & \ldots \\ s & r & 0 & 0 & \ldots \\ 0 & s & r & 0 & \ldots \\ . & . & . & . & \ldots \\ . & . & . & . & \ldots \end{bmatrix}.$$

If $T : \gamma \to \gamma$ is a bounded linear operator with the matrix A then its adjoint operator $T^* : \gamma^* \to \gamma^*$ is defined by transpose of the matrix A and γ is isomorphic to ℓ_1 with the norm $||x|| = \sum_{k=0}^{\infty} |x_k|$.

The following lemma determines the norm and verifies that $B(r, s) \in B(\gamma)$.

Lemma 9.14 *The matrix $A = (a_{nk})$ gives rise to a bounded linear operator $T \in B(\ell_1)$ from ℓ_1 to itself if and only if the supremum of ℓ_1 norms of the columns of A is bounded.*

Lemma 9.15 *$B(r, s) : \gamma \to \gamma$ is a bounded linear operator with $||B(r, s)||_{(\gamma,\gamma)} \leq |r| + |s|$.*

They established the following results.

Theorem 9.40 $\sigma_p(B(r, s), \gamma) = \{\alpha \in C : |\alpha - r| \leq s\}$.

Theorem 9.41 $\sigma_p(B(r, s), \gamma) = \emptyset$.

Theorem 9.42 $\sigma_p(B(r, s)^*, \gamma^*) = \{\alpha \in C : |\alpha - r| < s\}$.

Theorem 9.43 $\sigma_r(B(r, s), \gamma) = \{\alpha \in C : |\alpha - r| < s\}$.

Theorem 9.44 $\sigma_c(B(r, s), \gamma) = \{\alpha \in C : |\alpha - r| = s\}$.

Karakaya and Altun [25] determined the spectrum of the operator over the sequence spaces c_0 and c with the corresponding matrix as an upper triangular double-band matrix $U(r, s)$.

A triangle is defined as a lower triangular matrix with all of the principal diagonal elements as nonzero elements. The class ℓ_∞, c and c_0 represents the spaces of all bounded, convergent and null sequences, respectively. Let μ and γ be two classes of sequence and $A = (a_{nk})$ be an infinite matrix of real or complex numbers a_{nk}, where $n, k \in N$. Then, A defines a matrix mapping from μ into γ and denoted it by $A : \mu \to \gamma$ if, for every sequence $x = (x_k) \in \mu$, the sequence $Ax = \{(Ax)_n\}$, the A-transform of x, is in γ, where

$$(Ax)_n = \sum_k a_{nk} x_k (n \in N). \tag{9.4}$$

By $(\mu : \gamma)$ we denote the class of all matrices A such that $A : \mu \to \gamma$. Thus, $A \in (\mu : \gamma)$ if and only if the series on the right-hand side of (9.4) converges for each $n \in N$ and every $x \in \mu$ and we have $Ax = \{(Ax)_n\}_{n \in N} \in \gamma$ for all $x \in \mu$. An upper triangular double-band infinite matrix is of the form

$$U = U(r, S) = \begin{bmatrix} r & s & 0\,0\,0\,\ldots \\ 0 & r & s\,0\,0\,\ldots \\ 0\,1 - s & s & 0\,0\,\ldots \\ 0 & 0 & r\,s\,0\,0.. \\ . & . & .\,.\,\ldots \\ . & . & .\,.\,\ldots \end{bmatrix}.$$

Lemma 9.16 *The matrix $A = (a_{nk}$ derives a bounded linear operator $T \in B(c)$ defined from c to itself if and only if the following two conditions are satisfied.*

(1) the rows of A in ℓ_1 and their ℓ_1 norms are bounded,
(2) the columns of A are in c_0,
(3) the sequence of row sums of A is in c.

The operator norm of T is the supremum of the ℓ_1 norms of the rows.

Corollary 9.7 *$U(r, s) : c \to c$ is a bounded linear operator and $\|U(r, s)/|_{(c:c)} = |r| + |s|$.*

Lemma 9.17 *The matrix $A = (a_{nk})$ derives a bounded linear operator $T \in B(c_0)$ defined from c_0 to itself if and only if the following two conditions are satisfied.*

(1) the rows of A in ℓ_1 and their ℓ_1 norms are bounded,
(2) the columns of A are in c_0.

The operator norm of T is the supremum of the ℓ_1 norms of the rows.

Corollary 9.8 *The operator $U(r, s) : c_0 \to c_0$ is a bounded linear operator and $\|U(r, s)\|_{c_0:c_0} = |r| + |s|$.*

Lemma 9.18 Lemma 9.4 *Let (a_n) be a bounded sequence of complex numbers such that $lim_{n\to\infty}(a_{n+1} + \alpha a_n)$ exists and is finite, say b, for some $\alpha \in C$. Then the following two cases arise:*

(a) If $|\alpha| \neq 1$, the sequence (a_n) is convergent. Moreover, in this case $lim_n \to \infty a_n = b/1 + \alpha$.
(b) For each $|\alpha| = 1$, the sequence (a_n) can be divergent.

9.10 Spectra of Upper Triangular Matrices

The studies on the spectra of lower triangular matrices opened the mind on the spectra of upper triangular matrices. The following results are established by Karakaya and Altun [25]. They have denoted the upper triangular matrix by U.

Theorem 9.45 $\sigma_p(U, c_0) = \{\lambda \in C : |\lambda - r| < |s|\}$.

Lemma 9.19 *The operator T has a dense range if and only if T^* is one to one.*

Lemma 9.20 *The operator T has a bounded inverse if and only if T^* is onto.*

Theorem 9.46 *For any $\lambda \in C$, $U_\lambda : c_0 \to c_0$ has a dense range.*

Corollary 9.9 $\sigma_r(U, c_0) = \emptyset$.

Theorem 9.47 $\sigma_c(U, c_0) = \{\lambda \in C : |\lambda - r| = |s|\}$ *and* $\sigma(U, c_0) = \{|\lambda \in C : |\lambda - r| \leq |s|\}$.

Theorem 9.48 *If $|\lambda - r| < |s|$, then $\lambda \in I_3\sigma(U, c_0)$.*

Theorem 9.49 *If $|\lambda - r| = |s|$, then $\lambda \in II_2\sigma(U, c_0)$.*

Theorem 9.50 $\sigma_p(U, c) = \{\lambda \in C|\lambda - r| < |s|\} \cup \{r + s\}$.

Theorem 9.51 *$U_\lambda : c \to c$ has a dense range if and only if $\lambda \neq r + s$.*

Corollary 9.10 $\sigma_r(U, c) = \Phi$.

Theorem 9.52 $\sigma_r(U, c) = \{\lambda \in C : |\lambda - r| = |s|\} \setminus \{r + s\}$ *and* $\sigma(U, c) = \{\lambda \in C : |\lambda - r| \leq |s|\}$.

Corollary 9.11 $\sigma(U, \ell_\infty) = \{\lambda \in C : |\lambda - r| \leq |s|$.

Theorem 9.53 *If $|\lambda - r| < |s|$, then $\lambda \in I_3\sigma(U, c)$.*

Theorem 9.54 *If $|\lambda - r| = |s|$ and $\lambda \neq r + s$, then $\lambda \in II_2(U, c)$. If $\lambda = r + s$, then $\lambda \in II_3(U, c)$.*

Theorem 9.55 *Let us suppose that $\mu \in \{c_0, c\}$. The resolvent operator of $U(r, s)$ over the class μ exists if and only if $|r| > |s|$, where the domain of the resolvent operator is the whole space μ. In this case the resolvent operator is represented by the following Toeplitz matrix*

$$U^{-1}(r, S) = \frac{1}{r}\begin{bmatrix} 1 & -\frac{s}{r} & (-\frac{s}{r})^2 & (-\frac{s}{r})^3 & 0 & 0\ldots & & & \\ 0 & 1 & -\frac{s}{r} & (-\frac{s}{r})^2 & (-\frac{s}{r})^3 & 0 & 0\ldots & & \\ 0 & 0 & 1 & -\frac{s}{r} & (-\frac{s}{r})^2 & (-\frac{s}{r})^3 & 0 & 0\ldots & \\ . & . & . & . & . & \ldots. & & & \\ . & . & . & . & . & \ldots & & & \\ . & . & . & . & . & \ldots & & & \end{bmatrix}$$

which is continuous.

Theorem 9.56 *Let $\mu \in \{c_0, c\}$. Let P be a polynomial with $P(z) = a_0 + a_1 z + \ldots.. + a_n z^n = a_n(z - z_1)(z - z_2)\ldots..(z - z_n)$. Let T be the Toeplitz matrix associated with P represented by*

$$T = \begin{bmatrix} a_0 \; a_1 \; a_2 \ldots \; a_n & 0 & 0 \; \ldots & & \\ 0 \; a_0 \; a_1 & a_2 \ldots \; a_n & 0 \; 0 \; \ldots & & \\ 0 \; 0 \; a_0 & a_1 & a_2 \ldots \; a_n \; 0 \; 0 \; \ldots & & \\ . \; . \; . & . & . \; \ldots & & \\ . \; . \; . & . & \ldots & & \\ . \; . \; . & . & \ldots & & \end{bmatrix}.$$

The resolvent operator of T over μ exists if and only if all the roots of the polynomial are outside the unit disc $\{z \in C : |z| \leq 1\}$, where the domain of the resolvent operator is the whole space μ. In this case, the resolvent operator is given by the product of infinite banded Toeplitz matrices as

$$T^{-1} = \frac{1}{a_n} U^{-1}(-z_1, 1) U^{-1}(-z_2, 1) \ldots U^{-1}(-z_n, 1),$$

where T^{-1} is continuous.

9.11 Spectra of Some Matrix Operators Representing Difference Operators

In this section we present some results on the spectra of some generalized difference operators on sequences' spaces. The result of this section is new and we established the results.

The difference operator Δ is defined as $\Delta x_k = x_k - x_{k+1}$, for all $k \in N$. The spectra of the difference operator were investigated over different classes of sequence by several Authors. Altay and Başar [5–7] investigated on the spectra of the difference operator and generalized difference operator over the sequence spaces c_0, c and ℓ_p. Okutoyi [30] determined the spectrum of the Cesàro operator C_1 over the class of sequence bv_0. Here we investigate on the spectrum of the difference operator Δ over the sequence space bv_0. The difference operator Δ is represented by the following matrix

$$\Delta = \begin{bmatrix} 1 & 0 & 0 \; 0 \ldots \\ -1 & 1 & 0 \; 0 \ldots \\ 0 & -1 & 1 \; 0 \ldots \\ . & . & . \; \ldots. \\ . & . & . \; \ldots. \end{bmatrix}.$$

For the bounded linear operator $T : bv_0 \to bv_0$ with the matrix A, the adjoint operator $T^* : bv_0^* \to bv_0^*$ is defined by the transpose of the matrix A which is given by

$$\Delta^* = \begin{bmatrix} 1 & -1 & 0 & 0... \\ 0 & 1 & -1 & 0... \\ 0 & 0 & 1 & 0... \\ . & . & . & .... \\ . & . & . & .... \end{bmatrix}$$

We procure some results which will be useful for establishing the results of this article. The following result is well known.

Lemma 9.21 *The matrix $A = (a_{nk})$ derives a bounded linear operator $T \in B(c_0)$ defined from c_0 to itself if and only if the following two conditions are satisfied:*

(i) The rows of A in ℓ_1 and their ℓ_1 norms are bounded.
(ii) The columns of A in c_0. The operator norm of T is the supremum of the ℓ_1 norms of the rows.

Moreover,

$$\begin{aligned} \|\Delta x\|_{bv_0} &= \|x_{k+1} - x_k\|_{bv_0} \\ &= \sum_{k=0}^{\infty} |(x_{k+1} - x_k) - (x_k - x_{k-1})| \\ &\leq \sum_{k=0}^{\infty} |(x_{k+1} - x_k)| + \sum_{k=0}^{\infty} (x_k - x_{k-1})| = 2\|x\|_{bv_0}. \end{aligned}$$

Hence we have the following results.

Lemma 9.22 *The operator $\Delta : bv_0 \to bv_0$ is a bounded linear operator and $|\Delta| =$* 2.

Since for any operator T on a normed linear space X, $\|T\|_X = \|T^*\|_{X^*}$ so we can write $\|\Delta\|_{bv_0} = \|\Delta^*_{bv_0} = 2$.

The following lemmas are found in Goldberg [21].

Lemma 9.23 *Let $T \in B(X)$, where X is any Banach space. Then the spectrum of T^* is identical with the spectrum of T. Furthermore $R_\lambda(T) = (T - \lambda I)^{-1}$ and $\rho(T) = \{\lambda \in C : (T - \lambda I)^{-1} exist\}$.*

Lemma 9.24 *The operator T has a dense range if and only if T^* is one to one, where T^* denote the adjoint operator of T.*

The following characterization result is well known.

Lemma 9.25 *The matrix $A = (a_{nk})$ give rise to a bounded linear operator $T \in B(bv_0)$ from bv_0 to itself if and only if $\sup_\ell \sum_n |\sum_{k=0}^{\ell} (a_{nk} - a_{n-1,k})| < \infty$ and $limit_n a_{nk} = 0$, for all $k \in N$.*

They established the following results:

Theorem 9.57 $\sigma(\Delta, bv0) = \{\alpha \in C : |\alpha - 1| \leq 1\}$.

Proof It is sufficient to show that $(\Delta - \alpha I)^{-1}$ exist and is in $B(bv_0)$ for all α such that $|\alpha - 1| > 1$.

Since $(\Delta - \alpha I)$ is a lower triangular matrix, so $(\Delta - \alpha I)^{-1}$ exists. Now solving the equation $(\Delta - \alpha I)x = y$ for x in terms of y we get the following system of equations:

$$x_0 = \frac{1}{1-\alpha} y_0$$

$$x_1 = \frac{1}{(1-\alpha)^2} y_0 + \frac{1}{1-\alpha} y_1$$

$$x_2 = \frac{1}{(1-\alpha)^3} y_0 + \frac{1}{(1-\alpha)^2} y_1 + \frac{1}{1-\alpha} y_2$$

....

....

The nth row is given by

$$x_n = \sum_{i=0}^{n} \frac{1}{(1-\alpha)^{(n+1)-i}} y_i.$$

Therefore, the matrix of $(\Delta - \alpha I)^{-1} = (a_{nk})$ is defined by

$$a_{nk} = \begin{cases} \frac{(1-\alpha)^k}{(1-\alpha^{n+1})} & for\ n \leq k, \\ 0 & otherwise, \end{cases}$$

which can be put in the following form

$$(\Delta - \alpha I)^{-1} = \begin{bmatrix} \frac{1}{1-\alpha} & 0 & 0 & 0 & 0... & \\ \frac{1}{(1-\alpha)^2} & \frac{1}{1-\alpha} & 0 & 0... & & \\ \frac{1}{(1-\alpha)^3} & \frac{1}{(1-\alpha)^2} & \frac{1}{1-\alpha} & 0 & 0... & \\ . & . & . & . & . & .... \\ . & . & . & . & . & ... \\ . & . & . & . & . & ... \end{bmatrix}.$$

Thus $x = (\Delta - \alpha I)^{-1} y$ and we have

$$x_n - x_{n-1} = \sum_{i=0} n \frac{1}{(1-\alpha)^{(n+1)-i}} (y_i - y_{i-1});\ (n \in N).$$

We observe that $(\Delta - \alpha I)^{-1} \in B(bv_0)$ if and only if $|1 - \alpha| > 1$. This completes the proof.

Theorem 9.58 $\sigma_p(\Delta, bv_0) = \emptyset$.

Proof Suppose that $\Delta x = \alpha x$, for $x \neq \theta (= (0, 0, 0, ...))$ in bv_0. Now we get the following system of equations

$$x_0 = \alpha x_0$$

$$x_1 - x_0 = \alpha x_1$$

$$x_2 - x_1 = \alpha x_2$$

$$.......$$

$$.......$$

If x_{n_0} denote the first nonzero term of the sequence $x = (x_n) \in bv_0$, then $\alpha = 1$. Hence $x_{n_0+k} = 0$, for all $k \in N$, which is a contradiction. Thus $\alpha = 0$ that is $\sigma_p(\Delta, bv_0) = \phi$. Hence the proof.

Theorem 9.59 $\sigma(\Delta^*, bv_0^*) = \{\alpha \in C : |\alpha - 1| \leq 1\}$.

Proof We have $\sigma(\Delta^*) = \sigma(\Delta)$. The proof follows from Lemma 9.23 and Theorem 9.57.

Theorem 9.60 $\sigma_p(\Delta^*, bv_0^*) = \{\alpha \in C : |\alpha - 1| < 1\}$.

Proof We consider $\Delta^* y = \alpha y$, for $y \neq \theta \in bv_0^* (= c_0 \oplus bs)$. We get the following system of equations

$$y_0 - y_1 = \alpha y_0$$

$$y_1 - y_2 = \alpha y_1$$

$$y_2 - y_3 = \alpha y_2$$

$$.......$$

$$.......$$

Solving the above equations we get

$$y_n = (1 - \alpha)^n y_0$$

Therefore, $y_0 \in bv_0^*$ if and only if

$$\sup_n \sum_{i=0}^{n} |(1 - \alpha)^i| < \infty$$

$$\Rightarrow \sup_n \frac{1-(1-\alpha)^{n+1}}{1-(1-\alpha)} < \infty \text{ if and only if} |1-\alpha| < 1.$$

This completes the proof of the theorem.

Theorem 9.61 $\sigma_r(\Delta, bv_0) = \{\alpha \in C : |\alpha - 1| < 1\}$.

Proof For $|\alpha - 1| < 1$, the operator $(\Delta - \alpha I)$ is one to one and hence has an inverse. But theorem 4 implies that $(\Delta^* - \alpha I)$ is not onc to one for $|\alpha - 1| < 1$. Therefore using lemma 4 we can conclude that the range of $(\Delta - \alpha I)$ is not dense in bv_0, that is $\overline{R(\Delta - \alpha I)}$. This completes the proof.

Theorem 9.62 $\sigma_c(\Delta, bv_0) = 1$.

Proof The proof follows from the fact that the set of spectrum is the disjoint union of the point spectrum, residual spectrum and continuous spectrum, that is

$$\sigma(\Delta, bv_0) = \sigma_p(\Delta, bv_0) \cap \sigma_r(\Delta, bv_0) \cap \sigma_c(\Delta, bv_0).$$

Conclusion

Spectral theory is one of the developing branches of research in Mathematical Sciences and has many applications in other branches of science and engineering and thereby able to connect with other branches of science and technology. For instance, in numerical analysis the spectral values may predict whether a discretization of a differential equation is possible or not and how fast a conjugate gradient iteration will converge. In aeronautics engineering the spectral values may determine whether the flow over the wing is laminar or turbulent. In electrical engineering the spectral values are used to determine the frequency response of an amplifier. In quantum mechanics the spectral values may determine the atomic energy levels and the spectral signature of a star. In structural mechanics the spectral values may be helpful to determine whether an automobile can be too noisy or whether a building will collapse in an earthquake. In ecology the spectral values may determine the period over which a food web will be settled into a steady equilibrium.

References

1. Akhmedov, A.M., Başar, F.: On spectrum of the Cesàro operator. Proc. Inst. Math. Mech. Natl. Acad. Sci. Azerb. **19**, 3–8 (2003)
2. Akhmedov, A.M., Başar, F.: On the fine spectrum of the Cesàro operator in c_0. Math. J. Ibaraki Univ. **36**, 25–32 (2004)
3. Akhmedov, A.M., Başar, F.: On the fine spectra of the difference operator Δ over the sequence space ℓ_p, $(1 \leq p < \infty)$. Demons. Math. **39**(3), 585–595 (2006)

4. Akhmedov, A.M., Başar, F.: The fine spectra of the difference operator Δ over the sequence space bv_p, $(1 \leq p<\infty)$. Acta Math. Sin. Eng. Ser. **23** (2007)
5. Altay, B., Baüşar, F.: On the fine spectrum of the difference operator Δ on c_0 and c. Inform. Sci. **168**, 217–224 (2004)
6. Altay, B., Başar, F.: On the fine spectrum of the difference operator on c_0 and c. Inform. Sci. **168**, 217–224 (2004)
7. Altay, B., Başar, F.: The fine spectrum and the matrix domain of the difference operator Δ on the sequence space ℓ_p, $(0<p<1)$. Commun. Math. Anal. **2**(2), 1–11 (2007)
8. Altay, B., Karakuş, M.: On the spectrum and the fine spectrum of the Zweier matrix as an operator on some sequence spaces, **3**(2), 153–162 (2005)
9. Altun, M.: Fine spectra of tridiagonal symmetric matrices. Int. J. Math. Math. Sci. 161209 (2011). https://doi.org/10.1155/2011/161209
10. Altun, M., Karakaya, V.: Fine spectra of lacunary matrices. Commun. Math. Anal. **7**(1), 1–10 (2009)
11. Başar, F., Altay, B.: On the space of sequences of p-bounded variation and related matrix mappings. Ukrainian Math. J. **55**(1), 136–147 (2003)
12. Bilgic, H., Furkan, H.: On the fine spectrum of the operator $B(r, s, t)$ over the sequence spaces ℓ_1 and bv. Math. Comput. Model. **45**, 883–891 (2007)
13. Cartlidge, J.P.: Weighted mean matrices as operators on ℓ^p, Ph.D. dissertation, Indiana University, Indianapolis, Ind, USA (1978)
14. Coşkun, C.: The spectra and fine spectra for p-Cesàro operators. Turkish J. Math. **21**(2), 207–212 (1997)
15. Còlak, R., Et, M.: On some generalized difference sequence spaces and related matrix transformations. Hokkaido Math. J. **26**(3), 483–492 (1997)
16. Das, R., Tripathy, B.C.: Spectrum and fine spectrum of the lower triangular matrix $B(r, s, t)$ over the sequence space cs. Songklanakarin J. Sci. Techn. **38**(3), 265–274
17. Dutta, A.J., Tripathy, B.C.: Fine spectrum of the generalized difference operator $B(r, s)$ over the class of convergent series. Int. J. Anal. 1–4 (2013). https://doi.org/10.1155/2013/630436
18. Furkan, H., Bilgic, H., Altay, B.: On the fine spectrum of the operator $B(r, s, t)$ over c_0 and c. Comput. Math. Appl. **53**(6), 989–998 (2007)
19. Furkan, H., Bilgic, H., Kayaduman, K.: On the fine spectrum of the generalized difference operator $B(r, s)$ over the sequence spaces ℓ_1 and bv. Hokkaido Math. J. **35** (2006). https://doi.org/10.14492/hokmj/1285766434
20. Gonzàlez, M.: The fine spectrum of the Cesàro operator in $\ell_p(1<p<\infty)$. Arch. Math. **44**, 355–358 (1985)
21. Goldberg, S.: Unbounded Lineer Operators. Dover Publications, Inc., New York (1985)
22. Grosse-Erdmann, K.G.: Matrix transformation between the sequence spaces of Maddox. J. Math. Anal. Appl. **180**(2), 223–238 (1993)
23. Kayaduman, K., Furkan, H.: The fine spectra of the difference operator Δ over the sequence spaces ℓ_1 and bv. Int. Math. Forum **1**(21–24), 1153–1160 (2006)
24. Kizmaz, H.: On certain sequence spaces. Canad. Math. Bull. **24**(2), 169–176 (1981)
25. Karakaya, V., Altun, M.: Fine spectra of upper triangular double-band matrices. J. Comput. Appl. Math. **234**, 1387–1394 (2010)
26. Kreyszig, E.: Introductory Functional Analysis with Applications. Wiley, New York-Chichester-Brisbane-Toronto (1978)
27. Malafosse, B.D.: Properties of some sets of sequences and application to the spaces of bounded difference sequences of order μ. Hokkaido Math. J. **31**, 283–299 (2002)
28. Malkowsky, E.: Linear operators in certain BK-spaces. Bolyai Soc. Math. Stud. **5**, 259–273 (1996)
29. Malkowsky, E., Parashar, S.D.: Matrix transformations in space of bounded and convergent difference sequence of order m. Analysis **17**, 87–97 (1997)
30. Okutoyi, J.I.: On the spectrum of C_1 as an operator on bv_0. J. Austral. Math. Soc. Ser. A **48**, 79–86 (1990)

31. Okutoyi, J.T.: On the spectrum of C_1 as an operator on bv. Commun. Fac. Sci. Univ. Ank. Ser. **A1**(41), 197–207 (1992)
32. Rath, D., Tripathy, B.C.: Matrix maps on sequence spaces associated with sets of integers. Indian Jour. Pure Appl. Math. **27**(2), 197–206 (1996)
33. Reade, J.B.: On the spectrum of the Cesàro operator. Bull. Lond. Math. Soc. **17**, 263–267 (1985)
34. Rhoades, B.E.: The fine spectra for weighted mean operators. Pacific J. Math. **104**(1), 219–230 (1983)
35. Rhoades, B.E.: The spectrum of weighted mean operators. Canad. Math. Bull. **30**(4), 446–449 (1987)
36. Rhoades, B.E.: The fine spectra of some weighted mean operators in $B(\ell_p)$. Int. Equat. Oper. Th. **12**, 82–98 (1989)
37. Rhoades, B.E.: The spectra of weighted mean operators on bv_0. J. Austral. Math. Soc. (Ser. A) **52**, 242–250 (1992)
38. Rhoades, B.E., Yildirim, M.: Spectra and fine spectra for factorable matrices. Int. Equat. Oper. Th. **53**(1), 127–144 (2005)
39. Tripathy, B.C.: Matrix transformations between some classes of sequences. J. Math. Anal. Appl. **206**, 448–450 (1997)
40. Tripathy, B.C., Baruah, A.: Nörlund and Riesz mean of sequences of fuzzy real numbers. Appl. Math. Lett. **23**, 651–655 (2010)
41. Tripathy, B.C., Das, R.: Spectra of the Rhaly operator on the sequence space $\bar{bv_0} \cap \ell_\infty$. Bol. Soc. Para. Mat. **32**(1), 263–275 (2014)
42. Tripathy, B.C., Das, R.: Spectrum and fine spectrum of the upper triangular matrix $U(r, s)$ over the sequence space cs. Proyecciones J. Math. **34**(2), 107–125 (2015)
43. Tripathy, B.C., Paul, A.: The spectrum of the operator $D(r, 0, s, 0, t)$ over the sequence spaces c_0 and c. J. Math. **2013**, Article ID 430965, 7 pages (2013)
44. Tripathy, B.C., Paul, A.: Spectra of the operator $B(f, g)$ on the vector valued sequence space $c_0(X)$. Boletim da Sociedade Paranaense de Matemática **31**(1), 105–111 (2013)
45. Tripathy, B.C., Paul, A.: The spectrum of the operator $D(r, 0, 0, s)$ over the sequence space c_0 and c. Kyungpook Math. J. **53**(2), 247–256 (2013)
46. Tripathy, B.C., Paul, A.: The spectrum of the operator $D(r, 0, 0, s)$ over the sequence spaces ℓ_p and bv_p. Hacettepe J. Math. Stat. **43**(3), 425–434 (2014)
47. Tripathy, B.C., Paul, A.: The spectrum of the operator $D(r, 0, 0, s)$ over the sequence space bv_0. Georgian J. Math. **22**(3), 421–426 (2015)
48. Tripathy, B.C., Saikia, P.: On the spectrum of the Cesáro operator C_1 on $\bar{bv} \cap \ell_\infty$. Math. Slovaca **63**(3), 563–572 (2013)
49. Tripathy, B.C., Sen, M.: Characterization of some matrix classes involving paranormed sequence spaces. Tamkang J. Math. **37**(2), 155–162 (2006)
50. Wenger, R.B.: The fine spectra of Holder summability operators. Indian J. Pure Appl. Math. **6**, 695–712 (1975)
51. Wilansky, A.: Summability Through Functional Analysis, North-Holland Mathematics Studies, vol. 85. North-Holland, Amsterdam, The Netherlands (1984)
52. Yildirim, M.: On the spectrum and fine spectrum of the compact Rhally operators. Indian J. Pure Appl. Math. **27**(8), 779–784 (1996)

Chapter 10
Some New Fixed Point Theorems Via New Classes of Condensing Operators and Their Application

Anupam Das, Bipan Hazarika, and Mohsen Rabbani

Abstract In this chapter some new condensing operators are being introduced in Banach spaces and with the help of these condensing operators a number of new fixed point theorems have been established. Also with the help of these new results the existence of solutions of integrals equations is studied and with proper examples these findings are justified.

Keywords Condensing operators · Fixed point · Integral equations

10.1 Introduction and Preliminaries

Throughout the chapter $\mathbb{R}$:= the set of real numbers, $\mathbb{R}_+ := [0, +\infty)$, $\mathbb{N}$:= the set of natural numbers, and $\mathbb{N}^* := \mathbb{N} \cup \{0\}$.

The notion of measure of noncompactness (MNC, in short) was introduced by Kuratowski [14] in the year 1930 to provide a new direction of research combined with some algebraic arguments. It is very useful for solving different types of nonlinear real-life problems under certain conditions. The Kuratowskii and Hausdorff MNCs are well-known in the metric space. Fixed point theory has two main branches: Constructive fixed point theorems in the line of the Banach Contraction Principle and nonconstructive fixed point theorems, where results are obtained by using topological properties in the direction of Brouwer's/Schauder's, and Darbo's fixed point theorem. Schauder discussed the convexity of domains and the compactness of operators. Darbo relaxed the strong condition of compactness of operators with the use of MNC

A. Das
Department of Mathematics, Cotton University, Panbazar Guwahati-781001, Assam, India

B. Hazarika (✉)
Department of Mathematics, Gauhati University, Guwahati-781014 Assam, India
e-mail: bh_rgu@yahoo.co.in

M. Rabbani
Department of Mathematics, Sari Branch, Islamic Azad University, Sari, Iran
e-mail: mrabbani@iausari.ac.ir

B. Hazarika et al. (eds.), *Advances in Functional Analysis and Fixed-Point Theory*, Industrial and Applied Mathematics,
https://doi.org/10.1007/978-981-99-9207-2_10

and defined appropriate classes of operators [8] which is extended and generalized by many authors [1, 2, 4, 16–26]. Recently in papers [34–36] authors have studied Riemann-Liouville integral, Atangana-Baleanu integral operator and nonlinear Telegraph equation via fixed point method, novel coronavirus 2019- nCoV/SARS-CoV-2 modeling in the aspect of fractional derivatives and fixed points as well as using differential operators based on the generalized Mittag-Leffler Kernel and fixed points.

Notations:

- $(\mathcal{X}, \|\cdot\|)$ is a Banach space
- θ is the zero element of $\mathcal{X}$
- $\mathcal{B}(\vartheta, \zeta)$ or $\mathcal{B}_\zeta$ denote the closed ball with center ϑ are radius ζ
- $\mathfrak{M}_{\mathcal{X}}$ denote the family of nonempty bounded subsets of $\mathcal{X}$
- $\mathfrak{N}_{\mathcal{X}}$ denote the subfamily consisting of all relatively compact subsets of $\mathcal{X}$.

Definition 10.1 [7] A MNC is a function $\varkappa : \mathfrak{M}_{\mathcal{X}} \to \mathbb{R}^+$, if it satisfies ($\Gamma, \Gamma_1, \Gamma_2 \in \mathfrak{M}_{\mathcal{X}}$):

(1°) $ker\, \varkappa := \{\Gamma \in \mathfrak{M}_{\mathcal{X}} : \varkappa(\Gamma) = 0\} \neq \emptyset$ and $ker\, \varkappa \subset \mathfrak{N}_{\mathcal{X}}$,
(2°) $\Gamma_1 \subseteq \Gamma_2 \Rightarrow \varkappa(\Gamma_1) \leq \varkappa(\Gamma_2)$,
(3°) $\varkappa(\overline{\Gamma}) = \varkappa(\Gamma)$,
(4°) $\varkappa(\overline{conv\, \Gamma}) = \varkappa(\Gamma)$,
(5°) $\varkappa(\lambda\Gamma_1 + (1-\lambda)\Gamma_2) \leq \lambda\varkappa(\Gamma_1) + (1-\lambda)\varkappa(\Gamma_2)$ for $\lambda \in [0, 1]$,
(6°) $\varkappa(\Gamma_1 \cup \Gamma_2) = \max\{\varkappa(\Gamma_1), \varkappa(\Gamma_2)\}$,
(7°) If (Γ_n) is a decreasing sequence of nonempty closed sets in $\mathfrak{M}_{\mathcal{X}}$ and if $\lim_{n\to\infty} \varkappa(\Gamma_n) = 0$, then the set $\Gamma_\infty = \bigcap_{n=1}^{\infty} \Gamma_n$ is nonempty and compact.

The family $ker\varkappa$ defined in axiom (1^0) is called the kernel of the MNC $\varkappa$.

One of the properties of the MNC is $\mathcal{B}_\infty \in ker\varkappa$. Indeed, from the inequality $\varkappa(\mathcal{B}_\infty) \leq \varkappa(\mathcal{B}_n)$ for $n = 1, 2, 3, \ldots$, we infer that $\varkappa(\mathcal{B}_\infty) = 0$.

The Kuratowski MNC is a map $\chi : \mathfrak{M}_{\mathcal{X}} \to \mathbb{R}^+$ is defined by

$$\chi(\Gamma) = \inf\left\{\epsilon > 0 : \sigma \subset \bigcup_{k=1}^{n} \mathcal{U}_k, \mathcal{U}_k \subset \mathcal{X}, \operatorname{diam}(\mathcal{U}_k) < \epsilon \; (k \in \mathbb{N})\right\}. \tag{10.1}$$

We denote $\Lambda(\mathcal{X})$ a nonempty, bounded, closed and convex set on Banach space $\mathcal{X}$.

The space of real continuous functions $\mathbf{F} = C(I)$ on I, where $I = [0, T]$, becomes a Banach space with the norm

$$\| \xi \| = \sup\{|\xi(\vartheta)| : \vartheta \in I\}, \; \xi \in \mathbf{F}.$$

Let $X (\neq \emptyset) \subseteq \mathbf{F}$ be bounded. For $\xi \in X$ and $\epsilon > 0$, denote by $\omega(\xi, \epsilon)$ the modulus of the continuity of ξ, i.e.,

$$\omega(\xi, \epsilon) = \sup\{|\xi(\vartheta_1) - \xi(\vartheta_2)| : \vartheta_1, \vartheta_2 \in I, |\vartheta_1 - \vartheta_1| \leq \epsilon\}.$$

Further, we define

$$\omega(X, \epsilon) = \sup\{\omega(\xi, \epsilon) : \xi \in X\}; \ \omega_0(X) = \lim_{\epsilon \to 0} \omega(X, \epsilon).$$

It is clear that the function ω_0 is a MNC in $\mathbf{F}$ such that the Hausdorff MNC $\varkappa$ is given by $\varkappa(X) = \frac{1}{2}\omega_0(X)$ (see [7]).

The following extensions of topological Schauder fixed point theorem 10.1 and classical Banach fixed point theorem were proved by Darbo in 1955.

Theorem 10.1 *(Shauder fixed point theorem) Let σ be an unbounded subset of a Banach space $\mathcal{X}$. Then every compact, continuous function $\mathcal{D} : \sigma \to \Gamma$ has at least one fixed oint.*

Theorem 10.2 *[8] (Darbo's fixed point theorem) Let $\sigma \subset \mathcal{X}$ and $\mathcal{D} : \Gamma \to \Gamma$ be a continuous operator such that there exists a $\lambda \in [0, 1)$ with*

$$\varkappa(\mathcal{D}(\mathcal{U})) \leq \lambda\, \varkappa(\mathcal{U})$$

for any $\mathcal{U}(\neq \emptyset) \subset \Gamma$, where $\varkappa$ is the arbitrary MNC on $\mathcal{X}$. Then $\mathcal{D}$ has at least one fixed point.

Definition 10.2 [9] Let $\mathfrak{G}$ be the collection of functions $\mathbb{E} : \mathbb{R}_+ \times \mathbb{R}_+ \to \mathbb{R}_+$ satisfying:
(1) $\max\{u, v\} \leq \mathbb{E}(u, v)$ for $u, v \geq 0$.
(2) $\mathbb{E}$ is continuous and nondecreasing.
(3) $\mathbb{E}(u_1 + u_2, v_1 + v_2) \leq \mathbb{E}(u_1, v_1) + M(u_2, v_2)$.

For example, $\mathbb{E}(u, v) = u + v$.

In this work, we extend their work with defining new contraction condition and derive some new type fixed point results.

Definition 10.3 Let $\Gamma \in \Lambda(\mathcal{X})$ and Ψ is an MNC on $\mathcal{X}$. We call $\mathcal{D} : \Gamma \to \Gamma$ is said to be a generalized condensing operator and if there exists a continuous increasing mapping $\sigma : \mathbb{R}_+ \to \mathbb{R}_+$ and a mapping $\eta : \mathfrak{M}_{\mathcal{X}} \to [0, 1)$ satisfying

$$\mathbb{E}\left(\Psi(\mathcal{D}\mathcal{B}), \sigma(\Psi(\mathcal{D}\mathcal{B}))\right) \leq \eta(\mathcal{B})\left\{\mathbb{E}\left(\Psi(\mathcal{B}), \sigma(\Psi(\mathcal{B}))\right)\right\}$$

for all $\mathcal{B} \subseteq \Gamma$. Denote this class of condensing operators by $\mathcal{K}$.

Theorem 10.3 *Let $\Gamma \in \Lambda(\mathcal{X})$ and $\mathcal{D} : \Gamma \longrightarrow \Gamma$ be a continuous and $\mathcal{D} \in \mathcal{K}$. Then $\mathcal{D}$ has at least one fixed point in Γ.*

Proof Define a sequence $\{\Gamma_n\}$ with $\Gamma_1 = \Gamma$ and $\Gamma_{n+1} = Conv(\mathcal{D}\Gamma_n), n \geq 1$. $\mathcal{D}\Gamma_1 = \mathcal{D}\Gamma \subseteq \Gamma = \Gamma_1$, $\Gamma_1 = Conv(\mathcal{D}\Gamma_1) \subseteq \Gamma = \Gamma_1$, by continuing of this process, we obtain

$$\Gamma_1 \supseteq \Gamma_2 \supseteq \Gamma_3 \supseteq \cdots.$$

If there is an integer $N \geq 0$ such that $\mathbb{E}\,(\Psi(\Gamma_N), \sigma(\Psi(\Gamma_N))) = 0$, that is, $\Psi(\Gamma_N) = 0$, then Γ_N is relatively compact and since $\mathcal{D}\Gamma_N \subseteq Conv(\mathcal{D}\Gamma_N) = \Gamma_{N+1} \subseteq \Gamma_N$. Thus, $\mathcal{D}$ has a fixed point, by Theorem 10.1. So we assume that $\mathbb{E}\,(\Psi(\Gamma_n), \sigma(\Psi(\Gamma_n))) \neq 0$ for $n \geq 1$.

Without loss of generality, we assume for any $n \geq 1$ gives $\mathbb{E}\,(\Psi(\Gamma_n), \sigma(\Psi(\Gamma_n))) > 0$. Since $\{\Psi\,(\Gamma_n)\}_{n=1}^{\infty}$ is non-negative and non-increasing sequence, so

$$\{\mathbb{E}\,(\Psi(\Gamma_n), \sigma(\Psi(\Gamma_n)))\}_{n=1}^{\infty}$$

is also positive and non-increasing sequence, hence it converges to some $C \geq 0$.

If $C = 0$ then $\mathcal{D}$ has a fixed point.

If $C > 0$, then we have

$$\begin{aligned}
\mathbb{E}\,(\Psi(\Gamma_{n+1}), \sigma(\Psi(\Gamma_{n+1}))) &= \mathbb{E}\,(\Psi(Conv\mathcal{D}\Gamma_n), \sigma(\Psi(Conv\mathcal{D}\Gamma_n))) \\
&= \mathbb{E}\,(\Psi(\mathcal{D}\Gamma_n), \sigma(\Psi(\mathcal{D}\Gamma_n))) \\
&\leq \eta(\Gamma_n)\mathbb{E}\,(\Psi(\Gamma_n), \sigma(\Psi(\Gamma_n))) \\
&\leq \eta(\Gamma_n)\eta(\Gamma_{n-1})\mathbb{E}\,(\Psi(\Gamma_{n-1}), \sigma(\Psi(\Gamma_{n-1}))) \\
&\;\;\vdots \\
&\leq \eta(\Gamma_n)\eta(\Gamma_{n-1})\ldots\eta(\Gamma_1)\mathbb{F}\,(\Psi(\Gamma_1), \sigma(\Psi(\Gamma_1))) \quad (10.2)
\end{aligned}$$

From (10.2), we obtain $\lim_{n\to\infty} \eta(\Gamma_n)\eta(\Gamma_{n-1})\ldots\eta(\Gamma_1) = 0$. Now, we get

$$\lim_{n\to\infty} \mathbb{E}\,(\Psi(\mathcal{D}\Gamma_n), \sigma(\Psi(\mathcal{D}\Gamma_n))) = 0$$

therefore $\lim\limits_{n\to\infty} \Psi(\Gamma_n) = 0$. Since $\Gamma_{n+1} \subseteq \Gamma_n$ and $\mathcal{D}\Gamma_n \subseteq \Gamma_n$ for all $n \geq 1$, then from (7^0) of Definition 10.1, $\Gamma_\infty = \bigcap_{n=1}^{\infty} \Gamma_n$ is a nonempty convex closed set, invariant under $\mathcal{D}$. Therefore Theorem 10.1 gives the desired conclusion. □

Corollary 10.1 *Let $\Gamma \in \Lambda(\mathcal{X})$ and $\mathcal{D} : \Gamma \longrightarrow \Gamma$ be a continuous and*

$$\Psi\,(\mathcal{D}\mathcal{B}) + \sigma\,(\Psi\,(\mathcal{D}\mathcal{B})) \leq \eta(\mathcal{B})\,[\Psi\,(\mathcal{B}) + \sigma\,(\Psi\,(\mathcal{B}))]$$

for all $\mathcal{B} \subset \Gamma$ then $\mathcal{D}$ has at least one fixed point in Γ.

Proof These results can be obtained by taking $\mathbb{E}(u, v) = u + v$ in Theorem 10.3. □

Corollary 10.2 *Let $\Gamma \in \Lambda(\mathcal{X})$ and $\mathcal{D} : \Gamma \longrightarrow \Gamma$ be a continuous and*

$$\Psi\,(\mathcal{D}\mathcal{B}) \leq k\Psi\,(\mathcal{B})$$

for all $\mathcal{B} \subset \Gamma$ then $\mathcal{D}$ has a fixed point in Γ.

Proof For $\Gamma \equiv 0$ and $\eta(\mathcal{B}) = k \in [0, 1)$ in Corollary 10.1. □

This can be observed that Corollary 10.2 is Darbo's fixed point theorem.

Theorem 10.4 *Let $\Gamma \in \Lambda(\mathcal{X})$ and Ψ is an arbitrary MNC in $\mathcal{X}$. Also, let I be a index set and $\{\mathcal{D}_i\}_{i\in I}$, S be continuous self-mapping on Γ such that for any $i \in I$, $\mathcal{D}_i$ commutes with S and $\mathcal{D}_i(ConvA) \subseteq Conv(T_i A)$ for any $A \subseteq \Gamma$. If there exists a mapping $\eta : \mathfrak{M}_{\mathcal{X}} \to [0, 1)$ and $\sigma : \mathbb{R}_+ \to \mathbb{R}_+$ is a continuous and nondecreasing mapping such that*

$$\mathbb{E}\left(\Psi(SA), \sigma(\Psi(SA))\right) \leq \eta(A)\left\{\mathbb{E}\left(\Psi(\mathcal{D}_i A), \sigma(\Psi(\mathcal{D}_i A))\right)\right\}$$

where $\mathbb{E} \in \mathfrak{F}$ then S has a fixed point in Γ.

Proof We consider the sequence $\{\Gamma_n\}_{n=1}^{\infty}$ defined as $\Gamma_1 = \Gamma$ and $\Gamma_{n+1} = Conv(S\Gamma_n)$ for all $n = 1, 2, \ldots$.

Then we show that $\Gamma_{n+1} \subseteq \Gamma_n$, $\mathcal{D}_i(\Gamma_n) \subseteq \Gamma_n$ for all $n = 1, 2, \ldots$ and $i \in I$.

It is obvious that $\Gamma_2 \subseteq \Gamma_1$ and

$$\begin{aligned}
&\mathcal{D}_i(\Gamma_2) \\
&= \mathcal{D}_i\left(Conv\left(S\Gamma_1\right)\right) \\
&\subseteq Conv\left(\mathcal{D}_i\left(S(\Gamma_1)\right)\right) \\
&= Conv\left(S\left(\mathcal{D}_i(\Gamma_1)\right)\right) \\
&\subseteq Conv\left(S\left(\Gamma_1\right)\right) \\
&= \Gamma_2
\end{aligned}$$

For $n = 1$, $\Gamma_{n+1} \subseteq \Gamma_n$ holds. We assume that $\Gamma_{n+1} \subseteq \Gamma_n$ holds for some n.

Therefore $\Gamma_{n+2} = Conv(S\Gamma_{n+1}) \subseteq Conv(S\Gamma_n) = \Gamma_{n+1}$. Hence by induction $\Gamma_{n+1} \subseteq \Gamma_n$ holds for $n \geq 1$.

Again for $n = 1, 2$ the following holds: $\mathcal{D}_i(\Gamma_n) \subseteq \Gamma_n$. Assume that for some n the above expression hold. Therefore,

$$\begin{aligned}
&\mathcal{D}_i(\Gamma_{n+1}) \\
&= \mathcal{D}_i\left(Conv\left(S\Gamma_n\right)\right) \\
&\subseteq Conv\left(S\left(\mathcal{D}_i(\Gamma_n)\right)\right) \\
&\subseteq Conv\left(S\left(\Gamma_n\right)\right) \\
&= \Gamma_{n+1}.
\end{aligned}$$

Hence by induction for all n, $\mathcal{D}_i(\Gamma_n) \subseteq \Gamma_n$ holds.

If there exists an integer $N \geq 0$ such that $\mathbb{F}\left(\Psi(\Gamma_N), \sigma(\Psi(\Gamma_N))\right) = 0$, that is, $\Psi(\Gamma_N) = 0$, then Γ_N is relatively compact, because $S\Gamma_N \subseteq Conv(S\Gamma_N) = \Gamma_{N+1} \subseteq \Gamma_N$. Thus, S has a fixed point, by Theorem 10.1. Therefore, we assume $\mathbb{F}\left(\Psi(\Gamma_n), \sigma(\Psi(\Gamma_n))\right) \neq 0$ for $n \geq 1$.

Without loss of generality, we assume for any $n \geq 1$ gives $\mathbb{E}\left(\Psi(\Gamma_n), \sigma(\Psi(\Gamma_n))\right) > 0$. Since $\{\Psi\left(\Gamma_n\right)\}_{n=1}^{\infty}$ is non-negative and non-increasing sequence and so

$$\{\mathbb{E}\left(\Psi(\Gamma_n), \sigma(\Psi(\Gamma_n))\right)\}_{n=1}^{\infty}$$

is also positive and non-increasing sequence and hence convergent to some $C \geq 0$.

If $C = 0$ then the result will be trivial.

If $C > 0$ then

$$\begin{aligned}
&\mathbb{E}\left(\Psi(\Gamma_{n+1}), \sigma(\Psi(\Gamma_{n+1}))\right) \\
&= \mathbb{E}\left(\Psi(ConvS\Gamma_n), \sigma(\Psi(Conv\Gamma_n))\right) \\
&= \mathbb{E}\left(\Psi(S\Gamma_n), \sigma(\Psi(\Gamma_n))\right) \\
&\leq \eta(\Gamma_n)\mathbb{E}\left(\Psi(\mathcal{D}_i\Gamma_n), \sigma(\Psi(\mathcal{D}_i\Gamma_n))\right) \\
&\leq \eta(\Gamma_n)\eta(\Gamma_{n-1})\mathbb{E}\left(\Psi(\mathcal{D}_i\Gamma_{n-1}), \sigma(\Psi(\mathcal{D}_i\Gamma_{n-1}))\right) \\
&\vdots \\
&\leq \eta(\Gamma_n)\eta(\Gamma_{n-1})\ldots\eta(\Gamma_1)\mathbb{E}\left(\Psi(\mathcal{D}_i\Gamma_1), \sigma(\Psi(\mathcal{D}_i\Gamma_1))\right).
\end{aligned}$$

As $\lim_{n\to\infty} \eta(\Gamma_n)\eta(\Gamma_{n-1})\ldots\eta(\Gamma_1) = 0$, we get

$$\lim_{n\to\infty} \mathbb{E}\left(\Psi(\mathcal{D}\Gamma_n), \sigma(\Psi(\mathcal{D}\Gamma_n))\right) = 0$$

therefore $\lim_{n\to\infty} \Psi(\sigma_n) = 0$. Since $\sigma_{n+1} \subseteq \sigma_n$ and $S\sigma_n \subseteq \sigma_n \ \forall\, n \geq 1$, then from (7^0) of Definition 10.1, $\Gamma_\infty = \bigcap_{n=1}^{\infty} \Gamma_n$ is a nonempty convex closed set, invariant under S. Therefore Theorem 10.1 gives the desired conclusion. □

Corollary 10.3 *Let $\sigma \subseteq \mathcal{X}$ and Ψ is an arbitrary MNC in the Banach space $\mathcal{X}$. Also, let I be a index set and $\{\mathcal{D}_i\}_{i\in I}$, S be continuous self-mapping on Γ such that for any $i \in I$, $\mathcal{D}_i$ commutes with S and $\mathcal{D}_i(ConvA) \subseteq Conv(T_iA)$ for any $A \subseteq \Gamma$. Tf there exists a mapping $\eta : \mathfrak{M}_{\mathcal{X}} \to [0, 1)$ and $\sigma : \mathbb{R}_+ \to \mathbb{R}_+$ is a continuous and nondecreasing mapping such that*

$$\Psi(SA) + \sigma(\Psi(SA)) \leq \eta(A)\left\{\Psi(\mathcal{D}_iA) + \sigma(\Psi(\mathcal{D}_iA))\right\}$$

then S has a fixed point in Γ.

Proof This result follows from Theorem 10.4 by taking $\mathbb{E}(u, v) = u + v$. □

Definition 10.4 Let $\phi : \mathfrak{M}_{\mathcal{X}} \to \mathbb{R}_+$ be a mapping such that $\phi(Y) \geq c > 0$ where $Y \in \mathfrak{M}_{\mathcal{X}}$.

We denote this class by Φ.

Theorem 10.5 *Let $\Gamma \in \Lambda(\mathcal{X})$ and $\mathcal{D} : \Gamma \longrightarrow \Gamma$ be a continuous and*

$$\mathbb{E}\left(\Psi(\mathcal{D}\mathcal{B}), \sigma(\Psi(\mathcal{D}\mathcal{B}))\right) \leq \mathbb{E}\left(\Psi(\mathcal{B}), \sigma(\Psi(\mathcal{B}))\right) - \phi(\mathcal{B}),\ \mathcal{B} \subseteq \Gamma,$$

where Ψ is an arbitrary NMC, $\sigma : \mathbb{R}_+ \to \mathbb{R}_+$ be continuous increasing mapping and $\phi \in \Phi$, $\mathbb{E} \in \mathfrak{F}$. Then $\mathcal{D}$ has a fixed point in Γ.

Proof Consider a sequence $\{\Gamma_n\}_{n=1}^{\infty}$ such that $\Gamma_1 = \Gamma$ and $\Gamma_{n+1} = ConvT\Gamma_n,\ n \geq 1$.

Now $\mathcal{D}\Gamma_1 = \mathcal{D}\Gamma \subseteq \Gamma = \Gamma_1$, $\Gamma_1 = Conv(\mathcal{D}\Gamma_1) \subseteq \Gamma = \Gamma_1$, therefore by continuing this process we have

$$\Gamma_1 \supseteq \Gamma_2 \supseteq \Gamma_3 \supseteq \cdots.$$

If there exists an integer $N \geq 0$ such that $\mathbb{F}(\Psi(\Gamma_N), \sigma(\Psi(\Gamma_N))) = 0$, that is, $\Psi(\Gamma_N) = 0$, then Γ_N is relatively compact, because $\mathcal{D}\Gamma_N \subseteq Conv(\mathcal{D}\Gamma_N) = \Gamma_{N+1} \subseteq \Gamma_N$. Thus, Theorem 10.1 implies that $\mathcal{D}$ has a fixed point. We assume that $\mathbb{E}(\Psi(\Gamma_n), \sigma(\Psi(\Gamma_n))) \neq 0$ for $n \geq 1$.

Without loss of generality, we assume for any $n \geq 1$ gives $\mathbb{F}(\Psi(\Gamma_n), \sigma(\Psi(\Gamma_n))) > 0$. Since $\{\Psi(\Gamma_n)\}_{n=1}^{\infty}$ is non-negative and non-increasing sequence and so

$$\{\mathbb{E}(\Psi(\Gamma_n), \sigma(\Psi(\Gamma_n)))\}_{n=1}^{\infty}$$

is also positive and non-increasing sequence and hence convergent to some $b \geq 0$.

If $b = 0$ then the result is trivial.

If possible let $b > 0$. Then

$$\begin{aligned}
&\mathbb{E}(\Psi(\Gamma_{n+1}), \sigma(\Psi(\Gamma_{n+1}))) \\
&= \mathbb{E}(\Psi(Conv\mathcal{D}\Gamma_n), \sigma(\Psi(Conv\mathcal{D}\Gamma_n))) \\
&= \mathbb{E}(\Psi(\mathcal{D}\Gamma_n), \sigma(\Psi(\mathcal{D}\Gamma_n))) \\
&\leq \mathbb{E}(\Psi(\Gamma_n), \sigma(\Psi(\Gamma_n))) - \phi(\Gamma_n).
\end{aligned}$$

As $n \to \infty$ then

$$b \leq b - c \implies c \leq 0$$

which is a contradiction, hence $b = 0$.

Now, we get

$$\lim_{n\to\infty} \mathbb{E}(\Psi(\mathcal{D}\Gamma_n), \sigma(\Psi(\mathcal{D}\Gamma_n))) = 0$$

therefore $\lim_{n\to\infty} \Psi(\Gamma_n) = 0$. Since $\Gamma_{n+1} \subseteq \Gamma_n$ and $\mathcal{D}\Gamma_n \subseteq \Gamma_n$ for all $n \geq 1$, then from (7^0) of Definition 10.1, $\Gamma_\infty = \bigcap_{n=1}^{\infty} \Gamma_n$ is a nonempty convex closed set, invariant under $\mathcal{D}$. Therefore Theorem 10.1 gives the desired conclusion. □

Corollary 10.4 *Let $\Gamma \in \Lambda(\mathcal{X})$ and $\mathcal{D} : \Gamma \longrightarrow \Gamma$ be a continuous and*

$$\Psi(\mathcal{D}\mathcal{B}) + \sigma(\Psi(\mathcal{D}\mathcal{B})) \leq \Psi(\mathcal{B}) + \sigma(\Psi(\mathcal{B})) - \phi(\mathcal{B}),\ \mathcal{B} \subseteq \Gamma$$

where Ψ is an arbitrary NMC, $\sigma : \mathbb{R}_+ \to \mathbb{R}_+$ be continuous increasing mapping and $\phi \in \Phi$ Then $\mathcal{D}$ has at least one fixed point in Γ.

Proof This result is obtained by taking $\mathbb{E}(u, v) = u + v$ in Theorem 10.5. □

10.2 Application

We recall the following fractional integrals:

Let $w(x) \neq 0,\ w^{-1}(x) = \frac{1}{w(x)}$ and g is strictly increasing differentiable functions.

The weighted fractional integral of a continuous function f on $[a, \infty),\ a \in \mathbb{R}$ by another function g of order χ is defined by [10],

$$\left({}_{a+}J_w^{\chi} f\right)(x) = \frac{1}{w(x)\Gamma(\chi)} \int_a^x \frac{w(t)g'(t)f(t)}{(g(x) - g(t))^{1-\chi}} dt,\ \chi > 0,\ x > a.$$

In this part, we study the following fractional integral equation:

$$x(t) = F(t, x(t)) + \frac{Q(t, x(t))}{w(t)\Gamma(\chi)} \int_0^t \frac{w(s)g'(s)G(s, x(s))}{(g(t) - g(s))^{1-\chi}} ds \tag{10.3}$$

where $\chi > 0,\ t \in I = [0, T]$.
Let

$$B_{r_0} = \{x \in \mathbf{E} : \| x \| \leq r_0\}.$$

Assume that

(A) $F,\ Q,\ G : I \times \mathbb{R} \to \mathbb{R}$ be continuous and there exists constants $F_c,\ Q_c,\ G_c \geq 0$ satisfying for all $t \in I,\ x \in E$,

$$|F(t, x) - F(t, \bar{x})| \leq F_c \, |x - \bar{x}|,$$

$$|Q(t, x) - Q(t, \bar{x})| \leq Q_c \, |x - \bar{x}|,$$

and

$$|G(t, x) - G(t, \bar{x})| \leq G_c \, |x - \bar{x}|.$$

Also,

$$\max |F(t, 0)| = F_M,\ Q(t, 0) = G(t, 0) = 0.$$

(B) The functions $g : I \to \mathbb{R}_+,\ w : I \to \mathbb{R}$ is C^1 and nondecreasing. Also let

$$0 < l \leq |w(t)| \leq L$$

for all $x \in I$. Also, g' is non-negative.

(C) there exists a positive solution r_0 of the inequality

$$F_c r + \frac{L G_c Q_c r^2 (g(T) - g(0))^{\chi}}{l \Gamma(\chi + 1)} + F_M < r.$$

Theorem 10.6 *If conditions (A)–(C) hold, then Eq. (10.3) has a solution in* $\mathbf{F} = C(I)$.

Proof Define an operator $\mathcal{D} : \mathbf{F} \to \mathbf{F}$ as follows:

$$(\mathcal{D}x)(t) = F(t, x(t)) + \frac{Q(t, x(t))}{w(t)\Gamma(\chi)} \int_0^t \frac{w(s)g'(s)G(s, x(s))}{(g(t) - g(s))^{1-\chi}} ds.$$

Step 1: To show that $\mathcal{D}$ is self-mapping on B_{r_0}. We have

$$\begin{aligned}
&|(\mathcal{D}x)(t)| \\
&\leq |F(t, x(t))| + \frac{|Q(t, x(t))|}{|w(t)|\,\Gamma(\chi)} \int_0^t \frac{|w(s)|\,g'(s)\,|G(s, x(s))|}{(g(t) - g(s))^{1-\chi}} ds \\
&\leq |F(t, x(t)) - F(t, 0)| + |F(t, 0)| \\
&+ \frac{|Q(t, x(t)) - Q(t, 0)|\, L}{l\Gamma(\chi)} \int_0^t \frac{g'(s)\,|G(s, x(s)) - G(s, 0)|}{(g(t) - g(s))^{1-\chi}} ds \\
&\leq F_c\,|x(t)| + \frac{LQ_c\,|x(t)|}{l\Gamma(\chi)} \int_0^t \frac{g'(s)G_c\,|x(s)|}{(g(t) - g(s))^{1-\chi}} ds + F_M \\
&\leq F_c \parallel x \parallel + \frac{LQ_cG_c \parallel x \parallel^2}{l\Gamma(\chi)} \int_0^t \frac{g'(s)}{(g(t) - g(s))^{1-\chi}} ds + F_M \\
&\leq F_c \parallel x \parallel + \frac{LQ_cG_c \parallel x \parallel^2}{l\Gamma(\chi)} . \frac{(g(t) - g(0))^{\chi}}{\chi} + F_M \\
&\leq F_c \parallel x \parallel + \frac{LQ_cG_c \parallel x \parallel^2 (g(T) - g(0))^{\chi}}{l\Gamma(\chi + 1)} + F_M.
\end{aligned}$$

For $\parallel x \parallel \leq r_0$ and c condition concludes that,

$$\parallel T \parallel \leq F_c r_0 + \frac{LQ_cG_c r_0^2 (g(T) - g(0))^{\chi}}{l\Gamma(\chi + 1)} + F_M < r_0.$$

Due to the assumption (D) $\mathcal{D}$ maps B_{r_0} into B_{r_0}.

Step 2: We prove that $\mathcal{D}$ is continuous on B_{r_0}. Let $\epsilon > 0$ and $x, \bar{x} \in B_{r_0}$ such that $\parallel x - \bar{x} \parallel < \epsilon$. We have

$$\begin{aligned}
&|(\mathcal{D}x)(t) - (\mathcal{D}\bar{x})(t)| \\
&\leq |F(t, x(t) - F(t, \bar{x}(t))| \\
&+ \frac{1}{|w(t)|\,\Gamma(\chi)} \left| Q(t, x(t)) \int_0^t \frac{w(s)g'(s)G(s, x(s))}{(g(t) - g(s))^{1-\chi}} ds - Q(t, \bar{x}(t)) \int_0^t \frac{w(s)g'(s)G(s, \bar{x}(s))}{(g(t) - g(s))^{1-\chi}} ds \right| \\
&\leq F_c\,|x(t) - \bar{x}(t)| + \frac{1}{l\Gamma(\chi)} \left| Q(t, x(t)) \int_0^t \frac{w(s)g'(s)\,\{G(s, x(s)) - G(s, \bar{x}(s))\}}{(g(t) - g(s))^{1-\chi}} ds \right| \\
&+ \frac{1}{l\Gamma(\chi)} \left| \{Q(t, x(t)) - Q(t, \bar{x}(t))\} \int_0^t \frac{w(s)g'(s)\,\{G(s, \bar{x}(s))\}}{(g(t) - g(s))^{1-\chi}} ds \right|
\end{aligned}$$

$$\leq F_c |x(t) - \bar{x}(t)| + \frac{LQ_c |x(t)|}{l\Gamma(\chi)} \int_0^t \frac{g'(s)G_c |x(s) - \bar{x}(s)|}{(g(t) - g(s))^{1-\chi}} ds$$
$$+ \frac{LQ_c |x(t) - \bar{x}(t)|}{l\Gamma(\chi)} \int_0^t \frac{g'(s)G_c |\bar{x}(s)|}{(g(t) - g(s))^{1-\chi}} ds$$
$$\leq F_c \parallel x - \bar{x} \parallel + \frac{LQ_c G_c \parallel x \parallel \parallel x - \bar{x} \parallel}{l\Gamma(\chi)} \int_0^t \frac{g'(s)}{(g(t) - g(s))^{1-\chi}} ds$$
$$+ \frac{LQ_c G_c \parallel \bar{x} \parallel \parallel x - \bar{x} \parallel}{l\Gamma(\chi)} \int_0^t \frac{g'(s)}{(g(t) - g(s))^{1-\chi}} ds$$
$$< \epsilon + \frac{2LQ_c G_c r_0 \epsilon \, (g(T) - g(0))^{\chi}}{l\Gamma(\chi + 1)}.$$

As $\epsilon \to 0$ we get $|(\mathcal{D}x)(t) - (\mathcal{D}\bar{x})(t)| \to 0$. This shows that $\mathcal{D}$ is continuous on B_{r_0}.

Step 3: An estimate of $\mathcal{D}$ with respect to Γ_0: Assume that $\Gamma (\neq \emptyset) \subseteq B_{r_0}$. Let $\epsilon > 0$ be arbitrary and choose $x \in \Gamma$ and $t_1, t_2 \in I$ such that $|t_2 - t_1| \leq \epsilon$ and $t_2 \geq t_1$.

Now,

$$|(\mathcal{D}x)(t_2) - (\mathcal{D}x)(t_1)|$$
$$\leq |F(t_2, x(t_2) - F(t_1, x(t_1))|$$
$$+ \frac{1}{\Gamma(\chi)} \left| \frac{Q(t_2, x(t_2))}{w(t_2)} \int_0^{t_2} \frac{w(s)g'(s)G(s, x(s))}{(g(t_2) - g(s))^{1-\chi}} ds - \frac{Q(t_1, x(t_1))}{w(t_1)} \int_0^{t_1} \frac{w(s)g'(s)G(s, x(s))}{(g(t_1) - g(s))^{1-\chi}} ds \right|.$$

We have

$$\left| \frac{Q(t_2, x(t_2))}{w(t_2)} \int_0^{t_2} \frac{w(s)g'(s)G(s, x(s))}{(g(t_2) - g(s))^{1-\chi}} ds - \frac{Q(t_1, x(t_1))}{w(t_1)} \int_0^{t_1} \frac{w(s)g'(s)G(s, x(s))}{(g(t_1) - g(s))^{1-\chi}} ds \right|$$
$$\leq \left| \frac{Q(t_2, x(t_2))}{w(t_2)} \right| \left| \int_0^{t_2} \frac{w(s)g'(s)G(s, x(s))}{(g(t_2) - g(s))^{1-\chi}} ds - \int_0^{t_1} \frac{w(s)g'(s)G(s, x(s))}{(g(t_1) - g(s))^{1-\chi}} ds \right|$$
$$+ \left| \frac{Q(t_2, x(t_2))}{w(t_2)} - \frac{Q(t_1, x(t_1))}{w(t_1)} \right| \int_0^{t_1} \frac{|w(s)| \, g'(s) \, |G(s, x(s))|}{(g(t_1) - g(s))^{1-\chi}} ds$$
$$\leq \frac{Q_c \parallel x \parallel}{l} \left| \int_0^{t_2} \frac{w(s)g'(s)G(s, x(s))}{(g(t_2) - g(s))^{1-\chi}} ds - \int_0^{t_1} \frac{w(s)g'(s)G(s, x(s))}{(g(t_1) - g(s))^{1-\chi}} ds \right|$$
$$+ \left| \frac{Q(t_2, x(t_2))}{w(t_2)} - \frac{Q(t_1, x(t_1))}{w(t_1)} \right| \frac{G_c L \parallel x \parallel (g(T) - g(0))^{\chi}}{\chi}$$

and

$$|F(t_2, x(t_2)) - F(t_1, x(t_1))|$$
$$\leq |F(t_2, x(t_2)) - F(t_1, x(t_2))| + |F(t_1, x(t_2)) - F(t_1, x(t_1))|$$
$$\leq \gamma_{r_0}(F, \epsilon) + F_c |x(t_2) - x(t_1)|$$
$$\leq \gamma_{r_0}(F, \epsilon) + F_c w(x, \epsilon),$$

where

$$\gamma_{r_0}(F, \epsilon) = \sup \left\{ |F(t_2, x) - F(t_1, x)| : |t_2 - t_1| \leq \epsilon; t_1, t_2 \in I; \parallel x \parallel \leq r_0 \right\}.$$

Therefore

$$
\begin{aligned}
&|(Tx)(t_2) - (Tx)(t_1)| \\
&\leq \gamma_{r_0}(F, \epsilon) + F_c w(x, \epsilon) \\
&+ \frac{Q_c r_0}{l\Gamma(\chi)} \left| \int_0^{t_2} \frac{w(s)g'(s)G(s, x(s))}{(g(t_2) - g(s))^{1-\chi}} ds - \int_0^{t_1} \frac{w(s)g'(s)G(s, x(s))}{(g(t_1) - g(s))^{1-\chi}} ds \right| \\
&+ \left| \frac{Q(t_2, x(t_2))}{w(t_2)} - \frac{Q(t_1, x(t_1))}{w(t_1)} \right| \frac{G_c r_0 L \left(g(T) - g(0)\right)^{\chi}}{\Gamma(\chi + 1)}
\end{aligned}
$$

As $\epsilon \to 0$, then $t_2 \to t_1$ and so by the uniform continuity of F, Q w we have

$$\gamma_{r_0}(F, \epsilon) \to 0$$

and

$$\left| \frac{Q(t_2, x(t_2))}{w(t_2)} - \frac{Q(t_1, x(t_1))}{w(t_1)} \right| \to 0.$$

Also, as $\epsilon \to 0$ we get

$$\left| \int_0^{t_2} \frac{w(s)g'(s)G(s, x(s))}{(g(t_2) - g(s))^{1-\chi}} ds - \int_0^{t_1} \frac{w(s)g'(s)G(s, x(s))}{(g(t_1) - g(s))^{1-\chi}} ds \right| \to 0.$$

Now taking supremum over Γ and $\epsilon \to 0$ we get

$$\Gamma_0(\mathcal{D}\Gamma) \leq F_c \sigma(x, \epsilon).$$

Thus by assumption (C) and corollary 10.2, $\mathcal{D}$ has a fixed point in $\sigma \subseteq B_{T_0}$, i.e., Eq. (10.3) has a solution in **F**. □

Example 10.1 Consider the following equation:

$$x(t) = \frac{|x(t)| + t^2}{6} + \frac{|x(t)|}{4\Gamma(2)} \int_0^t \frac{2sx(s)\left(t^2 - s^2\right)}{e^s + 1} ds \tag{10.4}$$

for $t \in [0, 1] = I$.

Here

$$F(t, x) = \frac{|x| + t^2}{6}, \; Q(t, x) = \frac{|x|}{4}, \; G(t, x) = \frac{x}{e^t + 1},$$

$$\chi = 2, \; w(t) = 1,$$

and

$$g(t) = t^2.$$

Also,

$$|F(t,x) - F(t,\bar{x})| \leq \frac{1}{6}|x - \bar{x}|,$$

$$|Q(t,x) - Q(t,\bar{x})| \leq \frac{1}{4}|x - \bar{x}|,$$

$$|G(t,x) - G(t,\bar{x})| \leq \frac{1}{2}|x - \bar{x}|,$$

$$Max|F(t,0)| = F_M = \frac{1}{6}, t \in I, \ Q(t,0) = G(t,0) = 0,$$

$$F_c = \frac{1}{6}, \ Q_c = \frac{1}{4}, \ G_c = \frac{1}{2},$$

and

$$L = l = 1.$$

Clearly assumptions $(A)-(B)$ are satisfied.
Substituting these values in the inequality of assumption (C) we get

$$\frac{1}{6}r + \frac{1.\frac{1}{2}.\frac{1}{4}.r^2.1}{1.\Gamma(3)} + \frac{1}{6} < r$$
$$\implies 0.2 < r < 13.13$$

The assumption (C) is also satisfied for $r_0 = 10$. By Theorem 10.6, Eq. (10.4) has a solution in $\mathbf{F} = C(I)$.

10.3 An Iterative Algorithm Via Sinc Interpolation to Solve Fractional Volterra Integral Eq. (10.4)

Theoretical techniques to obtain the exact solution of Eq. (10.4) are usually difficult or impossible, therefore, numerical methods can be useful to approximate the solution. There are many numerical methods for approximating the solution of the integral equations, some of which are based on collocation and projection methods (see [15, 29, 30]). In the above methods, the nonlinear problems are converted to an algebraic system with unknown coefficients. In this section, we use Sinc interpolation with suitable collocation points to make an iterative algorithm to approximate the solution of Eq. (10.4). In the proposed algorithm, we don't need to convert the nonlinear problem to an algebraic system by expanding $x(t)$ in terms of Sinc functions with unknown coefficients. Moreover, the above algorithm contains a simple replacement process to approximate the solution. Also, for exponential accuracy, see [33]. Now, we give a brief review on Sinc function,

$$\mathrm{Sinc}(s) = \begin{cases} \dfrac{\sin(\pi s)}{\pi s}, & s \neq 0 \\ 1, & s = 0. \end{cases} \tag{10.5}$$

The k'th Sinc function with step size $h > 0$ is defined as:

$$S(k,h)(s) = \mathrm{Sinc}\left(\frac{s-kh}{h}\right), k \text{ is an integer} \tag{10.6}$$

thus

$$S(k,h)(jh) = \delta_{kj} = \begin{cases} 1, k = j \\ 0, k \neq j. \end{cases} \tag{10.7}$$

Whittaker cardinal expansion of function x on real line defined via in the series form,

$$C(x,h)(s) = \sum_{k=-\infty}^{\infty} x(kh)S(k,h)(s), \tag{10.8}$$

which this series converges [33]. By (10.7)–(10.8) the cardinal function interpolates function x at the points $\{kh\}_{k=-\infty}^{\infty}$ on real line. For interpolating on interval $[0, 1]$, we use a suitable combination of conformal map and Sinc function as follows:

$$\begin{aligned} \varphi : [0,1] &\longrightarrow (-\infty, \infty) \\ s &\longrightarrow \ln(\frac{s}{1-s}) \end{aligned} \tag{10.9}$$

in which $\lim_{s\to 0} \varphi(s) = -\infty$, $\lim_{s\to 1} \varphi(s) = \infty$. Therefore by considering (10.6) and (10.9) the composite function $S(k,h)o\phi$ is defined on interval $[0, 1]$. The integrand function of (10.4) can be approximated by $S(k,h)o\phi$ in the Sinc interpolation form. Let x be an integrand function by applying cardinal function (10.8) and the above process, we have

$$x_n(s) = \sum_{k=-N}^{N} x(kh)S(k,h)o\phi(s), \quad s \in [0,t], 0 \leq t \leq 1. \tag{10.10}$$

In (10.10), if $\phi(s) = kh$ for $k = -N, \ldots, N$, then $x_n(kh) = x(kh)$. This means that (10.10) is an interpolation of function x on interval $[0, 1]$ such that the interpolating points are as

$$\left\{ s_k = \phi^{-1}(kh) = \frac{e^{kh}}{1+e^{kh}}, \quad k = -N, \ldots, N \right. \tag{10.11}$$

Similar to [5, 6], we use Sinc function to approximate the integral on intervals $[0, t]$, for $0 \leqslant t \leqslant 1$.

$$\int_0^t x(s)ds \approx h \sum_{k=-N}^{N} \frac{x(s_k)}{\phi'(s_k)}, \quad \phi'(s_k) = \frac{1}{s_k(1-s_k)}. \tag{10.12}$$

According to (10.10)–(10.12), we construct an iterative algorithm to solve Eq. (10.4).

Algorithm

$$x_0(t) = 0, \tag{10.13}$$

$$x_n(t) = \frac{|x_{n-1}(t)| + t^2}{6} + \frac{|x_{n-1}(t)|}{4\Gamma(2)} h \sum_{k=-N}^{N} \frac{2s_k^2(1-s_k)(t^2-s_k^2)}{e^{s_k}+1} x_{n-1}(s_k)$$

$$n = 1, 2, 3, \ldots,$$

where s_k for $k = -N, \ldots, N$ are given by (10.11). Following are a few terms from the approximate solution sequence for $h = \frac{pi}{200}$ and $n = 10$

$$\begin{aligned}
x_0(t) &= 0, \\
x_1(t) &= 0.166667t^2, \\
x_2(t) &= 0.194431t^2 + 0.0000541843t^4, \qquad (10.14) \\
x_3(t) &= 0.199053t^2 + 0.0000827711t^4 + 2.05515 \times 10^{-8}t^6, \\
x_4(t) &= 0.199823t^2 + 0.0000910834t^4 + 3.5565 \times 10^{-8}t^6 + 7.98059 \times 10^{-12}t^8, \\
x_5(t) &= 0.199951t^2 + 0.0000930675t^4 + 4.14306 \times 10^{-8}t^6 + 1.51934 \times 10^{-11}t^8 \\
&\quad + 3.11104 \times 10^{-15}t^{10}, \\
x_6(t) &= 0.199972t^2 + 0.0000934981t^4 + 4.32045 \times 10^{-8}t^6 + 1.86919 \times 10^{-11}t^8 \\
&\quad + 6.4448 \times 10^{-15}t^{10} + 1.21354 \times 10^{-18}t^{12}.
\end{aligned}$$

According to Table.1 the above sequence is a Cauchy sequence in $C([0, 1])$, so it is convergent (Table 10.1).
We can get $x_6(t)$ as an approximate solution of Eq. (10.4) with the following graph (Fig. 10.1).

10.4 Conclusion

In this chapter, some novel condensing operators in Banach spaces are introduced. These condensing operators assist in establishing a number of new fixed point

Table 10.1 Cauchy sequence $\{x_n(t)\}$

n	$\| x_n(t) - x_{n-1}(t) \|$
1	1.6×10^{-1}
2	2.7×10^{-2}
3	4.6×10^{-3}
4	7.7×10^{-4}
5	1.2×10^{-4}
6	2.1×10^{-5}

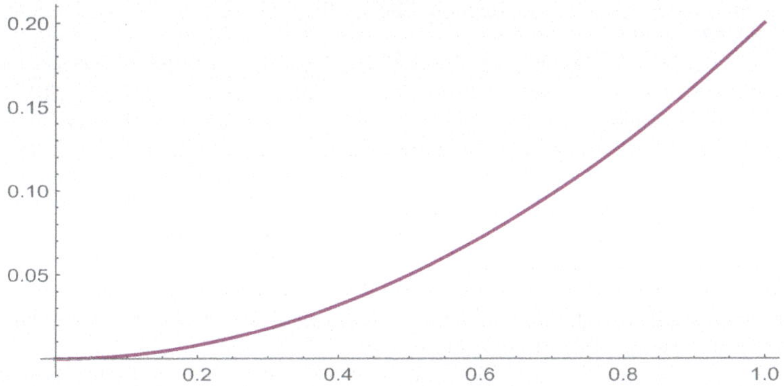

Fig. 10.1 Approximate solution $x_6(t)$

theorems. Additionally, we have used it to prove the existence of a solution in $C[0, T], T > 0$ for a weighted fractional integral equation. Iterative technique and Sinc interpolation were finally used to solve the fractional integral equation. Different types of fractional integral equations as well as differential equations can be tested for solvability using this fixed point theorem. Additionally, several sorts of equations can be solved using the iterative approach we have explained.

References

1. Aghajani, A., Allahyari, R., Mursaleen, M.: A generalization of Darbo's theorem with application to the solvability of systems of integral equations. J. Comput. Appl. Math. **260**, 68–77 (2014)
2. Aghajani, A., Banas, J., Sabzali, N.: Some generalizations of Darbofixed point theorem and applications. Bull. Belg. Math. Soc. Simon Stevin **20**, 345–358 (2013)
3. Aghajani, A., Sabzali, N.: A coupled fixed point theorem for condensing operators with application to system of integral equations. J. Nonlinear Convex Anal. **15**(5), 941–952 (2014)
4. Arab, R.: Some fixed point theorems in generalized Darbo fixed point theorem and the existence of solutions for system of integral equations. J. Korean Math. Soc. **52**(1), 125–139 (2015)

5. Arab, R., Rabbani, M., Mollapourasl, R.: On solution of nonlinear integral equations with deviating argument based on fixed point technique. Appl. Comput. Math. **14**(1), 38–49 (2015)
6. Arab, R., Rabbani, M.: Existence of solution of functional Volterra-Fredholm integral equations in space $L^{\infty}(R+)$ and Sinc interpolation to find solution. J. Integr. Eq. Appl. (2022 Inpress)
7. Banas, J., Goebel, K.: Measures of Noncompactness in Banach Spaces. Lecture Notes in Pure and Applied Mathematics, p. 60. Dekker, New York (1980)
8. Darbo, G.: Punti uniti in transformazioni a codominio non compatto (Italian). Rend. Sem. Math. Univ. Padova **24**, 84–92 (1955)
9. Das, A., Hazarika, B., Kumam, P.: Some new generalization of darbo's fixed point theorem and its application on integral equations. Mathematics **7**(3), 214 (2019). https://doi.org/10.3390/math7030214
10. Jarad, F., Abdeljawad, T., Shah, K.: On the weighted fractional operators of a function with respect to another function **28**(8), 2040011 (2020). https://doi.org/10.1142/S0218348X20400113
11. Falset, J.G., Latrach, K.: On Darbo-Sadovskii's fixed point theorems type for abstract measures of (weak) noncompactness. Bull. Belg. Math. Soc. Simon Stevin **22**, 797–812 (2015)
12. Guo, D., Lakshmikantham, V., Liu, X.: Nonlinear Integral Equations in Abstract Spaces. Mathematics and Its Applications, vol. 373. Kluwer Academic Publishers, Dordrecht, The Netherlands (1996)
13. Akmerov, R., Kamenski, M., Potapov, A., Rodkina, A., Sadovskii, B.: Measures of Noncompactness and Condensing Operators. Birkhauser-Verlag, Basel (1992)
14. Kuratowski, K.: Sur les espaces completes. Fund. Math. **15**, 301–309 (1930)
15. Marzban, H.R., Tabrizidooz, H.R., Razzaghi, M.: A composite collocation method for the nonlinear mixed Volterra-Fredholm-Hammerstein integral equations. Commun. Nonlinear Sci. Numer. Simul. **16**(3), 1186–1194 (2011)
16. Mursaleen, M., Mohiuddine, S.A.: Applications of measures of noncompactness to the infinite system of differential equations in l_p spaces. Nolinear Anal. (TMA) **75**, 2111–2115 (2012)
17. Mursaleen, M., Alotaibi, A.: Infinite system of differential equations in some spaces. Abstr. Appl. Anal. **20**863483 (2012) https://doi.org/10.1155/2012/863483
18. Mursaleen, M., Rizvi, S.M.H.: Solvability of infinite system of second order differential equations in c_0 and ℓ_1 by Meir-Keeler condensing operator. Proc. Amer. Math. Soc. **144**(10), 4279–4289 (2016)
19. Mursaleen, M., Bilalov, B., Rizvi, S.M.H.: Applications of measures of noncompactness to infinite system of fractional differential equations. Filomat **31**(11), 3421–3432 (2017)
20. Nashine, H.K., Arab, R.: Existence of solutions to nonlinear functional-integral equations via the measure of noncompactness. J. Fixed Point Theory Appl. **20**(6), (2018) https://doi.org/10.1007/s11784-018-0546-1
21. Nashine, H.K., Arab, R., Agarwal, R.P.: Ali Shole Haghighi, Darbo type fixed and coupled fixed point results and its application to integral equation. Period Math. Hung. **77**, 94–107 (2018)
22. Nashine, H.K., Arab, R., Agarwal, R.P., Sen, M., De, L.A.: Positive solutions of fractional integral equations by the technique of measure of noncompactness. J. Inequal. Appl. **2017**, 225 (2017)
23. Nashine, H.K., Arab, R.: Some new type generalization of Darbo's fixed point and its applications to integral equations. Indian J. Math. **61**(3), 395–420 (2019)
24. Nashine, H.K., Ibrahim, R.W.: Monotone solutions of iterative fractional equations found by modified Darbo-type fixed-point theorems. J. Fixed Point Theory Appl. **19**(14), 3217–3229 (2017)
25. Nashine, H.K., Ibrahim, R.W., Agarwal, R.P.: Moments solution of fractional evolution equation found by new krasnoselskii type fixed point theorems. Fixed Point Theory **22**(1), 263–278 (2021)
26. Nashine, H.K., Roshan, J.R.: Fixed point theorem via measure of noncompactness and application to Volterra integral equations in Banach algebras. J. Math. Ext. **13**(4), 91–116 (2019)

27. Rabbani, M., Das, A., Hazarika, B., Arab, R.: Measure of noncompactness of a new space of tempered sequences and its application on fractional differential equations. Chaos. Solitons Frac. **140**, 110221 (2020)
28. Rabbani, M., Das, A., Hazarika, B., Arab, R.: Existence of solution for two dimensional non-linear fractional integral equation by measure of noncompactness and iterative algorithm to solve it. J. Comput. Appl. Math. **370**, 1–17 (2020)
29. Rabbani, M., Kiasoltani, S.H.: Solving of nonlinear system of Fredholm-Volterra integro-differential equations by using discrete collocation method. J. Math. Comput. Sci. **3**(4), 382–389 (2011)
30. Rabbani, M., Maleknejad, K.: Using orthonormal wavelet basis in Petrov?Galerkin method for solving Fredholm integral equations of the second kind. Kybernetes **41**(3), 465–481 (2012)
31. Rezpecki, B.: On measure of noncompactness in topological vector spaces. Comment. Math. Univ. Carolinae **23**, 105–116 (1982)
32. Sadovskii, B.N.: Limit-compact and condensing operators (Russian). Uspehi Mat. Nauk. **27**(1, 163), 81–146 (1972)
33. Stenger, F.: Numerical Methods Based on Sinc and Analytic Functions. Springer, NewYork (1993)
34. Panda, S.K., Abdeljawad, T., Ravichandran, C.: A complex valued approach to the solutions of Riemann-Liouville integral, Atangana-Baleanu integral operator and non-linear Telegraph equation via fixed point method. Chaos Solitons Fract. **130**, 109439 (2020)
35. Panda, S.K., Atangana, A., Nieto, J.J.: New insights on novel coronavirus 2019-nCoV/SARS-CoV-2 modelling in the aspect of fractional derivatives and fixed points. Math. Biosci. Eng. **18**(6), 8683–8726 (2021). https://doi.org/10.3934/mbe.2021430
36. Panda, S.K., Atangana, A., Abdeljawad, T.: Existence results and numerical study on novel coronavirus 2019-NCOV/SARS-COV-2 model using differential operators based on the generalized Mittag-Leffler kernel and fixed points. https://doi.org/10.1142/S0218348X22402149
37. Wardowski, D.: Fixed points of new type of contractive mappings in com-plete metric spaces. Fixed Point Theory Appl. **2012**(94) (2012)

Chapter 11
Dual F-Suzuki, Triple F-Suzuki Contraction, and Related Fixed Point Results with an Application

Naeem Saleem, Shahid Bashir, Syed Muhammad Husnine, and Vahid Parvaneh

Abstract In this chapter, we extend Wardowski's idea of F-contraction by introducing the dual F-Suzuki type contraction and the triple F-Suzuki type contraction. We prove some fixed point theorems related to these contractions. Further, we apply our obtained results to find the existence of a solution to an integral equation.

Keywords F-Suzuki type contraction · Fixed point · Integral equation

11.1 Introduction

Fixed points are the points that remain unchanged upon the application of a mapping. In a plethora of real and theoretical world problems, the existence of a fixed point for a suitable mapping or operator ensures the existence of a solution to the problem. Fixed point theory, therefore, is of phenomenal importance in many areas of sciences and has become the subject of scientific research. The fixed point theory is a fabulous and fantastic blend of analysis (pure and applied), topology, and geometry. In the last five decades, the fixed point theory has been augmented as a versatile and dynamic tool in the analysis of nonlinear phenomena. Precisely, we use fixed point methodology in such diversified, pioneering, and multiple fields as physics, chemistry, biology,

N. Saleem
Department of Mathematics, University of Management and Technology, Lahore 54770, Pakistan

Department of Mathematics and Applied Mathematics, Sefako Makgatho Health Sciences University, Ga-Rankuwa, Pretoria, 0204 Medunsa, South Africa

S. Bashir · S. M. Husnine
Department of Sciences and Humanities, National University of Computer and Emerging Sciences, Lahore Campus 54700, Pakistan
e-mail: syed.husnine@nu.edu.pk

V. Parvaneh (✉)
Department of Mathematics, Gilan-E-Gharb Branch, Islamic Azad University, Gilan-E-Gharb, Iran
e-mail: zam.dalahoo@gmail.com

B. Hazarika et al. (eds.), *Advances in Functional Analysis and Fixed-Point Theory*, Industrial and Applied Mathematics,
https://doi.org/10.1007/978-981-99-9207-2_11

engineering, economics, and game theory, which endorses the fact that mathematics is the mother of all sciences. A fixed point problem provides conditions under which there exists a member x of the set X such that the equation $f(x) = x$, where f is a mapping from X into X, has a solution. The solution set of this equation may be empty, finite, countable, or uncountable. This theory provides a basic tool for solving problems in different branches of analysis. For example, solving problems such as variational inequalities, nonlinear optimization, equilibrium problems, and solving differential and integral equations are closely related to this theory. Banach's contraction principle, proved in 1922, is at the heart of fixed point theory in metric spaces and plays a fundamental role in many aspects of nonlinear functional analysis. This principle was initiated from the idea of successive approximation, which was proposed by Picard in 1890 to prove the existence and uniqueness of the solution to the equation $f(x) = x$. For more details on fixed point results in different spaces and their applications, we refer the reader to [4, 7, 10–12, 15, 16].

In 2012, Wardowski [21] introduced a new contraction, called F-contraction, and established a fixed point result as a generalization of the Banach contraction principle. Recently, Piri and Kumam [17] proved a fixed point theorem for an F-Suzuki contraction that generalizes the result of Wardowski. Several applications guarantee the existence of solutions for integral equations concerning the class of F-contractions and F-Suzuki contractions. We use multiple functions in the definition of an F-Suzuki contraction that will allow researchers to find the existence of solutions with an extensive range of integral equations.

Moreover, Alsulami et al. [1] introduced a kind of generalized F-Suzuki type contractions in b-metric spaces and investigated the existence of fixed points of such mappings only using the conditions $(F1)$ and $(F2)$. Recently, by adding the following four new values $\mathscr{D}(\Re^2 x, x)$, $\mathscr{D}(\Re^2 x, \Re x)$, $\mathscr{D}(\Re^2 x, y)$, and $\mathscr{D}(\Re^2 x, \Re y)$ to the contraction, Dung and Hang [5] proved some fixed point theorems for a new generalized F-Suzuki type contraction. Motivated by Dung and Hang [5], Gilić et al. [6] proved some fixed point theorems for a generalized $F1$-contraction using condition $(F1)$ only. For more on Wardowski's results, one can refer to [13] and so forth.

Zhou and Bashir [20] introduced a generalized F-contraction by using a new kind of auxiliary function F, for which one may have $F(\mathscr{D}(\Re x, \Re y)) > F(\mathscr{D}(x, y))$, for some $x, y \in X$.

Motivated by the above discussions, in this chapter, we define two types of dual F-Suzuki contractions. Furthermore, we introduce the triple F-Suzuki contraction. We prove some fixed point theorems for these contractions under certain conditions. Finally, we provide an application of our result which provides the existence and uniqueness of the solution to an integral equation.

11.2 Preliminaries

Throughout this chapter, $\mathbb{N}, \mathbb{N}_0, \mathbb{R}$, and $\mathbb{R}_+$ denote the set of natural numbers, $\mathbb{N} \cup \{0\}$, real numbers, and positive real numbers, respectively. Moreover, every set X taken into account is non-empty. In 2012, Wardowski [21] defined F-contraction as follows.

Definition 11.1 [21] Let $(X, \mathscr{D})$ be a metric space. A mapping $\mathscr{G} : X \mapsto X$ is said to be a F-contraction, if there exists a real number $\tau > 0$ such that for all $x, y \in X$,

$$[\mathscr{D}(\mathscr{G}x, \mathscr{G}y) > 0 \quad \Longrightarrow \quad \tau + F(\mathscr{D}(\mathscr{G}x, \mathscr{G}y)) \leq F(\mathscr{D}(x, y))],$$

where $F : \mathbb{R}_+ \mapsto \mathbb{R}$ is a mapping satisfying the following conditions:
$(F1)$ F is strictly increasing,
$(F2)$ For each sequence $\{\alpha_n\}_{n=1}^{\infty}$ of positive numbers, we have
$\lim_{n\to\infty}\alpha_n = 0$ if and only if $\lim_{n\to\infty} F(\alpha_n) = -\infty$,
$(F3)$ There exists $k \in (0, 1)$ such that $\lim_{\alpha\mapsto 0^+}\alpha^k F(\alpha) = 0$.
Let $\mathscr{F}$ be the set of all functions satisfying the conditions $(F1)$–$(F3)$.

Example 11.1 Let $F_i : \mathbb{R}_+ \mapsto \mathbb{R}$, $i = 1, 2, 3, 4$, be defined by
i. $F_1(t) = \ln t$.
ii. $F_2(t) = t + \ln t$.
iii. $F_3(t) = -\frac{1}{\sqrt{t}}$.
iv. $F_4(t) = \ln\left(t^2 + t\right)$.
Then, $F_1, F_2, F_3, F_4 \in \mathscr{F}$.

Remark 11.1 From the conditions of F-contraction, it is easy to conclude that every F-contraction is necessarily continuous.

Further, Wardowski [21] stated a modified version of Banach contraction principle via the below theorem.

Theorem 11.1 *Let $(X, \mathscr{D})$ be a complete metric space and $\mathscr{G} : X \mapsto X$ be a F-contraction. Then $\mathscr{G}$ has a unique fixed point $x^* \in X$ and for every $x \in X$ the sequence $\{\mathscr{G}^n x\}_{n\in N}$ converges to x^*.*

Definition 11.2 [1] Let $(X, \mathscr{D})$ be a metric space. A mapping $\mathscr{G} : X \mapsto X$ is said to be a F-Suzuki contraction if there exists a real number $\tau > 0$ such that for all $x, y \in X$,

$$\left[\frac{1}{2}\mathscr{D}(x, \mathscr{G}x) < \mathscr{D}(x, y) \quad \Longrightarrow \quad \tau + F(\mathscr{D}(\mathscr{G}x, \mathscr{G}y)) \leq F(\mathscr{D}(x, y))\right],$$

where $F \in \mathscr{F}$.

Definition 11.3 [20] Let (X, d) be a metric space. A mapping $G : X \to X$ is said to be a generalized F-contraction if there exists $\tau > 0$ and an integer $p > 1$ such that for all $x, y \in X$,

$$d(\mathscr{G}^i x, \mathscr{G}^i y) > 0 \quad \Longrightarrow \quad \tau + F(\min(d(\mathscr{G}^i x, \mathscr{G}^i y))) \leq F(d(x, y)), \quad i = 1, ..., p,$$

where $F : (0, +\infty) \mapsto \mathbb{R}$ is a mapping satisfying the following conditions:
(F_1^*) For some $t > 0$ one has $0 < F(\tau) < t$, if $0 < \alpha < \beta$, we have $F(\alpha) < F(\beta)$ and $F(\beta) \leq F(\alpha) + F(\tau)$.
(F_2^*) $\lim_{n\to\infty} F(\alpha_n) = -\infty \Leftrightarrow \lim_{n\to+\infty} \alpha_n = 0$, for each positive number sequence $\{\alpha_n\}$.
(F_3^*) F is continuous on $(0, \infty)$.
Denote by $\mathfrak{F}$ the set of all functions F satisfying the conditions (F_1^*), (F_2^*), (F_3^*).

11.3 Main Results

At first, we define two types of dual F-Suzuki contractions.

Definition 11.4 Let $(X, \mathscr{D})$ be a metric space. A mapping $\mathscr{G} : X \mapsto X$ is said to be a dual F-Suzuki contraction of type-I, if there exists a real number $\tau > 0$ and $F_1, F_2 \in \mathscr{F}$ such that for all $x, y \in X$,

$$\frac{1}{2}\mathscr{D}(x, \mathscr{G}x) < \mathscr{D}(x, y) \quad \to \quad \tau + \min\left\{F_2\left(\mathscr{D}\left(\mathscr{G}^2 x, \mathscr{G}^2 y\right)\right), F_1\left(\mathscr{D}\left(\mathscr{G}x, \mathscr{G}y\right)\right)\right\} \leq \alpha_2 F_2\left(\mathscr{D}\left(x, y\right)\right) + \alpha_1 F_1\left(\mathscr{D}\left(x, y\right)\right), \quad (11.1)$$

where $\begin{Bmatrix} \alpha_1 = 0, \ \alpha_2 = 1, \ \text{if}, & F_2\left(\mathscr{D}\left(\mathscr{G}^2 x, \mathscr{G}^2 y\right)\right) \leq F_1\left(\mathscr{D}\left(\mathscr{G}x, \mathscr{G}y\right)\right) \\ \alpha_1 = 1, \ \alpha_2 = 0, \ \text{if}, & F_2\left(\mathscr{D}\left(\mathscr{G}^2 x, \mathscr{G}^2 y\right)\right) > F_1\left(\mathscr{D}\left(\mathscr{G}x, \mathscr{G}y\right)\right) \end{Bmatrix}.$

Remark 11.2 For all $x, y \in X$, the conditions of Definition 11.4 yield, either $\tau + F_2\left(\mathscr{D}\left(\mathscr{G}^2 x, \mathscr{G}^2 y\right)\right) \leq F_2\left(\mathscr{D}\left(x, y\right)\right)$, or $\tau + F_1\left(\mathscr{D}\left(\mathscr{G}x, \mathscr{G}y\right)\right) \leq F_1\left(\mathscr{D}\left(x, y\right)\right)$.

Definition 11.5 Let $(X, \mathscr{D})$ be a metric space. A mapping $\mathscr{G} : X \mapsto X$ is said to be a dual F-Suzuki contraction of type-II, if there exists a real number $\tau > 0$ and $F_1, F_2 \in \mathscr{F}$ such that, for all $x, y \in X$,

$$\frac{1}{2}\mathscr{D}(x, \mathscr{G}x) < \mathscr{D}(x, y) \quad \to \quad \tau + \min\left\{F_2\left(\mathscr{D}\left(\mathscr{G}^2 x, \mathscr{G}^2 y\right)\right), F_1\left(\mathscr{D}\left(\mathscr{G}x, \mathscr{G}y\right)\right)\right\} \leq \alpha_2 F_2\left(\mathscr{D}\left(x, y\right)\right) + \alpha_1 F_1\left(\mathscr{D}\left(x, y\right)\right), \quad (11.2)$$

where either $\alpha_1 = 0$ or $\alpha_2 = 0$ and $\alpha_1 + \alpha_2 = 1$.

Remark 11.3 For all, $x, y \in X$, the dual F-Suzuki contraction of type-II copes with one of the following two cases:

$$\tau + \min\left\{F_2\left(\mathscr{D}\left(\mathscr{G}^2 x, \mathscr{G}^2 y\right)\right), F_1\left(\mathscr{D}\left(\mathscr{G}x, \mathscr{G}y\right)\right)\right\} \leq F_2\left(\mathscr{D}\left(x, y\right)\right), \qquad (R1)$$

or

$$\tau + \min\left\{F_2\left(\mathscr{D}\left(\mathscr{G}^2 x, \mathscr{G}^2 y\right)\right), F_1\left(\mathscr{D}\left(\mathscr{G}x, \mathscr{G}y\right)\right)\right\} \leq F_1\left(\mathscr{D}\left(x, y\right)\right). \qquad (R2)$$

For all $x, y \in X$, $(R1)$ yields either
$\tau + F_2\left(\mathscr{D}\left(\mathscr{G}^2 x, \mathscr{G}^2 y\right)\right) \leq F_2\left(\mathscr{D}\left(x, y\right)\right)$, or $\tau + F_1\left(\mathscr{D}\left(\mathscr{G}x, \mathscr{G}y\right)\right) \leq F_2\left(\mathscr{D}\left(x, y\right)\right)$.
Similarly, for all, $x, y \in X$, $(R2)$ yields either
$\tau + F_2\left(\mathscr{D}\left(\mathscr{G}^2 x, \mathscr{G}^2 y\right)\right) \leq F_1\left(\mathscr{D}\left(x, y\right)\right)$, or $\tau + F_1\left(\mathscr{D}\left(\mathscr{G}x, \mathscr{G}y\right)\right) \leq F_1\left(\mathscr{D}\left(x, y\right)\right)$.

Next, we will introduce the triple F-Suzuki contraction that can be considered as the most generalized form of the F-Suzuki contraction.

Definition 11.6 Let $(X, \mathscr{D})$ be a metric space. A mapping $\mathscr{G} : X \mapsto X$ is said to be a triple F-Suzuki contraction if there exists a real number $\tau > 0$ and $F, F_1, F_2 \in \mathscr{F}$ such that, for all $x, y \in X$, we have

$$\frac{1}{2}\mathscr{D}(x, \mathscr{G}x) < \mathscr{D}(x, y) \implies \tau + \min\left\{F_2\left(\mathscr{D}\left(\mathscr{G}^2 x, \mathscr{G}^2 y\right)\right), F_1\left(\mathscr{D}\left(\mathscr{G}x, \mathscr{G}y\right)\right)\right\} \leq F\left(\mathscr{D}\left(x, y\right)\right). \qquad (11.3)$$

Example 11.2 Let $F_1(\alpha) = \ln \alpha$, $F_2(\alpha) = \ln k\alpha$, $k > 0$, where $F_1, F_2 \in \mathscr{F}$. So the dual F-Suzuki contraction of type-I can be written as

$$\tau + \min\left\{\ln k\left(\mathscr{D}\left(\mathscr{G}^2 x, \mathscr{G}^2 y\right)\right), \ln\left(\mathscr{D}\left(\mathscr{G}x, \mathscr{G}y\right)\right)\right\} \leq \alpha_2 \ln k\left(\mathscr{D}\left(x, y\right)\right) + \alpha_1 \ln\left(\mathscr{D}\left(x, y\right)\right).$$

The condition $(F1)$ allows us to write

$$\tau + \ln \min\left\{k\mathscr{D}\left(\mathscr{G}^2 x, \mathscr{G}^2 y\right), \mathscr{D}\left(\mathscr{G}x, \mathscr{G}y\right)\right\} \leq \ln k^{\alpha_2}\left(\mathscr{D}\left(x, y\right)\right)^{\alpha_1 + \alpha_2}.$$

By the assumption of the definition, we have $\alpha_1 + \alpha_2 = 1$, which yields

$$\min\left\{k\mathscr{D}\left(\mathscr{G}^2 x, \mathscr{G}^2 y\right), \mathscr{D}\left(\mathscr{G}x, \mathscr{G}y\right)\right\} \leq e^{-\tau} k^{\alpha_2} \mathscr{D}\left(x, y\right).$$

It is interesting to note that if we suppose that for all $x, y \in X$, $\mathscr{D}\left(\mathscr{G}x, \mathscr{G}y\right) < k\mathscr{D}\left(\mathscr{G}^2 x, \mathscr{G}^2 y\right)$, we will have $\mathscr{D}\left(\mathscr{G}x, \mathscr{G}y\right) \leq e^{-\tau}\mathscr{D}\left(x, y\right)$ with $\alpha_2 = 0$. That is, $\mathscr{G}$ is a contraction. Further, if for all $x, y \in X$, $\mathscr{D}\left(\mathscr{G}x, \mathscr{G}y\right) > k\mathscr{D}\left(\mathscr{G}^2 x, \mathscr{G}^2 y\right)$, we will have $\mathscr{D}\left(\mathscr{G}^2 x, \mathscr{G}^2 y\right) \leq e^{-\tau}\mathscr{D}\left(x, y\right)$. Then, $\mathscr{G}$ is neither a contraction nor an expansive mapping and fixed point of dual F-Suzuki contraction of type-I will represent the fixed point of $\mathscr{G}$.

Example 11.3 Let $F_1(\alpha) = \ln \alpha$, $F_2(\alpha) = \ln k\alpha$, $k > 1$, where $F_1, F_2 \in \mathscr{F}$. Then by the definition of dual F-Suzuki contraction of type-II, for all $x, y \in X$, we have

$$\tau + \min\left\{F_2\left(\mathscr{D}\left(\mathscr{G}^2x, \mathscr{G}^2y\right)\right), F_1\left(\mathscr{D}\left(\mathscr{G}x, \mathscr{G}y\right)\right)\right\} \leq \alpha_2 F_2\left(\mathscr{D}\left(x, y\right)\right) + \alpha_1 F_1\left(\mathscr{D}\left(x, y\right)\right).$$

Conditions $(R1)$ and $(R2)$ allow us to write

$$\tau + \min\left\{F_2\left(\mathscr{D}\left(\mathscr{G}^2x, \mathscr{G}^2y\right)\right), F_1\left(\mathscr{D}\left(\mathscr{G}x, \mathscr{G}y\right)\right)\right\} \leq F_1\left(\mathscr{D}\left(x, y\right)\right) \tag{11.4}$$

and

$$\tau + \min\left\{F_2\left(\mathscr{D}\left(\mathscr{G}^2x, \mathscr{G}^2y\right)\right), F_1\left(\mathscr{D}\left(\mathscr{G}x, \mathscr{G}y\right)\right)\right\} \leq F_2\left(\mathscr{D}\left(x, y\right)\right). \tag{11.5}$$

Relation (11.4) yields

$$\tau + \min\left\{\ln\left(k\mathscr{D}\left(\mathscr{G}^2x, \mathscr{G}^2y\right)\right), \ln\left(\mathscr{D}\left(\mathscr{G}x, \mathscr{G}y\right)\right)\right\} \leq \ln\left(\mathscr{D}\left(x, y\right)\right).$$

The condition $(F1)$ allows us to write

$$\tau + \ln\min\left\{k\mathscr{D}\left(\mathscr{G}^2x, \mathscr{G}^2y\right), \mathscr{D}\left(\mathscr{G}x, \mathscr{G}y\right)\right\} \leq \ln\left(\mathscr{D}\left(x, y\right)\right),$$

so that

$$\min\left\{k\mathscr{D}\left(\mathscr{G}^2x, \mathscr{G}^2y\right), \mathscr{D}\left(\mathscr{G}x, \mathscr{G}y\right)\right\} \leq e^{-\tau}\left(\mathscr{D}\left(x, y\right)\right). \tag{11.6}$$

Relation (11.5) implies

$$\tau + \min\left\{\ln\left(k\mathscr{D}\left(\mathscr{G}^2x, \mathscr{G}^2y\right)\right), \ln\left(\mathscr{D}\left(\mathscr{G}x, \mathscr{G}y\right)\right)\right\} \leq \ln\left(k\mathscr{D}\left(x, y\right)\right),$$

which using the condition $(F1)$, we can write

$$\min\left\{k\mathscr{D}\left(\mathscr{G}^2x, \mathscr{G}^2y\right), \mathscr{D}\left(\mathscr{G}x, \mathscr{G}y\right)\right\} \leq e^{-\tau}\left(k\mathscr{D}\left(x, y\right)\right). \tag{11.7}$$

Note that, if we consider for all $x, y \in X$, $k\mathscr{D}\left(\mathscr{G}^2x, \mathscr{G}^2y\right) > \mathscr{D}\left(\mathscr{G}x, \mathscr{G}y\right)$, then relations (11.6) and (11.7) represent a k-Lipschitzian mapping under certain conditions and the fixed point of dual F-Suzuki contraction of type-II will represent the fixed point of $\mathscr{G}$.

Example 11.4 Suppose that $F_2(\alpha) = \ln k_2\alpha$, $F_1(\alpha) = \ln k_1\alpha$, $F(\alpha) = \ln k\alpha$, where $k, k_1, k_2 > 0$, and $F, F_1, F_2 \in \mathscr{F}$. Then by the definition of a triple F-Suzuki contraction, for all $x, y \in X$, we have

$$\tau + \min\left\{F_2\left(\mathscr{D}\left(\mathscr{G}^2x, \mathscr{G}^2y\right)\right), F_1\left(\mathscr{D}\left(\mathscr{G}x, \mathscr{G}y\right)\right)\right\} \leq F\left(\mathscr{D}\left(x, y\right)\right),$$

or

$$\tau+\min\left\{\ln\left(k_2\mathscr{D}\left(\mathscr{G}^2x,\mathscr{G}^2y\right)\right),\ln\left(k_1\mathscr{D}\left(\mathscr{G}x,\mathscr{G}y\right)\right)\right\}\le\ln\left(k\mathscr{D}\left(x,y\right)\right).$$

The condition $(F1)$ allows us to write

$$\tau+\ln\min\left\{k_2\mathscr{D}\left(\mathscr{G}^2x,\mathscr{G}^2y\right),k_1\mathscr{D}\left(\mathscr{G}x,\mathscr{G}y\right)\right\}\le\ln k\mathscr{D}\left(x,y\right),$$

or

$$\min\left\{k_2\mathscr{D}\left(\mathscr{G}^2x,\mathscr{G}^2y\right),k_1\mathscr{D}\left(\mathscr{G}x,\mathscr{G}y\right)\right\}\le e^{-\tau}k\mathscr{D}\left(x,y\right).$$

Then either

$$k_2\left(\mathscr{D}\left(\mathscr{G}^2x,\mathscr{G}^2y\right)\right)\le ke^{-\tau}\left(\mathscr{D}\left(x,y\right)\right),$$

or

$$k_1\left(\mathscr{D}\left(\mathscr{G}x,\mathscr{G}y\right)\right)\le ke^{-\tau}\left(\mathscr{D}\left(x,y\right)\right).$$

We can define $k_2=2\alpha_2, k_1=2\alpha_1, \frac{k_1}{2}+\frac{k_2}{2}=1,\quad$ where $\alpha_1+\alpha_2=1$.
So,

$$\alpha_2\left(\mathscr{D}\left(\mathscr{G}^2x,\mathscr{G}^2y\right)\right)\le\frac{1}{2}ke^{-\tau}\left(\mathscr{D}\left(x,y\right)\right),$$

or

$$\alpha_1\left(\mathscr{D}\left(\mathscr{G}x,\mathscr{G}y\right)\right)\le\frac{1}{2}ke^{-\tau}\left(\mathscr{D}\left(x,y\right)\right).$$

Therefore,

$$\alpha_2\left(\mathscr{D}\left(\mathscr{G}^2x,\mathscr{G}^2y\right)\right)+\alpha_1\left(\mathscr{D}\left(\mathscr{G}x,\mathscr{G}y\right)\right)\le ke^{-\tau}\left(\mathscr{D}\left(x,y\right)\right),$$

which is a mean Lipschitzian mapping and the fixed point of triple F-Suzuki contraction will represent the fixed point of $\mathscr{G}$.

Theorem 11.2 *Let $(X,\mathscr{D})$ be a complete metric space. Suppose that the continuous mapping $\mathscr{G}:X\mapsto X$ is a dual F-Suzuki contraction of type-I, and for all $t_1,t_2\in\mathbb{R}_+$, there exists $\upsilon>0$ and $\tau>2\upsilon$, such that*

$$F_2\left(t_2\right)<F_1\left(t_1\right)\le F_2\left(t_2\right)+\upsilon. \qquad (A)$$
$$F_1\left(t_1\right)<F_2\left(t_2\right)\le F_1\left(t_1\right)+\upsilon. \qquad (B)$$

Then $\mathscr{G}$ has a unique fixed point in X and for every $x_0\in X$, the sequence $\{\mathscr{G}^mx_0\}_{m=1}^{\infty}$ converges to x^*.

Proof Consider a sequence $\{x_1,x_2,\ldots.\}$, $x_{m+1}=\mathscr{G}x_m=\mathscr{G}^{m+1}x_0$ for all $m\in\mathbb{N}_0$. If for some $m\in\mathbb{N}$, $\mathscr{D}\left(x_m,\mathscr{G}x_m\right)=0$, $\mathscr{G}$ will admit a fixed point. Let $\mathscr{D}\left(x_m,\mathscr{G}x_m\right)=\mathscr{D}\left(\mathscr{G}x_{m-1},\mathscr{G}x_m\right)>0,\quad$ for all $m\in\mathbb{N}$.
We will prove that $\lim\limits_{m\to\infty}\mathscr{D}\left(x_m,\mathscr{G}x_m\right)=0$.

We know that $\frac{1}{2}\mathcal{D}(x_{m-1}, \mathcal{G}x_{m-1}) < (x_{m-1}, \mathcal{G}x_{m-1})$ for all $m \in N$. Therefore, for any $m \in \mathbb{N}$, we can write

$$\begin{aligned}&\tau + \min\left\{F_2\left(\mathcal{D}\left(\mathcal{G}^2x_{m-1}, \mathcal{G}^2x_m\right)\right), F_1\left(\mathcal{D}\left(\mathcal{G}x_{m-1}, \mathcal{G}x_m\right)\right)\right\}\\ &\leq \alpha_2 F_2\left(\mathcal{D}\left(x_{m-1}, x_m\right)\right) + \alpha_1 F_1\left(\mathcal{D}\left(x_{m-1}, x_m\right)\right).\end{aligned}$$

So,

$$\begin{aligned}&\min\left\{F_2\left(\mathcal{D}\left(\mathcal{G}^2x_{m-1}, \mathcal{G}^2x_m\right)\right), F_1\left(\mathcal{D}\left(\mathcal{G}x_{m-1}, \mathcal{G}x_m\right)\right)\right\}\\ &\leq \alpha_2 F_2\left(\mathcal{D}\left(x_{m-1}, x_m\right)\right) + \alpha_1 F_1\left(\mathcal{D}\left(x_{m-1}, x_m\right)\right) - \tau.\end{aligned} \tag{11.8}$$

Now, we will discuss the two possible cases (C) and (D).

$$F_1\left(\mathcal{D}\left(\mathcal{G}x_{m-1}, \mathcal{G}x_m\right)\right) = \min\left\{F_2\left(\mathcal{D}\left(\mathcal{G}^2x_{m-1}, \mathcal{G}^2x_m\right)\right), F_1\left(\mathcal{D}\left(\mathcal{G}x_{m-1}, \mathcal{G}x_m\right)\right)\right\}, \tag{C}$$

or

$$F_2\left(\mathcal{D}\left(\mathcal{G}^2x_{m-1}, \mathcal{G}^2x_m\right)\right) = \min\left\{F_2\left(\mathcal{D}\left(\mathcal{G}^2x_{m-1}, \mathcal{G}^2x_m\right)\right), F_1\left(\mathcal{D}\left(\mathcal{G}x_{m-1}, \mathcal{G}x_m\right)\right)\right\}. \tag{D}$$

If (C) holds, then by the conditions of Definition 11.4, inequality (11.8) will take the form

$$\min\left\{F_2\left(\mathcal{D}\left(\mathcal{G}^2x_{m-1}, \mathcal{G}^2x_m\right)\right), F_1\left(\mathcal{D}\left(\mathcal{G}x_{m-1}, \mathcal{G}x_m\right)\right)\right\} \leq F_1\left(\mathcal{D}\left(x_{m-1}, x_m\right)\right) - \tau. \tag{11.9}$$

Then, either

$$F_1\left(\mathcal{D}\left(x_{m-1}, x_m\right)\right) = \min\left\{F_2\left(\mathcal{D}\left(\mathcal{G}x_{m-1}, \mathcal{G}x_m\right)\right), F_1\left(\mathcal{D}\left(x_{m-1}, x_m\right)\right)\right\}, \tag{11.10}$$

or

$$F_2\left(\mathcal{D}\left(\mathcal{G}x_{m-1}, \mathcal{G}x_m\right)\right) = \min\left\{F_2\left(\mathcal{D}\left(\mathcal{G}x_{m-1}, \mathcal{G}x_m\right)\right), F_1\left(\mathcal{D}\left(x_{m-1}, x_m\right)\right)\right\}. \tag{11.11}$$

If (11.10) is true, the inequality (11.9) will take the form

$$\begin{aligned}&\min\left\{F_2\left(\mathcal{D}\left(\mathcal{G}^2x_{m-1}, \mathcal{G}^2x_m\right)\right), F_1\left(\mathcal{D}\left(\mathcal{G}x_{m-1}, \mathcal{G}x_m\right)\right)\right\}\\ &\leq \min\left\{F_2\left(\mathcal{D}\left(\mathcal{G}x_{m-1}, \mathcal{G}x_m\right)\right), F_1\left(\mathcal{D}\left(x_{m-1}, x_m\right)\right)\right\} - \tau.\end{aligned} \tag{11.12}$$

If (11.11) is true, we have

$$F_2\left(\mathcal{D}\left(\mathcal{G}x_{m-1}, \mathcal{G}x_m\right)\right) < F_1\left(\mathcal{D}\left(x_{m-1}, x_m\right)\right).$$

Condition (A) allows us to write

$$F_2(\mathscr{D}(\mathscr{G}x_{m-1}, \mathscr{G}x_m)) < F_1(\mathscr{D}(x_{m-1}, x_m)) \leq F_2(\mathscr{D}(\mathscr{G}x_{m-1}, \mathscr{G}x_m)) + \upsilon. \tag{11.13}$$

Using (11.13) in (11.9), we have

$$\begin{aligned}&\min\left\{F_2\left(\mathscr{D}\left(\mathscr{G}^2x_{m-1}, \mathscr{G}^2x_m\right)\right), F_1(\mathscr{D}(\mathscr{G}x_{m-1}, \mathscr{G}x_m))\right\}\\ &\leq F_2(\mathscr{D}(\mathscr{G}x_{m-1}, \mathscr{G}x_m)) + \upsilon - \tau.\end{aligned}$$

Moreover, with the use of condition (11.11) the above inequality takes the following form:

$$\begin{aligned}&\min\left\{F_2\left(\mathscr{D}\left(\mathscr{G}^2x_{m-1}, \mathscr{G}^2x_m\right)\right), F_1(\mathscr{D}(\mathscr{G}x_{m-1}, \mathscr{G}x_m))\right\}\\ &\leq \min\left\{F_2(\mathscr{D}(\mathscr{G}x_{m-1}, \mathscr{G}x_m)), F_1(\mathscr{D}(x_{m-1}, x_m))\right\} + \upsilon - \tau.\end{aligned} \tag{11.14}$$

Combining both possible cases of (11.12) and (11.14), we can write

$$\begin{aligned}&\min\left\{F_2\left(\mathscr{D}\left(\mathscr{G}^2x_{m-1}, \mathscr{G}^2x_m\right)\right), F_1(\mathscr{D}(\mathscr{G}x_{m-1}, \mathscr{G}x_m))\right\}\\ &\leq \min\left\{F_2(\mathscr{D}(\mathscr{G}x_{m-1}, \mathscr{G}x_m)), F_1(\mathscr{D}(x_{m-1}, x_m))\right\} + \delta_m\upsilon - \tau,\end{aligned} \tag{11.15}$$

where

$$\delta_m = \begin{cases}0 & if,\ F_2(t_2) > F_1(t_1)\\ 1 & if,\ F_2(t_2) < F_1(t_1)\end{cases}, \quad t_1, t_2 \in \mathbb{R}_+, t_1 \neq t_2.$$

Similarly, with the possible existence of (D), inequality (11.8) takes the following form:

$$\min\left\{F_2\left(\mathscr{D}\left(\mathscr{G}^2x_{m-1}, \mathscr{G}^2x_m\right)\right), F_1(\mathscr{D}(\mathscr{G}x_{m-1}, \mathscr{G}x_m))\right\} \leq F_2(\mathscr{D}(x_{m-1}, x_m)) - \tau. \tag{11.16}$$

If $F_2(\mathscr{D}(x_{m-1}, x_m)) \leq F_1(\mathscr{D}(x_{m-1}, x_m))$, then (11.16) can be written as

$$\min\left\{F_2\left(\mathscr{D}\left(\mathscr{G}^2x_{m-1}, \mathscr{G}^2x_m\right)\right), F_1(\mathscr{D}(\mathscr{G}x_{m-1}, \mathscr{G}x_m))\right\} \leq F_1(\mathscr{D}(x_{m-1}, x_m)) - \tau. \tag{11.17}$$

If $F_2(\mathscr{D}(x_{m-1}, x_m)) > F_1(\mathscr{D}(x_{m-1}, x_m))$, then using (B) we can write

$$\min\left\{F_2\left(\mathscr{D}\left(\mathscr{G}^2x_{m-1}, \mathscr{G}^2x_m\right)\right), F_1(\mathscr{D}(\mathscr{G}x_{m-1}, \mathscr{G}x_m))\right\} \leq F_1(\mathscr{D}(x_{m-1}, x_m)) + \upsilon - \tau. \tag{11.18}$$

Combining the inequalities (11.17) and(11.18), we will have

$$\min\left\{F_2\left(\mathscr{D}\left(\mathscr{G}^2x_{m-1}, \mathscr{G}^2x_m\right)\right), F_1(\mathscr{D}(\mathscr{G}x_{m-1}, \mathscr{G}x_m))\right\} \leq F_1(\mathscr{D}(x_{m-1}, x_m)) + \eta_m\upsilon - \tau, \tag{11.19}$$

where

$$\eta_m = \begin{cases} 1 & if, \quad F_2(t) > F_1(t) \\ 0 & if, \quad F_2(t) < F_1(t) \end{cases}, \quad t \in \mathbb{R}_+.$$

By following the procedure of inequalities (11.9)–(11.15), the inequality (11.19) will take the form

$$\begin{aligned} &\min\left\{F_2\left(\mathscr{D}\left(\mathscr{G}^2 x_{m-1}, \mathscr{G}^2 x_m\right)\right), F_1\left(\mathscr{D}\left(\mathscr{G}x_{m-1}, \mathscr{G}x_m\right)\right)\right\} \\ &\le \min\left\{F_2\left(\mathscr{D}\left(\mathscr{G}x_{m-1}, \mathscr{G}x_m\right)\right), F_1\left(\mathscr{D}\left(x_{m-1}, x_m\right)\right)\right\} + \left(\delta_m + \eta_m\right)\upsilon - \tau. \end{aligned} \tag{11.20}$$

The inequalities (11.15) and (11.20) can be combined as follows:

$$\begin{aligned} &\min\left\{F_2\left(\mathscr{D}\left(\mathscr{G}^2 x_{m-1}, \mathscr{G}^2 x_m\right)\right), F_1\left(\mathscr{D}\left(\mathscr{G}x_{m-1}, \mathscr{G}x_m\right)\right)\right\} \\ &\le \min\left\{F_2\left(\mathscr{D}\left(\mathscr{G}x_{m-1}, \mathscr{G}x_m\right)\right), F_1\left(\mathscr{D}\left(x_{m-1}, x_m\right)\right)\right\} + \left(\delta_m + \varsigma_m\eta_m\right)\upsilon - \tau, \end{aligned}$$

where ς_m is either 0 or 1.
Equivalently,

$$\begin{aligned} &\min\left\{F_2\left(\mathscr{D}\left(\mathscr{G}^2 x_{m-1}, \mathscr{G}^2 x_m\right)\right), F_1\left(\mathscr{D}\left(\mathscr{G}x_{m-1}, \mathscr{G}x_m\right)\right)\right\} \\ &\le \min\left\{F_2\left(\mathscr{D}\left(\mathscr{G}^2 x_{m-2}, \mathscr{G}^2 x_{m-1}\right)\right), F_1\left(\mathscr{D}\left(\mathscr{G}x_{m-2}, \mathscr{G}x_{m-1}\right)\right)\right\} + \left(\delta_m + \varsigma_m\eta_m\right)\upsilon - \tau. \end{aligned}$$

Repeating this process, we have

$$\begin{aligned} &\min\left\{F_2\left(\mathscr{D}\left(\mathscr{G}^2 x_{m-1}, \mathscr{G}^2 x_m\right)\right), F_1\left(\mathscr{D}\left(\mathscr{G}x_{m-1}, \mathscr{G}x_m\right)\right)\right\} \\ &\le \min\left\{F_2\left(\mathscr{D}\left(\mathscr{G}^2 x_{m-3}, \mathscr{G}^2 x_{m-2}\right)\right), F_1\left(\mathscr{D}\left(\mathscr{G}x_{m-3}, \mathscr{G}x_{m-2}\right)\right)\right\} \\ &+ \left(\delta_m + \varsigma_m\eta_m\right)\upsilon + \left(\delta_{m-1} + \varsigma_{m-1}\eta_{m-1}\right)\upsilon - 2\tau \\ &\vdots \\ &\min\left\{F_2\left(\mathscr{D}\left(\mathscr{G}^2 x_{m-1}, \mathscr{G}^2 x_m\right)\right), F_1\left(\mathscr{D}\left(\mathscr{G}x_{m-1}, \mathscr{G}x_m\right)\right)\right\} \\ &\le \min\left\{F_2\left(\mathscr{D}\left(\mathscr{G}^2 x_1, \mathscr{G}^2 x_0\right)\right), F_1\left(\mathscr{D}\left(\mathscr{G}x_1, \mathscr{G}x_0\right)\right)\right\} + \upsilon\sum_{j=1}^{m}\left(\delta_j + \varsigma_j\eta_j\right) - m\tau. \end{aligned}$$

Since $\tau > 2\upsilon$ and $\upsilon \sum_{j=1}^{m} \left(\delta_j + \varsigma_j\eta_j\right) < m\tau$, therefore,

$$\lim_{m\to\infty} \upsilon \sum_{j=1}^{m} \left(\delta_j + \varsigma_j\eta_j\right) - m\tau = -\infty.$$

So,

$$\lim_{m\to\infty} \min\left\{F_2\left(\mathscr{D}\left(\mathscr{G}^2 x_{m-1}, \mathscr{G}^2 x_m\right)\right), F_1\left(\mathscr{D}\left(\mathscr{G}x_{m-1}, \mathscr{G}x_m\right)\right)\right\} = -\infty. \tag{11.21}$$

Now, from Eq. (11.21) we have the following two cases:

$$\lim_{m\to\infty} F_2\left(\mathscr{D}\left(\mathscr{G}^2 x_{m-1}, \mathscr{G}^2 x_m\right)\right) = -\infty. \tag{E}$$

or

$$\lim_{m\to\infty} F_1\left(\mathscr{D}\left(\mathscr{G} x_{m-1}, \mathscr{G} x_m\right)\right) = -\infty. \tag{F}$$

Using condition (F2), then (E) yields

$$\lim_{m\to\infty} \mathscr{D}\left(\mathscr{G}^2 x_{m-1}, \mathscr{G}^2 x_m\right) = 0,$$

or equivalently

$$\lim_{m\to\infty} \mathscr{D}\left(\mathscr{G}^2 x_{m-1}, \mathscr{G}^2 x_m\right) = \lim_{m\to\infty} \mathscr{D}\left(x_{m+1}, \mathscr{G} x_{m+1}\right) = \lim_{m+1\to\infty} \mathscr{D}\left(x_m, \mathscr{G} x_m\right) = 0.$$

Using condition (F2), then (F) yields $\lim_{m\to\infty} \mathscr{D}\left(\mathscr{G} x_{m-1}, \mathscr{G} x_m\right) = 0$ that takes the form $\lim_{m\to\infty} \mathscr{D}\left(x_m, \mathscr{G} x_m\right) = 0$. Therefore, from (11.21), we get

$$\lim_{m\to\infty} \mathscr{D}\left(x_m, \mathscr{G} x_m\right) = 0. \tag{11.22}$$

Next in order to prove that the sequence $\{x_m\}_{n=1}^{\infty}$ is Cauchy, we suppose on contrary that there exists $\varepsilon > 0$ and sequences $\{\mathfrak{g}(m)\}_{n=1}^{\infty}$ and $\{\mathfrak{h}(m)\}_{n=1}^{\infty}$ of natural numbers such that

$$\mathfrak{g}(m) > \mathfrak{h}(m) > m, \quad \mathscr{D}\left(x_{\mathfrak{g}(m)}, x_{\mathfrak{h}(m)}\right) \geq \varepsilon, \quad \mathscr{D}\left(x_{\mathfrak{g}(m)-1}, x_{\mathfrak{h}(m)}\right) < \varepsilon, \quad \text{for all } m \in \mathbb{N}. \tag{11.23}$$

Therefore, we can write

$$\begin{aligned} \varepsilon &\leq \mathscr{D}\left(x_{\mathfrak{g}(m)}, x_{\mathfrak{h}(m)}\right) \\ &\leq \mathscr{D}\left(x_{\mathfrak{g}(m)-1}, x_{\mathfrak{g}(m)}\right) + \mathscr{D}\left(x_{\mathfrak{g}(m)-1}, x_{\mathfrak{h}(m)}\right) \\ &< \mathscr{D}\left(x_{\mathfrak{g}(m)-1}, x_{\mathfrak{g}(m)}\right) + \varepsilon \\ &= \mathscr{D}\left(x_{\mathfrak{g}(m)-1}, \mathscr{G} x_{\mathfrak{g}(m)-1}\right) + \varepsilon. \end{aligned}$$

That is,

$$\varepsilon \leq \mathscr{D}\left(x_{\mathfrak{g}(m)}, x_{\mathfrak{h}(m)}\right) < \mathscr{D}\left(x_{\mathfrak{g}(m)-1}, \mathscr{G} x_{\mathfrak{g}(m)-1}\right) + \varepsilon. \tag{11.24}$$

Therefore, (11.22) and (11.24) yield

$$\lim_{m\to\infty} \mathscr{D}\left(x_{\mathfrak{g}(m)}, x_{\mathfrak{h}(m)}\right) = \varepsilon.$$

Further, from (11.22) there exists $N \in \mathbb{N}$, such that

$$\mathscr{D}\left(x_{\mathfrak{g}(m)}, \mathscr{G}x_{\mathfrak{g}(m)}\right) < \frac{\varepsilon}{4} \quad and \quad \mathscr{D}\left(x_{\mathfrak{h}(m)}, \mathscr{G}x_{\mathfrak{h}(m)}\right) < \frac{\varepsilon}{4}, \quad \text{for all } m \geq N. \tag{11.25}$$

Next, we shall prove that

$$\frac{1}{2}\mathscr{D}\left(x_{\mathfrak{g}(m)}, x_{\mathfrak{g}(m)+1}\right) < \mathscr{D}\left(x_{\mathfrak{g}(m)}, x_{\mathfrak{h}(m)}\right), \qquad \text{for all } m \geq N. \tag{11.26}$$

For this purpose we suppose on contrary that there exists $r \geq N$, such that

$$\frac{1}{2}\mathscr{D}\left(x_{\mathfrak{g}(r)+1}, x_{\mathfrak{g}(r)+2}\right) \geq \mathscr{D}\left(x_{\mathfrak{g}(r)+1}, x_{\mathfrak{h}(r)+1}\right), \tag{11.27}$$

$$\frac{1}{2}\mathscr{D}\left(x_{\mathfrak{g}(r)+2}, x_{\mathfrak{g}(r)+3}\right) \geq \mathscr{D}\left(x_{\mathfrak{g}(r)+2}, x_{\mathfrak{h}(r)+1}\right). \tag{11.28}$$

The inequality (11.27) yields

$$2\mathscr{D}\left(x_{\mathfrak{g}(r)+1}, x_{\mathfrak{h}(r)+1}\right) \leq \mathscr{D}\left(x_{\mathfrak{g}(r)+1}, x_{\mathfrak{g}(r)+2}\right).$$

That can be written as

$$2\mathscr{D}\left(x_{\mathfrak{g}(r)+1}, x_{\mathfrak{h}(r)+1}\right) \leq \mathscr{D}\left(x_{\mathfrak{g}(r)+1}, x_{\mathfrak{h}(r)+1}\right) + \mathscr{D}\left(x_{\mathfrak{g}(r)+2}, x_{\mathfrak{h}(r)+1}\right)$$

so that

$$\mathscr{D}\left(x_{\mathfrak{g}(r)+1}, x_{\mathfrak{h}(r)+1}\right) \leq \mathscr{D}\left(x_{\mathfrak{g}(r)+2}, x_{\mathfrak{h}(r)+1}\right). \tag{11.29}$$

The inequalities (11.28) and (11.29) can be written in the combined form as follows:

$$\mathscr{D}\left(x_{\mathfrak{g}(r)+1}, x_{\mathfrak{h}(r)+1}\right) \leq \mathscr{D}\left(x_{\mathfrak{g}(r)+2}, x_{\mathfrak{h}(r)+1}\right) \leq \frac{1}{2}\mathscr{D}\left(x_{\mathfrak{g}(r)+2}, x_{\mathfrak{g}(r)+3}\right) = \frac{1}{2}\mathscr{D}\left(x_{\mathfrak{g}(r)+2}, \mathscr{G}x_{\mathfrak{g}(r)+2}\right). \tag{11.30}$$

It follows from (11.23), (11.25), and (11.30) that

$$\begin{aligned}
\varepsilon &\leq \mathscr{D}\left(x_{\mathfrak{g}(r)}, x_{\mathfrak{h}(r)}\right) \leq \mathscr{D}\left(x_{\mathfrak{g}(r)}, x_{\mathfrak{g}(r)+1}\right) + \mathscr{D}\left(x_{\mathfrak{g}(r)+1}, x_{\mathfrak{h}(r)}\right) \\
&\leq \mathscr{D}\left(x_{\mathfrak{g}(r)}, x_{\mathfrak{g}(r)+1}\right) + \mathscr{D}\left(x_{\mathfrak{g}(r)+1}, x_{\mathfrak{h}(r)+1}\right) + \mathscr{D}\left(x_{\mathfrak{h}(r)+1}, x_{\mathfrak{h}(r)}\right) \\
&= \mathscr{D}\left(x_{\mathfrak{g}(r)}, \mathscr{G}x_{\mathfrak{g}(r)}\right) + \mathscr{D}\left(x_{\mathfrak{g}(r)+1}, x_{\mathfrak{h}(r)+1}\right) + \mathscr{D}\left(x_{\mathfrak{h}(r)}, \mathscr{G}x_{\mathfrak{h}(r)}\right) \\
&< \frac{\varepsilon}{4} + \frac{\varepsilon}{8} + \frac{\varepsilon}{4} = \frac{5\varepsilon}{8},
\end{aligned}$$

which is a contradiction, therefore, relation (11.26) with the assumption of the theorem gives

$$\tau + \min\left\{F_2\left(\mathcal{D}\left(\mathcal{G}^2 x_{\mathfrak{g}(m)}, \mathcal{G}^2 x_{\mathfrak{h}(m)}\right)\right), F_1\left(\mathcal{D}\left(\mathcal{G} x_{\mathfrak{g}(m)}, \mathcal{G} x_{\mathfrak{h}(m)}\right)\right)\right\}$$
$$\leq \alpha_2 F_2\left(x_{\mathfrak{g}(m)}, x_{\mathfrak{h}(m)}\right) + \alpha_1 F_1\left(\mathcal{D}\left(x_{\mathfrak{g}(m)}, x_{\mathfrak{h}(m)}\right)\right). \tag{11.31}$$

From (11.22) and (11.28), we have

$$\tau + \min\{F_2(\varepsilon), F_1(\varepsilon)\} \leq \alpha_2 F_2(\varepsilon) + \alpha_1 F_1((\varepsilon)). \tag{11.32}$$

Relation (11.32) has two possible cases, that is, either $\tau + F_2(\varepsilon) \leq F_2(\varepsilon)$, or $\tau + F_1(\varepsilon) \leq F_1(\varepsilon)$.

Both of the above relations are contradictions. Therefore, $\{x_m\}_{n=1}^{\infty}$ is a Cauchy sequence. The completeness of $(X, \mathcal{D})$ proves that $\{x_m\}_{n=1}^{\infty}$ converges to some point x^* in X, and

$$\lim_{m\to\infty} \mathcal{D}\left(x_m, x^*\right) = 0. \tag{11.33}$$

Next, we claim that

$$\frac{1}{2}\mathcal{D}(x_m, \mathcal{G}x_m) < \mathcal{D}\left(x_m, x^*\right), \qquad (i)$$
$$\frac{1}{2}\mathcal{D}\left(\mathcal{G}x_m, \mathcal{G}^2 x_m\right) < \mathcal{D}\left(\mathcal{G}x_m, x^*\right), \qquad (ii)$$
$$\frac{1}{2}\mathcal{D}\left(\mathcal{G}^2 x_m, \mathcal{G}^3 x_m\right) < \mathcal{D}\left(\mathcal{G}^2 x_m, x^*\right). \qquad (iii)$$

On the contrary, suppose that for some $m \in N$, we have the following three inequalities:

$$\frac{1}{2}\mathcal{D}(x_m, \mathcal{G}x_m) \geq \mathcal{D}\left(x_m, x^*\right). \tag{11.34}$$

$$\frac{1}{2}\mathcal{D}\left(\mathcal{G}x_m, \mathcal{G}^2 x_m\right) \geq \mathcal{D}\left(\mathcal{G}x_m, x^*\right). \tag{11.35}$$

$$\frac{1}{2}\mathcal{D}\left(\mathcal{G}^2 x_m, \mathcal{G}^3 x_m\right) \geq \mathcal{D}\left(\mathcal{G}^2 x_m, x^*\right). \tag{11.36}$$

Now, (11.34) along with the triangular inequality will give

$$2\mathcal{D}\left(x_m, x^*\right) \leq \mathcal{D}(x_m, \mathcal{G}x_m) \leq \mathcal{D}\left(x_m, x^*\right) + \mathcal{D}\left(\mathcal{G}x_m, x^*\right)$$

that takes the form

$$\mathcal{D}\left(x_m, x^*\right) \leq \mathcal{D}\left(\mathcal{G}x_m, x^*\right). \tag{11.37}$$

The inequality (11.35) along with (11.37) will give

$$\mathcal{D}\left(x_m, x^*\right) \leq \mathcal{D}\left(\mathcal{G}x_m, x^*\right) \leq \frac{1}{2}\mathcal{D}\left(\mathcal{G}x_m, \mathcal{G}^2 x_m\right). \tag{11.38}$$

Similarly (11.35) implies that

$$2\mathscr{D}\left(\mathscr{G}x_m, x^*\right) \leq \mathscr{D}\left(\mathscr{G}x_m, \mathscr{G}^2x_m\right) \leq \mathscr{D}\left(\mathscr{G}x_m, x^*\right) + \mathscr{D}\left(\mathscr{G}^2x_m, x^*\right).$$

That yields

$$\mathscr{D}\left(\mathscr{G}x_m, x^*\right) \leq \mathscr{D}\left(\mathscr{G}^2x_m, x^*\right). \tag{11.39}$$

The inequality (11.39) along with (11.36) will take the following form:

$$\mathscr{D}\left(\mathscr{G}x_m, x^*\right) \leq \mathscr{D}\left(\mathscr{G}^2x_m, x^*\right) \leq \frac{1}{2}\mathscr{D}\left(\mathscr{G}^2x_m, \mathscr{G}^3x_m\right). \tag{11.40}$$

Now, by our assumption

$$\begin{aligned}&\tau + \min\left\{F_2\left(\mathscr{D}\left(\mathscr{G}^2x_m, \mathscr{G}^3x_m\right)\right), F_1\left(\mathscr{D}\left(\mathscr{G}x_m, \mathscr{G}^2x_m\right)\right)\right\}\\ &\leq \alpha_2 F_2\left(\mathscr{D}\left(x_m, \mathscr{G}x_m\right)\right) + \alpha_1 F_1\left(\mathscr{D}\left(x_m, \mathscr{G}x_m\right)\right).\end{aligned}$$

If

$$\min\left\{F_2\left(\mathscr{D}\left(\mathscr{G}^2x_m, \mathscr{G}^3x_m\right)\right), F_1\left(\mathscr{D}\left(\mathscr{G}x, \mathscr{G}^2x_m\right)\right)\right\} = F_1\left(\mathscr{D}\left(\mathscr{G}x, \mathscr{G}^2x_m\right)\right),$$

we have

$$\tau + F_1\left(\mathscr{D}\left(\mathscr{G}x_m, \mathscr{G}^2x_m\right)\right) \leq F_1\left(\mathscr{D}\left(x_m, \mathscr{G}x_m\right)\right).$$

As $\tau > 0$, we have

$$F_1\left(\mathscr{D}\left(\mathscr{G}x_m, \mathscr{G}^2x_m\right)\right) < F_1\left(\mathscr{D}\left(x_m, \mathscr{G}x_m\right)\right).$$

Using the condition $(F1)$, we can write

$$\begin{aligned}\mathscr{D}\left(\mathscr{G}x_m, \mathscr{G}^2x_m\right) &< \mathscr{D}\left(x_m, \mathscr{G}x_m\right)\\ &\leq \mathscr{D}\left(x_m, x^*\right) + \mathscr{D}\left(\mathscr{G}x_m, x^*\right)\\ &\leq \frac{1}{2}\mathscr{D}\left(\mathscr{G}x_m, \mathscr{G}^2x_m\right) + \frac{1}{2}\mathscr{D}\left(\mathscr{G}x_m, \mathscr{G}^2x_m\right) \qquad \text{using (11.38)}\\ &= \mathscr{D}\left(\mathscr{G}x_m, \mathscr{G}^2x_m\right),\end{aligned}$$

which yields a contradiction,
Similarly, if

$$\min\left\{F_2\left(\mathscr{D}\left(\mathscr{G}^2x_m, \mathscr{G}^3x_m\right)\right), F_1\left(\mathscr{D}\left(Tx, \mathscr{G}^2x_m\right)\right)\right\} = F_2\left(\mathscr{D}\left(\mathscr{G}^2x, \mathscr{G}^3x_m\right)\right),$$

we have

$$\tau + F_2\left(\mathscr{D}\left(\mathscr{G}^2x, \mathscr{G}^3x_m\right)\right) \leq F_2\left(\mathscr{D}\left(x_m, \mathscr{G}x_m\right)\right).$$

As $\tau > 0$, we have

$$F_2\left(\mathscr{D}\left(\mathscr{G}^2 x, \mathscr{G}^3 x_m\right)\right) < F_2\left(\mathscr{D}\left(x_m, \mathscr{G} x_m\right)\right).$$

Using condition $(F1)$, we can write

$$\begin{aligned}\mathscr{D}\left(\mathscr{G}^2 x, \mathscr{G}^3 x_m\right) &< \mathscr{D}\left(x_m, \mathscr{G} x_m\right)\\ &\leq \mathscr{D}\left(x_m, x^*\right) + \mathscr{D}\left(\mathscr{G} x_m, x^*\right)\\ &\leq \frac{1}{2}\mathscr{D}\left(\mathscr{G}^2 x_m, \mathscr{G}^3 x_m\right) + \frac{1}{2}\mathscr{D}\left(\mathscr{G}^2 x_m, \mathscr{G}^3 x_m\right) \quad \text{using (11.37), (11.40)}\\ &= \mathscr{D}\left(\mathscr{G}^2 x_m, \mathscr{G}^3 x_m\right),\end{aligned}$$

again a contradiction. Therefore, (i), (ii), and (iii) hold.

Now, we can write

$$\tau + \min\left\{F_2\left(\mathscr{D}\left(\mathscr{G}^2 x_m, \mathscr{G}^2 x^*\right)\right), F_1\left(\mathscr{D}\left(\mathscr{G} x_m, \mathscr{G} x^*\right)\right)\right\} \leq \alpha_2 F_2\left(\mathscr{D}\left(x_m, x^*\right)\right) + \alpha_1 F_1\left(\mathscr{D}\left(x_m, x^*\right)\right).$$

Using (11.21), (E), and (F), we have either $\lim_{m\to\infty} F_2\left(\mathscr{D}\left(\mathscr{G}^2 x_m, \mathscr{G}^2 x^*\right)\right) = -\infty$, or $\lim_{m\to\infty} F_1\left(\mathscr{D}\left(\mathscr{G} x_m, \mathscr{G} x^*\right)\right) = -\infty$.
Using condition $(F2)$ gives $\lim_{m\to\infty} \mathscr{D}\left(\mathscr{G}^2 x_m, \mathscr{G}^2 x^*\right) = 0$, or $\lim_{m\to\infty} \mathscr{D}\left(\mathscr{G} x_m, \mathscr{G} x^*\right) = 0$.
So, relation (11.33) implies $\mathscr{D}\left(x^*, \mathscr{G} x^*\right) = \lim_{m\to\infty} \mathscr{D}\left(x_{m+1}, \mathscr{G} x^*\right) = \lim_{m\to\infty} \mathscr{D}\left(\mathscr{G} x_m, \mathscr{G} x^*\right) = 0$. Therefore, $\mathscr{G}$ has a fixed point x^*. Now, for the uniqueness of fixed point, let us suppose that $\mathscr{G}$ has more than one fixed point. That is, there exist two distinct $x, y \in X$ such that $\mathscr{G} x = x \neq y = \mathscr{G} y$.
Since $0 = \frac{1}{2}\mathscr{D}(x, \mathscr{G} x) < \mathscr{D}(x, y)$, so, using the assumption of the theorem, we can write

$$\tau + \min\left\{F_2\left(\mathscr{D}\left(\mathscr{G}^2 x, \mathscr{G}^2 y\right)\right), F_1\left(\mathscr{D}\left(\mathscr{G} x, \mathscr{G} y\right)\right)\right\} \leq \alpha_2 F_2\left(\mathscr{D}(x, y)\right) + \alpha_1 F_1\left(\mathscr{D}(x, y)\right).$$

That yields either

$$F_1\left(\mathscr{D}(x, y)\right) = F_1\left(\mathscr{D}\left(\mathscr{G} x, \mathscr{G} y\right)\right) < \tau + F_1\left(\mathscr{D}\left(\mathscr{G} x, \mathscr{G} y\right)\right) \leq F_1\left(\mathscr{D}(x, y)\right), \tag{11.41}$$

or

$$F_2\left(\mathscr{D}(x, y)\right) = F_2\left(\mathscr{D}\left(\mathscr{G}^2 x, \mathscr{G}^2 y\right)\right) < \tau + F_2\left(\mathscr{D}\left(\mathscr{G}^2 x, \mathscr{G}^2 y\right)\right) \leq F_2\left(\mathscr{D}(x, y)\right). \tag{11.42}$$

Both relations (11.41) and (11.42) are contradictions that prove the existence of a unique fixed point.

Theorem 11.3 *Let $(X, \mathscr{D})$ be a complete metric space. Suppose that a continuous mapping $\mathscr{G} : X \mapsto X$ be a dual F-Suzuki contraction of type-II, and for all $t_1, t_2 \in \mathbb{R}_+$, there exist $\upsilon > 0$, $\tau > 2\upsilon$, such that*

$$F_2(t_2) < F_1(t_1) \le F_2(t_2) + \upsilon. \qquad (A)$$
$$F_1(t_1) < F_2(t_2) \le F_1(t_1) + \upsilon. \qquad (B)$$

Then $\mathscr{G}$ has a unique fixed point in X and for every $x_0 \in X$, $\{\mathscr{G}^m x_0\}_{m=1}^{\infty}$ converges to x^*.

Proof Consider a sequence $\{x_1, x_2, \ldots.\}$ such that $x_{m+1} = \mathscr{G}x_m = \mathscr{G}^{m+1}x_0$ for any $x_0 \in X$ and for all $m \in \mathbb{N}_0$. If for some $m \in \mathbb{N}$, $\mathscr{D}(x_m, \mathscr{G}x_m) = 0$, $\mathscr{G}$ will admit a fixed point. Let for all $m \in \mathbb{N}$, $\mathscr{D}(x_m, \mathscr{G}x_m) = \mathscr{D}(\mathscr{G}x_{m-1}, \mathscr{G}x_m) > 0$, so that $\frac{1}{2}$ $\mathscr{D}(x_m, \mathscr{G}x_m) < (x_m, \mathscr{G}x_m)$.
We will prove that $\lim_{m\to\infty} \mathscr{D}(x_m, \mathscr{G}x_m) = 0$, for both possible cases $(R1)$, $(R2)$.
If, for all $x, y \in X$, the relation $(R2)$ holds, then the procedure of inequalities (11.9) to (11.15) yields

$$\begin{aligned}&\min\left\{F_2\left(\mathscr{D}\left(\mathscr{G}^2x_{m-1}, \mathscr{G}^2x_m\right)\right), F_1\left(\mathscr{D}\left(\mathscr{G}x_{m-1}, \mathscr{G}x_m\right)\right)\right\}\\&\le \min\left\{F_2\left(\mathscr{D}\left(\mathscr{G}x_{m-1}, \mathscr{G}x_m\right)\right), F_1\left(\mathscr{D}\left(x_{m-1}, x_m\right)\right)\right\} + \delta_m\upsilon - \tau,\end{aligned}$$

where

$$\delta_m = \begin{cases}0 & if,\ F_2(t_2) > F_1(t_1)\\ 1 & if,\ F_2(t_2) < F_1(t_1)\end{cases}, \quad t_1, t_2 \in \mathbb{R}_+, t_1 \ne t_2.$$

The above inequality can be written as

$$\begin{aligned}&\min\left\{F_2\left(\mathscr{D}\left(\mathscr{G}^2x_{m-1}, \mathscr{G}^2x_m\right)\right), F_1\left(\mathscr{D}\left(\mathscr{G}x_{m-1}, \mathscr{G}x_m\right)\right)\right\}\\&\le \min\left\{F_2\left(\mathscr{D}\left(\mathscr{G}^2x_{m-2}, \mathscr{G}^2x_{m-1}\right)\right), F_1\left(\mathscr{D}\left(\mathscr{G}x_{m-2}, \mathscr{G}x_{m-1}\right)\right)\right\} + \delta_m\upsilon - \tau.\end{aligned}$$

Repeating this process, we have

$$\begin{aligned}&\min\left\{F_2\left(\mathscr{D}\left(\mathscr{G}^2x_{m-1}, \mathscr{G}^2x_m\right)\right), F_1\left(\mathscr{D}\left(\mathscr{G}x_{m-1}, \mathscr{G}x_m\right)\right)\right\}\\&\le \min\left\{F_2\left(\mathscr{D}\left(\mathscr{G}^2x_{m-3}, \mathscr{G}^2x_{m-2}\right)\right), F_1\left(\mathscr{D}\left(\mathscr{G}x_{m-3}, \mathscr{G}x_{m-2}\right)\right)\right\} + \delta_m\upsilon + \delta_{m-1}\upsilon - 2\tau,\\&\vdots\\&\min\left\{F_2\left(\mathscr{D}\left(\mathscr{G}^2x_{m-1}, \mathscr{G}^2x_m\right)\right), F_1\left(\mathscr{D}\left(\mathscr{G}x_{m-1}, \mathscr{G}x_m\right)\right)\right\}\\&\le \min\left\{F_2\left(\mathscr{D}\left(\mathscr{G}^2x_1, \mathscr{G}^2x_0\right)\right), F_1\left(\mathscr{D}\left(\mathscr{G}x_1, \mathscr{G}x_0\right)\right)\right\} + \sum_{j=1}^{m}\delta_j\upsilon - m\tau,\end{aligned}$$

or

$$\min\left\{F_2\left(\mathscr{D}\left(\mathscr{G}^2x_{m-1},\mathscr{G}^2x_m\right)\right),F_1\left(\mathscr{D}\left(\mathscr{G}x_{m-1},\mathscr{G}x_m\right)\right)\right\}$$
$$\leq F_1\left(\mathscr{D}\left(x_1,x_0\right)\right)+\sum_{j=1}^{m}\delta_j\upsilon-(m+1)\tau.$$

Since $\tau>2\upsilon$ and $\sum_{j=1}^{m}\delta_j<2(m+1)$, therefore,

$$\lim_{m\to\infty}\upsilon\sum_{j=1}^{m}\delta_j-(m+1)\tau=-\infty.$$

So,

$$\lim_{m\to\infty}\min\left\{F_2\left(\mathscr{D}\left(\mathscr{G}^2x_{m-1},\mathscr{G}^2x_m\right)\right),F_1\left(\mathscr{D}\left(\mathscr{G}x_{m-1},\mathscr{G}x_m\right)\right)\right\}=-\infty.$$

If, for all $x,y\in X$, relation $(R1)$ holds, then the procedure of inequalities (11.16) to (11.19) yields

$$\min\left\{F_2\left(\mathscr{D}\left(\mathscr{G}^2x_{m-1},\mathscr{G}^2x_m\right)\right),F_1\left(\mathscr{D}\left(\mathscr{G}x_{m-1},\mathscr{G}x_m\right)\right)\right\}\leq F_1\left(\mathscr{D}\left(x_{m-1},x_m\right)\right)+\eta_m\upsilon-\tau,$$

where

$$\eta_m=\begin{cases}1 & if,\quad F_2(t)>F_1(t)\\ 0 & if,\quad F_2(t)<F_1(t)\end{cases},\quad t\in\mathbb{R}_+.$$

Following the procedure of inequalities (11.9)–(11.15), the above inequality will take the form

$$\min\left\{F_2\left(\mathscr{D}\left(\mathscr{G}^2x_{m-1},\mathscr{G}^2x_m\right)\right),F_1\left(\mathscr{D}\left(\mathscr{G}x_{m-1},\mathscr{G}x_m\right)\right)\right\}$$
$$\leq\min\left\{F_2\left(\mathscr{D}\left(\mathscr{G}x_{m-1},\mathscr{G}x_m\right)\right),F_1\left(\mathscr{D}\left(x_{m-1},x_m\right)\right)\right\}+\left(\delta_m+\eta_m\right)\upsilon-\tau.$$

That can be written as

$$\min\left\{F_2\left(\mathscr{D}\left(\mathscr{G}^2x_{m-1},\mathscr{G}^2x_m\right)\right),F_1\left(\mathscr{D}\left(\mathscr{G}x_{m-1},\mathscr{G}x_m\right)\right)\right\}$$
$$\leq\min\left\{F_2\left(\mathscr{D}\left(\mathscr{G}^2x_{m-2},\mathscr{G}^2x_{m-1}\right)\right),F_1\left(\mathscr{D}\left(\mathscr{G}x_{m-2},\mathscr{G}x_{m-1}\right)\right)\right\}+\left(\delta_m+\eta_m\right)\upsilon-\tau.$$

Repeating this process, we have

$$\begin{aligned}
&\min\left\{F_2\left(\mathscr{D}\left(\mathscr{G}^2x_{m-1},\mathscr{G}^2x_m\right)\right),F_1\left(\mathscr{D}\left(\mathscr{G}x_{m-1},\mathscr{G}x_m\right)\right)\right\}\\
&\leq\min\left\{F_2\left(\mathscr{D}\left(\mathscr{G}^2x_{m-3},\mathscr{G}^2x_{m-2}\right)\right),F_1\left(\mathscr{D}\left(\mathscr{G}x_{m-3},\mathscr{G}x_{m-2}\right)\right)\right\}\\
&+\left(\delta_m+\eta_m\right)\upsilon+\left(\delta_{m-1}+\eta_{m-1}\right)\upsilon-2\tau\\
&\vdots\\
&\min\left\{F_2\left(\mathscr{D}\left(\mathscr{G}^2x_{m-1},\mathscr{G}^2x_m\right)\right),F_1\left(\mathscr{D}\left(\mathscr{G}x_{m-1},\mathscr{G}x_m\right)\right)\right\}\\
&\leq\min\left\{F_2\left(\mathscr{D}\left(\mathscr{G}^2x_1,\mathscr{G}^2x_0\right)\right),F_1\left(\mathscr{D}\left(\mathscr{G}x_1,\mathscr{G}x_0\right)\right)\right\}+\upsilon\sum_{j=1}^{m}\left(\delta_j+\eta_j\right)-m\tau.
\end{aligned}$$

Since $\tau>2\upsilon$ and $\upsilon\sum_{j=1}^{m}\left(\delta_j+\eta_j\right)<m\tau$, therefore,

$$\upsilon_{\lim_{m\to\infty}}\sum_{j=1}^{m}\left(\delta_j+\eta_j\right)-m\tau=-\infty.$$

So,

$$\lim_{m\to\infty}\min\left\{F_2\left(\mathscr{D}\left(\mathscr{G}^2x_{m-1},\mathscr{G}^2x_m\right)\right),F_1\left(\mathscr{D}\left(\mathscr{G}x_{m-1},\mathscr{G}x_m\right)\right)\right\}=-\infty.$$

Therefore, for both cases $(R1)$, $(R2)$, we have

$$\lim_{m\to\infty}\mathscr{D}\left(x_m,\mathscr{G}x_m\right)=0. \tag{11.43}$$

Next in order to prove the sequence $\{x_m\}_{n=1}^{\infty}$ is Cauchy, we suppose on contrary that there exists $\varepsilon>0$ and sequences $\{\mathfrak{g}(m)\}_{n=1}^{\infty}$ and $\{\mathfrak{h}(m)\}_{n=1}^{\infty}$ of natural numbers such that

$$\mathfrak{g}(m)>\mathfrak{h}(m)>m,\quad\mathscr{D}\left(x_{\mathfrak{g}(m)},x_{\mathfrak{h}(m)}\right)\geq\varepsilon,\quad\mathscr{D}\left(x_{\mathfrak{g}(m)-1},x_{\mathfrak{h}(m)}\right)<\varepsilon,\quad\text{for all } m\in N.$$

Now, the analysis similar to the previous theorem yields

$$\begin{aligned}
&\tau+\min\left\{F_2\left(\mathscr{D}\left(\mathscr{G}^2x_{\mathfrak{g}(m)},\mathscr{G}^2x_{\mathfrak{h}(m)}\right)\right),F_1\left(\mathscr{D}\left(\mathscr{G}x_{\mathfrak{g}(m)},\mathscr{G}x_{\mathfrak{h}(m)}\right)\right)\right\}\\
&\leq\alpha_2F_2\left(x_{\mathfrak{g}(m)},x_{\mathfrak{h}(m)}\right)+\alpha_1F_1\left(\mathscr{D}\left(x_{\mathfrak{g}(m)},x_{\mathfrak{h}(m)}\right)\right).
\end{aligned} \tag{11.44}$$

From (11.43), the assumptions of the theorem, and the conditions of F-contraction, we will have

$$\tau+\min\left\{F_2\left(\varepsilon\right),F_1\left(\varepsilon\right)\right\}\leq\alpha_2F_2\left(\varepsilon\right)+\alpha_1F_1\left(\left(\varepsilon\right)\right\}.$$

For both the cases $(R1)$ and $(R2)$, the above inequality yields the following possibilities:

$$\tau + F_2(\varepsilon) \leq F_2(\varepsilon). \tag{11.45}$$

$$\tau + F_1(\varepsilon) \leq F_1(\varepsilon). \tag{11.46}$$

$$\tau + F_2(\varepsilon) \leq F_1(\varepsilon). \tag{11.47}$$

$$\tau + F_1(\varepsilon) \leq F_2(\varepsilon). \tag{11.48}$$

As $\tau > 0$, so the relations (11.45) and (11.46) are contradictions.
Now we will consider the inequality (11.47)

$$\tau + F_2(\varepsilon) \leq F_1(\varepsilon). \tag{11.49}$$

As $\tau > 2\upsilon$, then (11.49) will take the form

$$2\upsilon + F_2(\varepsilon) < F_1(\varepsilon).$$

That is,

$$\upsilon + (\upsilon + F_2(\varepsilon)) < F_1(\varepsilon).$$

Moreover, condition (A) allows us to write

$$\upsilon + F_1(\varepsilon) < F_1(\varepsilon).$$

Therefore, (11.47) yields a contradiction.
Similarly, $\tau + F_1(\varepsilon) \leq F_2(\varepsilon)$ represents the contradiction $\upsilon + F_2(\varepsilon) < F_2(\varepsilon)$.
Contradictions of relations (11.45)–(11.48) prove that $\{x_m\}_{m=1}^{\infty}$ is a Cauchy sequence. The completeness of $(X, \mathscr{D})$ proves that $\{x_m\}_{m=1}^{\infty}$ converges to some point x^* in X. Next, we claim that

$$\frac{1}{2}\mathscr{D}(x_m, \mathscr{G}x_m) < \mathscr{D}\left(x_m, x^*\right). \tag{i}$$

$$\frac{1}{2}\mathscr{D}\left(\mathscr{G}x_m, \mathscr{G}^2x_m\right) < \mathscr{D}\left(\mathscr{G}x_m, x^*\right). \tag{ii}$$

$$\frac{1}{2}\mathscr{D}\left(\mathscr{G}^2x_m, \mathscr{G}^3x_m\right) < \mathscr{D}\left(\mathscr{G}^2x_m, x^*\right). \tag{iii}$$

On the contrary, suppose that for some $m \in \mathbb{N}$, we have the following three inequalities:

$$\frac{1}{2}\mathscr{D}(x_m, \mathscr{G}x_m) \geq \mathscr{D}\left(x_m, x^*\right). \tag{11.50}$$

$$\frac{1}{2}\mathscr{D}\left(\mathscr{G}x_m, \mathscr{G}^2x_m\right) \geq \mathscr{D}\left(\mathscr{G}x_m, x^*\right). \tag{11.51}$$

$$\frac{1}{2}\mathscr{D}\left(\mathscr{G}^2 x_m, \mathscr{G}^3 x_m\right) \geq \mathscr{D}\left(\mathscr{G}^2 x_m, x^*\right). \tag{11.52}$$

Now, (11.50) along with the triangular inequality gives

$$2\mathscr{D}\left(x_m, x^*\right) \leq \mathscr{D}\left(x_m, \mathscr{G}x_m\right) \leq \mathscr{D}\left(x_m, x^*\right) + \mathscr{D}\left(\mathscr{G}x_m, x^*\right)$$

that takes the form

$$\mathscr{D}\left(x_m, x^*\right) \leq \mathscr{D}\left(\mathscr{G}x_m, x^*\right). \tag{11.53}$$

(11.53) along with (11.51) gives

$$\mathscr{D}\left(x_m, x^*\right) \leq \mathscr{D}\left(\mathscr{G}x_m, x^*\right) \leq \frac{1}{2}\mathscr{D}\left(\mathscr{G}x_m, \mathscr{G}^2 x\right). \tag{11.54}$$

Similarly (11.51) implies that

$$2\mathscr{D}\left(\mathscr{G}x_m, x^*\right) \leq \mathscr{D}\left(\mathscr{G}x_m, \mathscr{G}^2 x_m\right) \leq \mathscr{D}\left(\mathscr{G}x_m, x^*\right) + \mathscr{D}\left(\mathscr{G}^2 x_m, x^*\right).$$

That yields

$$\mathscr{D}\left(\mathscr{G}x_m, x^*\right) \leq \mathscr{D}\left(\mathscr{G}^2 x_m, x^*\right). \tag{11.55}$$

(11.55) along with (11.52) will take the following form:

$$\mathscr{D}\left(\mathscr{G}x_m, x^*\right) \leq \mathscr{D}\left(\mathscr{G}^2 x_m, x^*\right) \leq \frac{1}{2}\mathscr{D}\left(\mathscr{G}^2 x_m, \mathscr{G}^3 x_m\right). \tag{11.56}$$

Now, by the assumption of the theorem,

$$\begin{aligned} &\tau + \min\left\{F_2\left(\mathscr{D}\left(\mathscr{G}^2 x_m, \mathscr{G}^3 x_m\right)\right), F_1\left(\mathscr{D}\left(\mathscr{G}x_m, \mathscr{G}^2 x_m\right)\right)\right\} \\ &\leq \alpha_2 F_2\left(\mathscr{D}\left(x_m, \mathscr{G}x_m\right)\right) + \alpha_1 F_1\left(\mathscr{D}\left(x_m, \mathscr{G}x_m\right)\right). \end{aligned}$$

For both the cases $(R1)$, $(R2)$, the above relation yields four possible inequalities (11.57)–(11.60)

$$\tau + F_1\left(\mathscr{D}\left(\mathscr{G}x_m, \mathscr{G}^2 x_m\right)\right) \leq F_1\left(\mathscr{D}\left(x_m, \mathscr{G}x_m\right)\right). \tag{11.57}$$

$$\tau + F_2\left(\mathscr{D}\left(\mathscr{G}^2 x_m, \mathscr{G}^3 x_m\right)\right) \leq F_2\left(\mathscr{D}\left(x_m, \mathscr{G}x_m\right)\right). \tag{11.58}$$

$$\tau + F_2\left(\mathscr{D}\left(\mathscr{G}^2 x_m, \mathscr{G}^3 x_m\right)\right) \leq F_1\left(\mathscr{D}\left(x_m, \mathscr{G}x_m\right)\right). \tag{11.59}$$

$$\tau + F_1\left(\mathscr{D}\left(\mathscr{G}x_m, \mathscr{G}^2 x_m\right)\right) \leq F_2\left(\mathscr{D}\left(x_m, \mathscr{G}x_m\right)\right). \tag{11.60}$$

As $\tau > 0$, inequality (11.57) can be written as

$$F_1\left(\mathscr{D}\left(\mathscr{G}x_m, \mathscr{G}^2x_m\right)\right) < F_1\left(\mathscr{D}\left(x_m, \mathscr{G}x_m\right)\right).$$

Using the condition $(F1)$, the triangular inequality and relation (11.54), we can write

$$\begin{aligned}\mathscr{D}\left(\mathscr{G}x_m, \mathscr{G}^2x_m\right) &< \mathscr{D}\left(x_m, \mathscr{G}x_m\right)\\ &\leq \mathscr{D}\left(x_m, x^*\right) + \mathscr{D}\left(\mathscr{G}x_m, x^*\right)\\ &\leq \frac{1}{2}\mathscr{D}\left(\mathscr{G}x_m, \mathscr{G}^2x_m\right) + \frac{1}{2}\mathscr{D}\left(\mathscr{G}x_m, \mathscr{G}^2x_m\right)\\ &= \mathscr{D}\left(\mathscr{G}x_m, \mathscr{G}^2x_m\right).\end{aligned}$$

That yields a contradiction.
Similarly, using condition $(F1)$, and relations (11.53), (11.57), the inequality (11.58) can be written as

$$\begin{aligned}\mathscr{D}\left(\mathscr{G}^2x, \mathscr{G}^3x_m\right) &< \mathscr{D}\left(x_m, \mathscr{G}x_m\right)\\ &\leq \mathscr{D}\left(x_m, x^*\right) + \mathscr{D}\left(\mathscr{G}x_m, x^*\right)\\ &\leq \frac{1}{2}\mathscr{D}\left(\mathscr{G}^2x_m, \mathscr{G}^3x_m\right) + \frac{1}{2}\mathscr{D}\left(\mathscr{G}^2x_m, \mathscr{G}^3x_m\right)\\ &= \mathscr{D}\left(\mathscr{G}^2x_m, \mathscr{G}^3x_m\right),\end{aligned}$$

again a contradiction.
Now, using condition (B), relations (11.53) and (11.57), the inequality (11.59) will take the form

$$2\upsilon + F_2\left(\mathscr{D}\left(\mathscr{G}^2x_m, \mathscr{G}^3x_m\right)\right) < F_1\left(\mathscr{D}\left(x_m, \mathscr{G}x_m\right)\right),$$

$$\upsilon + F_1\left(\mathscr{D}\left(\mathscr{G}^2x_m, \mathscr{G}^3x_m\right)\right) < F_1\left(\mathscr{D}\left(x_m, \mathscr{G}x_m\right)\right),$$

or

$$F_1\left(\mathscr{D}\left(\mathscr{G}^2x_m, \mathscr{G}^3x_m\right)\right) < F_1\left(\mathscr{D}\left(x_m, \mathscr{G}x_m\right)\right),$$

or

$$\begin{aligned}\mathscr{D}\left(\mathscr{G}^2x_m, \mathscr{G}^3x_m\right) &< \mathscr{D}\left(x_m, \mathscr{G}x_m\right)\\ &\leq \mathscr{D}\left(x_m, x^*\right) + \mathscr{D}\left(\mathscr{G}x_m, x^*\right)\\ &\leq \frac{1}{2}\mathscr{D}\left(\mathscr{G}^2x_m, \mathscr{G}^3x_m\right) + \frac{1}{2}\mathscr{D}\left(\mathscr{G}^2x_m, \mathscr{G}^3x_m\right)\\ &= \mathscr{D}\left(\mathscr{G}^2x_m, \mathscr{G}^3x_m\right),\end{aligned}$$

which is again a contradiction. Therefore, relations (i), (ii), and (iii) hold.
Finally, using condition (B), relation (11.54), and inequality (11.60), we have

$$\upsilon + F_2\left(\mathscr{D}\left(\mathscr{G}x_m, \mathscr{G}^2x_m\right)\right) < F_2\left(\mathscr{D}\left(x_m, \mathscr{G}x_m\right)\right),$$

or

$$\begin{aligned}\mathscr{D}\left(\mathscr{G}x_m, \mathscr{G}^2x_m\right) &< \mathscr{D}\left(x_m, \mathscr{G}x_m\right)\\ &\leq \mathscr{D}\left(x_m, x^*\right) + \mathscr{D}\left(\mathscr{G}x_m, x^*\right)\\ &\leq \frac{1}{2}\left(\mathscr{D}\left(\mathscr{G}x_m, \mathscr{G}^2x_m\right)\right) + \left(\mathscr{D}\left(\mathscr{G}x_m, \mathscr{G}^2x_m\right)\right)\\ &= \mathscr{D}\left(\mathscr{G}x_m, \mathscr{G}^2x_m\right).\end{aligned}$$

That yields a contradiction as well.
Therefore, the contradictions of (11.57)–(11.60) prove that (11.49) holds.
Now, using relation (11.49), and the assumption of the theorem, we can write

$$\tau + \min\left\{F_2\left(\mathscr{D}\left(\mathscr{G}^2x_m, \mathscr{G}^2x^*\right)\right), F_1\left(\mathscr{D}\left(\mathscr{G}x_m, \mathscr{G}x^*\right)\right)\right\} \leq \alpha_2 F_2\left(\mathscr{D}\left(x_m, x^*\right)\right) + \alpha_1 F_1\left(\mathscr{D}\left(x_m, x^*\right)\right).$$

By (11.43), and the analysis of the previous theorem, we have
either $\lim_{m\to\infty} F_2\left(\mathscr{D}\left(\mathscr{G}^2x_m, \mathscr{G}^2x^*\right)\right) = -\infty$, or $\lim_{m\to\infty} F_1\left(\mathscr{D}\left(\mathscr{G}x_m, \mathscr{G}x^*\right)\right) = -\infty$.
The use of condition $(F2)$ yields
$\lim_{m\to\infty} \mathscr{D}\left(\mathscr{G}^2x_m, \mathscr{G}^2x^*\right) = 0$, or $\lim_{m\to\infty} \mathscr{D}\left(\mathscr{G}x_m, \mathscr{G}x^*\right) = 0$.
So, $\mathscr{D}\left(x^*, \mathscr{G}x^*\right) = \lim_{m\to\infty} \mathscr{D}\left(x_{m+1}, \mathscr{G}x^*\right) = \lim_{m\to\infty} \mathscr{D}\left(\mathscr{G}x_m, \mathscr{G}x^*\right) = 0$.
Therefore, $\mathscr{G}$ has a fixed point x^*. Now, for the uniqueness, let us suppose that $\mathscr{G}$ has more than one fixed point. That is, there exist two distinct $x, y \in X$ such that $\mathscr{G}x = x \neq y = \mathscr{G}y$.
Since $0 = \frac{1}{2}\mathscr{D}\left(x, \mathscr{G}x\right) < \mathscr{D}\left(x, y\right)$ and from the assumption of the theorem, we can write

$$\tau + \min\left\{F_2\left(\mathscr{D}\left(\mathscr{G}^2x, \mathscr{G}^2y\right)\right), F_1\left(\mathscr{D}\left(\mathscr{G}x, \mathscr{G}y\right)\right)\right\} \leq \alpha_2 F_2\left(\mathscr{D}\left(x, y\right)\right) + \alpha_1 F_1\left(\mathscr{D}\left(x, y\right)\right).$$

That will give rise to four possible cases

$$F_1\left(\mathscr{D}\left(x, y\right)\right) = F_1\left(\mathscr{D}\left(\mathscr{G}x, \mathscr{G}y\right)\right) < \tau + F_1\left(\mathscr{D}\left(\mathscr{G}x, \mathscr{G}y\right)\right) \leq F_1\left(\mathscr{D}\left(x, y\right)\right). \tag{11.61}$$

$$F_2\left(\mathscr{D}\left(x, y\right)\right) = F_2\left(\mathscr{D}\left(\mathscr{G}^2x, \mathscr{G}^2y\right)\right) < \tau + F_2\left(\mathscr{D}\left(\mathscr{G}^2x, \mathscr{G}^2y\right)\right) \leq F_2\left(\mathscr{D}\left(x, y\right)\right). \tag{11.62}$$

$$F_2\left(\mathscr{D}\left(x, y\right)\right) = F_2\left(\mathscr{D}\left(\mathscr{G}^2x, \mathscr{G}^2y\right)\right) < \tau + F_2\left(\mathscr{D}\left(\mathscr{G}^2x, \mathscr{G}^2y\right)\right) \leq F_1\left(\mathscr{D}\left(x, y\right)\right). \tag{11.63}$$

$$F_1\left(\mathscr{D}\left(x, y\right)\right) = F_1\left(\mathscr{D}\left(\mathscr{G}x, \mathscr{G}y\right)\right) < \tau + F_1\left(\mathscr{D}\left(\mathscr{G}x, \mathscr{G}y\right)\right) \leq F_2\left(\mathscr{D}\left(x, y\right)\right). \tag{11.64}$$

Both the relations (11.61) and (11.62) are contradictions.
Relation (11.63) along with condition (B) yields a contradiction as follows:

$$\tau + F_2\left(\mathscr{D}\left(\mathscr{G}^2 x, \mathscr{G}^2 y\right)\right) \leq F_1\left(\mathscr{D}(x, y)\right).$$

As $\tau > 2\upsilon$, we can write

$$2\upsilon + F_2\left(\mathscr{D}\left(\mathscr{G}^2 x, \mathscr{G}^2 y\right)\right) < F_1\left(\mathscr{D}(x, y)\right),$$

or

$$\upsilon + \left(\upsilon + F_2\left(\mathscr{D}\left(\mathscr{G}^2 x, \mathscr{G}^2 y\right)\right)\right) < F_1\left(\mathscr{D}(x, y)\right),$$

or

$$\upsilon + F_1\left(\mathscr{D}\left(\mathscr{G}^2 x, \mathscr{G}^2 y\right)\right) < F_1\left(\mathscr{D}(x, y)\right).$$

As $\mathscr{G}^2 x = x$ and $\mathscr{G}^2 y = y$, we have

$$\upsilon + F_1\left(\mathscr{D}(x, y)\right) < F_1\left(\mathscr{D}(x, y)\right),$$

a contradiction.
Similarly, inequality (11.64) with the use of condition (A) implies a contradiction as follows:

$$\tau + F_1\left(\mathscr{D}(\mathscr{G}x, \mathscr{G}y)\right) \leq F_2\left(\mathscr{D}(x, y)\right).$$

As $\tau > 2\upsilon$, we can write

$$2\upsilon + F_1\left(\mathscr{D}(\mathscr{G}x, \mathscr{G}y)\right) < F_2\left(\mathscr{D}(x, y)\right),$$

or

$$\upsilon + (\upsilon + F_1\left(\mathscr{D}(\mathscr{G}x, \mathscr{G}y)\right)\} < F_2\left(\mathscr{D}(x, y)\right),$$

or

$$\upsilon + F_2\left(\mathscr{D}(\mathscr{G}x, \mathscr{G}y)\right) < F_2\left(\mathscr{D}(x, y)\right).$$

As $\mathscr{G}x = x$, $\mathscr{G}y = y$, we have

$$\upsilon + F_2\left(\mathscr{D}(x, y)\right) < F_2\left(\mathscr{D}(x, y)\right),$$

a contradiction. This proves the existence of a unique fixed point. □

Theorem 11.4 *Let $(X, \mathscr{D})$ be a complete metric space. Suppose that a continuous mapping $\mathscr{G} : X \mapsto X$ is a triple F-Suzuki contraction and for all $t, t_1, t_2 \in \mathbb{R}_+$, there exist $\upsilon > 0$, $\tau > 2\upsilon$, such that*

$$F_2(t_2) < F_1(t_1) \leq F_2(t_2) + \upsilon. \tag{A}$$

$$F_1(t_1) < F_2(t_2) \leq F_1(t_1) + \upsilon. \tag{B}$$

$$F_i(t_1) < F(t_2) \leq F_i(t_1) + \upsilon, \quad i = 1, 2. \tag{C}$$

Then $\mathscr{G}$ has a unique fixed point in X and for every $x_0 \in X$, the sequence $\{\mathscr{G}^m x_0\}_{m=1}^{\infty}$ converges to x^*.

Proof Consider a sequence $\{x_1, x_2, \ldots .\}$ such that $x_{m+1} = \mathscr{G}x_m = \mathscr{G}^{m+1}x_0$ for any $x_0 \in X$ and for all $m \in \mathbb{N}_0$. If for some $m \in \mathbb{N}$, $\mathscr{D}(x_m, Gx_m) = 0$, G will admit a fixed point:
$\mathscr{D}(x_m, \mathscr{G}x_m) = \mathscr{D}(\mathscr{G}x_{m-1}, \mathscr{G}x_m) > 0, \quad$ for all $m \in \mathbb{N}$.
We will prove that $\lim_{m\to\infty} \mathscr{D}(x_m, \mathscr{G}x_m) = 0$.
For any $m \in \mathbb{N}$, we have $\frac{1}{2}\mathscr{D}(x_m, \mathscr{G}x_m) < (x_m, \mathscr{G}x_m)$, and

$$\tau + \min\left\{F_2\left(\mathscr{D}\left(\mathscr{G}^2 x_{m-1}, \mathscr{G}^2 x_m\right)\right), F_1\left(\mathscr{D}\left(\mathscr{G}x_{m-1}, \mathscr{G}x_m\right)\right)\right\} \leq F\left(\mathscr{D}\left(x_{m-1}, x_m\right)\right).$$

So,

$$\min\left\{F_2\left(\mathscr{D}\left(\mathscr{G}^2 x_{m-1}, \mathscr{G}^2 x_m\right)\right), F_1\left(\mathscr{D}\left(\mathscr{G}x_{m-1}, \mathscr{G}x_m\right)\right)\right\} \leq F\left(\mathscr{D}\left(x_{m-1}, x_m\right)\right) - \tau. \tag{11.65}$$

Let $F(d(x_{m-1}, x_m)) \leq F_1(d(x_{m-1}, x_m))$, we have

$$\min\left(F_2\left(d\left(\mathscr{G}^2 x_{m-1}, \mathscr{G}^2 x_m\right)\right), F_1\left(d\left(\mathscr{G}x_{m-1}, \mathscr{G}x_m\right)\right)\right) \leq F_1\left(d\left(x_{m-1}, x_m\right)\right) - \tau.$$

Let $F(d(x_{m-1}, x_m)) > F_1(d(x_{m-1}, x_m))$.

Using condition (C), we can write (11.65) as follows:

$$\min\{F_2(d(\mathscr{G}^2 x_{m-1}, \mathscr{G}^2 x_m)), F_1(d(\mathscr{G}x_{m-1}, \mathscr{G}x_m))\} \leq F_1(d(x_{m-1}, x_m)) + \upsilon - \tau. \tag{11.66}$$

Then, either

$$F_1\left(\mathscr{D}\left(x_{m-1}, x_m\right)\right) = \min\left\{F_2\left(\mathscr{D}\left(\mathscr{G}x_{m-1}, \mathscr{G}x_m\right)\right), F_1\left(\mathscr{D}\left(x_{m-1}, x_m\right)\right)\right\}, \tag{11.67}$$

or

$$F_2\left(\mathscr{D}\left(\mathscr{G}x_{m-1}, \mathscr{G}x_m\right)\right) = \min\left\{F_2\left(\mathscr{D}\left(\mathscr{G}x_{m-1}, \mathscr{G}x_m\right)\right), F_1\left(\mathscr{D}\left(x_{m-1}, x_m\right)\right)\right\}. \tag{11.68}$$

If (11.67) is true, the inequality (11.66) will take the form

$$\begin{aligned} &\min\left\{F_2\left(\mathscr{D}\left(\mathscr{G}^2 x_{m-1}, \mathscr{G}^2 x_m\right)\right), F_1\left(\mathscr{D}\left(\mathscr{G}x_{m-1}, \mathscr{G}x_m\right)\right)\right\} \\ &\leq \min\left\{F_2\left(\mathscr{D}\left(\mathscr{G}x_{m-1}, \mathscr{G}x_m\right)\right), F_1\left(\mathscr{D}\left(x_{m-1}, x_m\right)\right)\right\} + \upsilon - \tau. \end{aligned} \tag{11.69}$$

If (11.68) is true, we have $F_2(\mathscr{D}(\mathscr{G}x_{m-1}, \mathscr{G}x_m)) < F_1(\mathscr{D}(x_{m-1}, x_m))$.
Condition (A) allows us to write

$$F_2(\mathscr{D}(\mathscr{G}x_{m-1}, \mathscr{G}x_m)) < F_1(\mathscr{D}(x_{m-1}, x_m)) \leq F_2(\mathscr{D}(\mathscr{G}x_{m-1}, \mathscr{G}x_m)) + \upsilon. \tag{11.70}$$

Using (11.70) in (11.66), we have

$$\min\left\{F_2\left(\mathscr{D}\left(\mathscr{G}^2x_{m-1}, \mathscr{G}^2x_m\right)\right), F_1(\mathscr{D}(\mathscr{G}x_{m-1}, \mathscr{G}x_m))\right\} \leq F_2(\mathscr{D}(\mathscr{G}x_{m-1}, \mathscr{G}x_m)) + 2\upsilon - \tau.$$

Moreover, condition (11.68) will change the above inequality in the following form:

$$\begin{aligned} &\min\left\{F_2\left(\mathscr{D}\left(\mathscr{G}^2x_{m-1}, \mathscr{G}^2x_m\right)\right), F_1(\mathscr{D}(\mathscr{G}x_{m-1}, \mathscr{G}x_m))\right\} \\ &\leq \min\{F_2(\mathscr{D}(\mathscr{G}x_{m-1}, \mathscr{G}x_m)), F_1(\mathscr{D}(x_{m-1}, x_m))\} + 2\upsilon - \tau. \end{aligned} \tag{11.71}$$

Combining both possible cases (11.69) and (11.71), we have

$$\begin{aligned} &\min\left\{F_2\left(\mathscr{D}\left(\mathscr{G}^2x_{m-1}, \mathscr{G}^2x_m\right)\right), F_1(\mathscr{D}(\mathscr{G}x_{m-1}, \mathscr{G}x_m))\right\} \\ &\leq \min\{F_2(\mathscr{D}(\mathscr{G}x_{m-1}, \mathscr{G}x_m)), F_1(\mathscr{D}(x_{m-1}, x_m))\} + \delta_m\upsilon - \tau, \end{aligned} \tag{11.72}$$

where $\delta_m = \begin{cases} 1 & if,\ F_2(t_2) \geq F_1(t_1) \\ 2 & if,\ F_2(t_2) \leq F_1(t_1) \end{cases}$.
Relation (11.72) can be written as

$$\begin{aligned} &\min\left\{F_2\left(\mathscr{D}\left(\mathscr{G}^2x_{m-1}, \mathscr{G}^2x_m\right)\right), F_1(\mathscr{D}(\mathscr{G}x_{m-1}, \mathscr{G}x_m))\right\} \\ &\leq \min\left\{F_2\left(\mathscr{D}\left(\mathscr{G}^2x_{m-2}, \mathscr{G}^2x_{m-1}\right)\right), F_1(\mathscr{D}(\mathscr{G}x_{m-2}, \mathscr{G}x_{m-1}))\right\} + \delta_m\upsilon - \tau. \end{aligned}$$

Repeating this process, we have

$$\begin{aligned} &\min\left\{F_2\left(\mathscr{D}\left(\mathscr{G}^2x_{m-1}, \mathscr{G}^2x_m\right)\right), F_1(\mathscr{D}(\mathscr{G}x_{m-1}, \mathscr{G}x_m))\right\} \\ &\leq \min\left\{F_2\left(\mathscr{D}\left(\mathscr{G}^2x_{m-3}, \mathscr{G}^2x_{m-2}\right)\right), F_1(\mathscr{D}(\mathscr{G}x_{m-3}, \mathscr{G}x_{m-2}))\right\} + \delta_{m-1}\upsilon + \delta_m\upsilon - 2\tau \\ &\vdots \\ &\min\left\{F_2\left(\mathscr{D}\left(\mathscr{G}^2x_{m-1}, \mathscr{G}^2x_m\right)\right), F_1(\mathscr{D}(\mathscr{G}x_{m-1}, \mathscr{G}x_m))\right\} \\ &\leq \min\left\{F_2\left(\mathscr{D}\left(\mathscr{G}^2x_1, \mathscr{G}^2x_0\right)\right), F_1(\mathscr{D}(\mathscr{G}x_1, \mathscr{G}x_0))\right\} + \upsilon\sum_{j=1}^{m}\delta_j - m\tau, \end{aligned}$$

or

$$\begin{aligned} &\min\left\{F_2\left(\mathscr{D}\left(\mathscr{G}^2x_{m-1}, \mathscr{G}^2x_m\right)\right), F_1(\mathscr{D}(\mathscr{G}x_{m-1}, \mathscr{G}x_m))\right\} \\ &\leq F(\mathscr{D}(x_1, x_0)) + \upsilon\sum_{j=1}^{m}\delta_j - (m+1)\tau. \end{aligned}$$

Since $\tau > 2\upsilon$ and $\sum_{j=1}^{m} \delta_j < m+1$, we have

$$\upsilon \lim_{m\to\infty} \sum_{j=1}^{m} \delta_j - (m+1)\tau = -\infty.$$

So,

$$\lim_{m\to\infty} \min\left\{F_2\left(\mathscr{D}\left(\mathscr{G}^2 x_{m-1}, \mathscr{G}^2 x_m\right)\right), F_1\left(\mathscr{D}\left(\mathscr{G} x_{m-1}, \mathscr{G} x_m\right)\right)\right\} = -\infty. \tag{11.73}$$

Now, we have two possible cases

$$\lim_{m\to\infty} F_2\left(\mathscr{D}\left(\mathscr{G}^2 x_{m-1}, \mathscr{G}^2 x_m\right)\right) = -\infty, \tag{E}$$

or

$$\lim_{m\to\infty} F_1\left(\mathscr{D}\left(\mathscr{G} x_{m-1}, \mathscr{G} x_m\right)\right) = -\infty. \tag{F}$$

Condition (F2) among (E) yields $\lim_{m\to\infty} \mathscr{D}\left(\mathscr{G}^2 x_{m-1}, \mathscr{G}^2 x_m\right) = 0$,
or equivalently
$\lim_{m\to\infty} \mathscr{D}\left(\mathscr{G}^2 x_{m-1}, \mathscr{G}^2 x_m\right) = \lim_{m\to\infty} \mathscr{D}\left(x_{m+1}, \mathscr{G} x_{m+1}\right) = \lim_{m+1\to\infty} \mathscr{D}\left(x_m, \mathscr{G} x_m\right) = 0.$
Condition (F2) among (F) yields $\lim_{m\to\infty} \mathscr{D}\left(\mathscr{G} x_{m-1}, \mathscr{G} x_m\right) = 0$,
or equivalently $\lim_{m\to\infty} \mathscr{D}\left(x_m, \mathscr{G} x_m\right) = 0$.
Therefore, from (11.73), we get

$$\lim_{m\to\infty} \mathscr{D}\left(x_m, \mathscr{G} x_m\right) = 0. \tag{11.74}$$

Next in order to prove the sequence $\{x_m\}_{n=1}^{\infty}$ is Cauchy, we suppose on contrary that there exists $\varepsilon > 0$ and sequences $\{\mathfrak{g}(m)\}_{n=1}^{\infty}$ and $\{\mathfrak{h}(m)\}_{n=1}^{\infty}$ of natural numbers such that

$$\mathfrak{g}(m) > \mathfrak{h}(m) > m, \quad \mathscr{D}\left(x_{\mathfrak{g}(m)}, x_{\mathfrak{h}(m)}\right) \geq \varepsilon, \quad \mathscr{D}\left(x_{\mathfrak{g}(m)-1}, x_{\mathfrak{h}(m)}\right) < \varepsilon, \quad \text{for all } m \in N.$$

Now, an analysis similar to the previous theorem yields

$$\tau + \min\left\{F_2\left(\mathscr{D}\left(\mathscr{G}^2 x_{\mathfrak{g}(m)}, \mathscr{G}^2 x_{\mathfrak{h}(m)}\right)\right), F_1\left(\mathscr{D}\left(\mathscr{G} x_{\mathfrak{g}(m)}, \mathscr{G} x_{\mathfrak{h}(m)}\right)\right)\right\} \leq F\left(\mathscr{D}\left(x_{\mathfrak{g}(m)}, x_{\mathfrak{h}(m)}\right)\right). \tag{11.75}$$

So,

$$\tau + \min\left\{F_2(\varepsilon), F_1(\varepsilon)\right) \leq F(\varepsilon).$$

That yields either $\tau + F_2(\varepsilon) \leq F(\varepsilon)$ or $\tau + F_1(\varepsilon) \leq F(\varepsilon)$.
If $\tau + F_2(\varepsilon) \leq F(\varepsilon)$, the condition (C) allows us to write $\tau + F_2(\varepsilon) \leq F_2(\varepsilon) + \upsilon$.
That yields a contradiction as $\tau > 2\upsilon$. Likewise, $\tau + F_1(\varepsilon) \leq F(\varepsilon)$ gives a contradiction $\tau + F_2(\varepsilon) \leq F_2(\varepsilon) + \upsilon$.
This contradiction shows that $\{x_m\}_{m=1}^{\infty}$ is a Cauchy sequence. The completeness of $(X, \mathscr{D})$ proves that $\{x_m\}_{m=1}^{\infty}$ converges to some point x^* in X.
Next, we claim that

$$\frac{1}{2}\mathscr{D}(x_m, \mathscr{G}x_m) < \mathscr{D}\left(x_m, x^*\right), \qquad \frac{1}{2}\mathscr{D}\left(\mathscr{G}x_m, \mathscr{G}^2x_m\right) < \mathscr{D}\left(\mathscr{G}x_m, x^*\right),$$
$$\frac{1}{2}\mathscr{D}\left(\mathscr{G}^2x_m, \mathscr{G}^3x_m\right) < \mathscr{D}\left(\mathscr{G}^2x_m, x^*\right). \tag{11.76}$$

Now, following the procedure of the previous theorem from inequality (11.49) to (11.57), we can write

$$\tau + \min\left\{F_2\left(\mathscr{D}\left(\mathscr{G}^2x_m, \mathscr{G}^3x_m\right)\right), F_1\left(\mathscr{D}\left(\mathscr{G}x_m, \mathscr{G}^2x_m\right)\right)\right\} \leq F(\mathscr{D}(x_m, \mathscr{G}x_m)).$$

If

$$\min\left\{F_2\left(\mathscr{D}\left(\mathscr{G}^2x_m, \mathscr{G}^3x_m\right)\right), F_1\left(\mathscr{D}\left(\mathscr{G}x, \mathscr{G}^2x_m\right)\right)\right\} = F_1\left(\mathscr{D}\left(\mathscr{G}x, \mathscr{G}^2x_m\right)\right),$$

we have

$$\tau + F_1\left(\mathscr{D}\left(\mathscr{G}x_m, \mathscr{G}^2x_m\right)\right) \leq F(\mathscr{D}(x_m, \mathscr{G}x_m)).$$

As $\tau > 2\upsilon$, we can write

$$2\upsilon + F_1\left(\mathscr{D}\left(\mathscr{G}x_m, \mathscr{G}^2x_m\right)\right) < F(\mathscr{D}(x_m, \mathscr{G}x_m)),$$

so,

$$\upsilon + F\left(\mathscr{D}\left(\mathscr{G}x_m, \mathscr{G}^2x_m\right)\right) < F(\mathscr{D}(x_m, \mathscr{G}x_m)).$$

Further, using the condition $(F1)$, we can write

$$\begin{aligned}
\mathscr{D}\left(\mathscr{G}x_m, \mathscr{G}^2x_m\right) &< \mathscr{D}(x_m, \mathscr{G}x_m) \\
&\leq \mathscr{D}\left(x_m, x^*\right) + \mathscr{D}\left(\mathscr{G}x_m, x^*\right) \\
&\leq \frac{1}{2}\mathscr{D}\left(\mathscr{G}x_m, \mathscr{G}^2x_m\right) + \frac{1}{2}\mathscr{D}\left(\mathscr{G}x_m, \mathscr{G}^2x_m\right) \quad \text{Using (11.54)}, \\
&= \mathscr{D}\left(\mathscr{G}x_m, \mathscr{G}^2x_m\right).
\end{aligned}$$

That yields a contradiction.
Similarly, if

$$\min\left\{F_2\left(\mathscr{D}\left(\mathscr{G}^2x_m, \mathscr{G}^3x_m\right)\right), F_1\left(\mathscr{D}\left(Tx, \mathscr{G}^2x_m\right)\right)\right\} = F\left(\mathscr{D}\left(\mathscr{G}^2x, \mathscr{G}^3x_m\right)\right),$$

we have

$$\tau + F_2\left(\mathscr{D}\left(\mathscr{G}^2 x, \mathscr{G}^3 x_m\right)\right) \leq F\left(\mathscr{D}\left(x_m, \mathscr{G} x_m\right)\right).$$

As $\tau > 2\upsilon$, we can write

$$2\upsilon + F_2\left(\mathscr{D}\left(\mathscr{G}^2 x_m, \mathscr{G}^3 x_m\right)\right) < F\left(\mathscr{D}\left(x_m, \mathscr{G} x_m\right)\right),$$

or

$$\upsilon + F\left(\mathscr{D}\left(\mathscr{G}^2 x_m, \mathscr{G}^3 x_m\right)\right) < F\left(\mathscr{D}\left(x_m, \mathscr{G} x_m\right)\right),$$

or

$$F\left(\mathscr{D}\left(\mathscr{G}^2 x, \mathscr{G}^3 x_m\right)\right) < F\left(\mathscr{D}\left(x_m, \mathscr{G} x_m\right)\right).$$

Using the condition $(F1)$, we can write

$$\begin{aligned}
\mathscr{D}\left(\mathscr{G}^2 x, \mathscr{G}^3 x_m\right) &< \mathscr{D}\left(x_m, \mathscr{G} x_m\right) \\
&\leq \mathscr{D}\left(x_m, x*\right) + \mathscr{D}\left(\mathscr{G} x_m, x*\right) \\
&\leq \frac{1}{2}\mathscr{D}\left(\mathscr{G}^2 x_m, \mathscr{G}^3 x_m\right) + \frac{1}{2}\mathscr{D}\left(\mathscr{G}^2 x_m, \mathscr{G}^3 x_m\right) \qquad \text{Using (11.53), (11.56),} \\
&= \mathscr{D}\left(\mathscr{G}^2 x_m, \mathscr{G}^3 x_m\right),
\end{aligned}$$

again a contradiction. Therefore, (11.76) holds.
Now, using the relations (11.76), we can write

$$\tau + \min\left\{F_2\left(\mathscr{D}\left(\mathscr{G}^2 x_m, \mathscr{G}^2 x^*\right)\right), F_1\left(\mathscr{D}\left(\mathscr{G} x_m, \mathscr{G} x^*\right)\right)\right\} \leq F\left(\mathscr{D}\left(x_m, x^*\right)\right).$$

By (11.73) and condition $(F2)$, we have
either $\lim_{m\to\infty} F_2\left(\mathscr{D}\left(\mathscr{G}^2 x_m, \mathscr{G}^2 x^*\right)\right) = -\infty$ or $\lim_{m\to\infty} F_1\left(\mathscr{D}\left(\mathscr{G} x_m, \mathscr{G} x^*\right)\right) = -\infty$.
So,
$\lim_{m\to\infty} \mathscr{D}\left(\mathscr{G}^2 x_m, \mathscr{G}^2 x^*\right) = 0$ or $\lim_{m\to\infty} \mathscr{D}\left(\mathscr{G} x_m, \mathscr{G} x^*\right) = 0$.
or $\mathscr{D}\left(x^*, \mathscr{G} x^*\right) = \lim_{m\to\infty} \mathscr{D}\left(x_{m+1}, \mathscr{G} x^*\right) = \lim_{m\to\infty} \mathscr{D}\left(\mathscr{G} x_m, \mathscr{G} x^*\right) = 0$.
Therefore, $\mathscr{G}$ has a fixed point x^*. Now, for the uniqueness, let us suppose that $\mathscr{G}$ has more than one fixed point. That is, there exist two distinct $x, y \in X$ such that $\mathscr{G} x = x \neq y = \mathscr{G} y$.
Since $0 = \frac{1}{2}\mathscr{D}(x, \mathscr{G} x) < \mathscr{D}(x, y)$ and from the assumption of the theorem, we can write

$$\tau + \min\left\{F_2\left(\mathscr{D}\left(\mathscr{G}^2 x, \mathscr{G}^2 y\right)\right), F_1\left(\mathscr{D}\left(\mathscr{G} x, \mathscr{G} y\right)\right)\right\} \leq F\left(\mathscr{D}(x, y)\right),$$

so, we have either

$$\tau + F_1\left(\mathscr{D}\left(\mathscr{G}x, \mathscr{G}y\right)\right) \leq F\left(\mathscr{D}\left(x, y\right)\right) \tag{11.77}$$

or

$$\tau + F_2\left(\mathscr{D}\left(\mathscr{G}^2x, \mathscr{G}^2y\right)\right) \leq F\left(\mathscr{D}\left(x, y\right)\right). \tag{11.78}$$

The inequality (11.77) yields a contradiction as follows:

$$2\upsilon + F_1\left(\mathscr{D}\left(x, y\right)\right) < F\left(\mathscr{D}\left(x, y\right)\right)$$

or

$$\upsilon + F\left(\mathscr{D}\left(x, y\right)\right) < F\left(\mathscr{D}\left(x, y\right)\right).$$

The inequality (11.78) yields a contradiction as follows:

$$2\upsilon + F_2\left(\mathscr{D}\left(x, y\right)\right) < F\left(\mathscr{D}\left(x, y\right)\right)$$

or

$$\upsilon + F\left(\mathscr{D}\left(x, y\right)\right) < F\left(\mathscr{D}\left(x, y\right)\right).$$

That proves the existence of a unique fixed point. □

Example 11.5 Consider a closed unit ball $\mathscr{B}$ in the ℓ_1 space of all absolutely summable sequences $u = (u_1, u_2, \ldots)$, with a metric inherited from the standard norm $\|u\| = \sum_{i=1}^{\infty} |u_i|$. Let $\mathfrak{h} : [-1, 1] \mapsto [-1, 1]$ be the function given by

$$\mathfrak{h}\left(w\right) = \begin{cases} 1 + 2w, & if \quad -1 \leq w \leq -1/2, \\ 0, & if \quad -1/2 \leq w \leq 1/2, \\ -1 + 2w, & if \quad 1/2 \leq w \leq 1. \end{cases}$$

Observe that, for all $s, w \in [-1, 1]$, we have $|\mathfrak{h}(s) - \mathfrak{h}(w)| \leq 2|s - w|$, and $|\mathfrak{h}(w)| \leq |w|$. Define a continuous mapping $\mathscr{G} : \mathscr{B} \to \mathscr{B}$ as

$$\mathscr{G}u = \mathscr{G}(u_1, u_2, \ldots) = e^{-\kappa}\left(\mathfrak{h}(u_2), \frac{2}{3}u_3, u_4, u_5, \ldots\right),$$

where $\kappa > 0$ is a real number. So,

$$\mathscr{G}^2u = e^{-\kappa}\left(\mathfrak{h}\left(\frac{2}{3}u_3\right), \frac{2}{3}u_4, u_5, u_6, \ldots\right).$$

For each $u = (u_1, u_2, \ldots)$ and $v = (v_1, v_2, \ldots)$ in $\mathscr{B}$, we have

$$
\begin{aligned}
e^{\kappa}\left\|\mathscr{G}u-\mathscr{G}v\right\| &= |\mathfrak{h}(u_2)-\mathfrak{h}(v_2)|+\frac{2}{3}|u_3-v_3|+\sum_{k=4}^{\infty}|u_k-v_k| \\
&\leq 2|u_2-v_2|+\frac{2}{3}|u_3-v_3|+\sum_{k=4}^{\infty}|u_k-v_k| \\
&\leq 2\left\|u-v\right\|.
\end{aligned}
$$

That is

$$\frac{e^{\kappa}}{2}\left\|\mathscr{G}u-\mathscr{G}v\right\| \leq \left\|u-v\right\|.$$

The above inequality can be written as

$$\kappa+\ln\left(\frac{1}{2}\left\|\mathscr{G}u-\mathscr{G}v\right\|\right) \leq \ln\left\|u-v\right\|. \tag{I}$$

Likewise,

$$
\begin{aligned}
e^{\kappa}\left\|\mathscr{G}^2u-\mathscr{G}^2v\right\| &= \left|\mathfrak{h}\left(\frac{2}{3}u_3\right)-\mathfrak{h}\left(\frac{2}{3}v_3\right)\right|+\frac{2}{3}|u_3-v_3| \\
&\quad+\sum_{k=5}^{\infty}|u_k-v_k| \\
&\leq \frac{4}{3}|u_3-v_3|+\frac{2}{3}|u_3-v_3|+\sum_{k=5}^{\infty}|u_k-v_k| \\
&\leq \frac{4}{3}\left\|u-v\right\|.
\end{aligned}
$$

That is

$$\frac{3e^{\kappa}}{4}\left\|\mathscr{G}^2u-\mathscr{G}^2v\right\| \leq \left\|u-v\right\|.$$

The above inequality can be written as

$$\kappa+\ln\left(\frac{3}{4}\left\|\mathscr{G}^2u-\mathscr{G}^2v\right\|\right) \leq \ln\left\|u-v\right\|. \tag{J}$$

Now, we define $F_1(t)=\ln\left(\frac{t}{2}\right)$, $F_2(t)=\ln\left(\frac{3t}{4}\right)$, and $F(t)=\ln(t)$. As $\frac{e^{\kappa}}{2}\left\|\mathscr{G}u-\mathscr{G}v\right\| \leq \left\|u-v\right\|$ implies $\frac{1}{2}\left\|\mathscr{G}u-\mathscr{G}^2u\right\| \leq \left\|u-\mathscr{G}u\right\|$, therefore, by combining the inequalities (I) and (J), we can write the triple F-Suzuki contraction as follows:

$$\kappa+\min\left\{F_1\left(\left\|\mathscr{G}u-\mathscr{G}^2u\right\|\right), F_2\left(\left\|\mathscr{G}^2u-\mathscr{G}^3u\right\|\right)\right\} \leq \min\{F\left(\left\|u-\mathscr{G}u\right\|\right)\}.$$

Further, the assumptions, $F_1(t) = \ln\left(\frac{t}{2}\right)$, $F_2(t) = \ln\left(\frac{3t}{4}\right)$ and $F(t) = \ln(t)$, yield $F_1(t) < F_2(t)$, so that we have $ln(\frac{t}{2}) < \ln(\frac{3t}{4})$. Now, we can define $0 < \upsilon \leq \ln(\frac{3}{2})$, such that $\ln(\frac{t}{2}) < \ln(\frac{3t}{4}) \leq \ln(\frac{t}{2}) + \upsilon$. That is, $F_1(t) < F_2(t) \leq F_1(t) + \upsilon$. Therefore, we conclude that $\mathscr{G} : \mathscr{B} \mapsto \mathscr{B}$ represents a triple F-contraction mapping. Next, we will show that for every $x_0 \in \mathscr{B}$, the sequence $\{\mathscr{G}^m x_0\}_{m=1}^{\infty}$ converges to a unique fixed point $(0, 0, ...)$.
For some fixed $i \in \mathbb{N}$, consider the absolutely summable sequences $a = (a_1, a_2, ..., a_i, 0, 0, ...) \in \mathscr{B}$, with a metric inherited from the standard norm. As $\mathscr{B}$ is a closed unit ball, we have $\sum_{k=1}^{i} |a_k| \leq 1$.
Now,

$$\mathscr{G}a = e^{-\kappa}\left(\mathfrak{h}(a_2), \frac{2}{3}a_3, ...a_i, 0, 0, ...\right)$$

implies that

$$\mathscr{G}^2 a = e^{-\kappa}\left(\mathfrak{h}\left(\frac{2}{3}a_3\right), a_4..., a_i, 0, 0, ...\right).$$

Then, for all $m > i$, we have $\mathscr{G}^m a = (0, 0, ...)$. That $(0, 0, ...)$ is a unique fixed point.

11.4 Application

As an application of our work, we will consider an engineering problem in which the transformed mathematical model of a problem represents the following integral equation:

$$y(x) = \int_0^x H(u, y(u))\, G(x, u) du, \quad x \in [0, I], \tag{11.79}$$

where Green function $G(x, u)$ is defined by

$$G(x, u) = \begin{cases} (x-u)e^{\kappa(x-u)} & if\ 0 \leq u \leq x \leq I \\ 0 & if\ 0 \leq x \leq u \leq I' \end{cases} \tag{11.80}$$

with constant $\kappa > 0$.

Let X be the set of all continuous functions $f : [0, I] \mapsto \mathbb{R}^+$ equipped with the norm,

$$\|f\|_\kappa = \sup_{x\in[0,I]} \left\{|f(x)|\, e^{-2\kappa x}\right\}.$$

Define a distance $\mathscr{D} : X \times X \to [0, \infty)$ by

$$\mathscr{D}(f, g) = \max\left\{\|f\|_\kappa, \|g\|_\kappa\right\},$$

for all $f, g \in X$. Now, to find the existence of the solution of integral equation (11.79), we consider a function $\mathscr{G} : X \mapsto X$ defined by

$$\mathscr{G}(y)(x) = \int_0^x H(u, y(u))\, G(x, u) du,$$

for all $y \in X$ and $x \in [0, I]$. We will prove that there exists some $v \in X$ such that $\mathscr{G}(v)(x) = v(x)$. That is, the fixed point of triple F-contraction will represent the solution to the problem (11.79).

Nonlinear integral equation (11.65) has a solution if the following conditions hold:

(a) $H(u, y(u))$ is an increasing function.

(b) There , $k > 1$, $\kappa > 0$, such that $ke^{-\kappa} < 2$, and

$$|H(u, y)| \leq k\kappa^2 e^{-\kappa} y,$$

where $u \in [0, I]$ and $y \in \mathbb{R}_+$.

Proof For all $v, w \in X$, we have,

$$|\mathscr{G}(v)(u)| = \int_0^x |H(u, v(u))|\, G(x, u) du.$$

Now, using the conditions (a) and (b), we can write

$$\begin{aligned}
|\mathcal{G}(v)(u)| &\leq k \int_0^x \kappa^2 e^{-\kappa} |v(u)| G(x,u) du \\
&= k \int_0^x \kappa^2 e^{-\kappa} |v(u)| (x-u) e^{\kappa(x-u)} du \\
&= k \int_0^x \kappa^2 e^{-\kappa} e^{2\kappa u} e^{-2\kappa u} |v(u)| (x-u) e^{\kappa(x-u)} du \\
&\leq k \int_0^x \kappa^2 e^{-\kappa} e^{2\kappa u} \|v\|_\kappa (x-u) e^{\kappa(x-u)} du \\
&= k\kappa^2 e^{-\kappa} e^{x\kappa} \|v\|_\kappa \int_0^x e^{2\kappa u} (x-u) e^{-\kappa u} du \\
&= k\kappa^2 e^{x\kappa-\kappa} \|v\|_\kappa \int_0^x (x-u) e^{\kappa u} du
\end{aligned}$$

$$\begin{aligned}
&= k\kappa^2 e^{x\kappa-\kappa} \|v\|_\kappa \left(-\frac{x}{\kappa} + \frac{e^{\kappa x}}{\kappa^2} - \frac{1}{\kappa^2}\right) \\
&= k e^{-\kappa} \|v\|_\kappa e^{\kappa x} \left(-x\kappa + e^{\kappa x} - 1\right).
\end{aligned}$$

Therefore,

$$e^{-2\kappa x} |\mathcal{G}(v)(u)| = k e^{-\kappa} \|v\|_\kappa \left(-x\kappa e^{-\kappa x} + 1 - e^{-\kappa x}\right).$$

Since $-x\kappa e^{-\kappa x} + 1 - e^{-\kappa x} \leq 1$, we have

$$\|\mathcal{G}(v)(u)\|_\kappa \leq k e^{-\kappa} \|v\|_\kappa.$$

Likewise, we can find that

$$\|\mathcal{G}(w)(u)\|_\kappa \leq k e^{-\kappa} \|w\|_\kappa.$$

Since

$$\begin{aligned}
\max\{\|\mathcal{G}v\|_\kappa, \|\mathcal{G}w\|_\kappa\} &\leq \max\left\{k e^{-\kappa} \|v\|_\kappa, k e^{-\kappa} \|w\|_\kappa\right\}, \\
&= k e^{-\kappa} \max\{\|v\|_\kappa, \|w\|_\kappa\}.
\end{aligned}$$

Therefore, for all $v, w \in X$, we have,

$$\mathscr{D}(\mathscr{G}v, \mathscr{G}w) \leq ke^{-\kappa} \max\{\|v\|_\kappa, \|w\|_\kappa\}.$$

or

$$\mathscr{D}(\mathscr{G}v, \mathscr{G}w) \leq ke^{-\kappa} \mathscr{D}(v, w). \tag{11.81}$$

Therefore,

$$\mathscr{D}\left(\mathscr{G}^2 v, \mathscr{G}^2 w\right) \leq k^2 e^{-2\kappa} \mathscr{D}(v, w). \tag{11.82}$$

So, relation (11.81) yields

$$\frac{1}{k}\mathscr{D}(\mathscr{G}v, \mathscr{G}w) \leq e^{-\kappa} \mathscr{D}(v, w),$$

$$\ln\left(\frac{1}{k}\mathscr{D}(\mathscr{G}v, \mathscr{G}w)\right) \leq \ln\left(e^{-\kappa} \mathscr{D}(v, w)\right),$$

$$\kappa + \ln\left(\frac{1}{k}\mathscr{D}(\mathscr{G}v, \mathscr{G}w)\right) \leq \ln(\mathscr{D}(v, w)). \tag{11.83}$$

Similarly, relation (11.82) yields

$$\frac{1}{k^2}\mathscr{D}\left(\mathscr{G}^2 v, \mathscr{G}^2 w\right) \leq e^{-2\kappa} \mathscr{D}(v, w),$$

$$\ln\left(\frac{1}{k^2}\mathscr{D}\left(\mathscr{G}^2 v, \mathscr{G}^2 w\right)\right) \leq \ln\left(e^{-2\kappa} \mathscr{D}(v, w)\right),$$

$$2\kappa + \ln\left(\frac{1}{k^2}\mathscr{D}\left(\mathscr{G}^2 v, \mathscr{G}^2 w\right)\right) \leq \ln(\mathscr{D}(v, w)). \tag{11.84}$$

Define $F_1(x) = \ln \frac{x}{k}$, $F_2(x) = \ln \frac{x}{k^2}$ and $F(x) = \ln x$. So, the inequalities (11.83) and (11.84) take the form,

$$\kappa + F_1(\mathscr{D}(\mathscr{G}v, \mathscr{G}w)) \leq F(\mathscr{D}(v, w)) \text{ and } \kappa + \mathrm{F}_2\left(\mathscr{D}\left(\mathscr{G}^2 v, \mathscr{G}^2 w\right)\right) < F(\mathscr{D}(v, w)).$$

Both of the above inequalities can be written as

$$\kappa + \min\left\{F_1(\mathscr{D}(\mathscr{G}v, \mathscr{G}w)), \mathrm{F}_2\left(\mathscr{D}\left(\mathscr{G}^2 v, \mathscr{G}^2 w\right)\right)\right\} \leq F(\mathscr{D}(v, w)).$$

Now, $\mathscr{D}(\mathscr{G}v, w) \leq \max\left\{ke^{-\kappa}\|v\|_\kappa, \|w\|_\kappa\right\} < \max\{2\|v\|_\kappa, \|w\|_\kappa\}$, as $ke^{-\kappa} < 2$. The above inequality is equivalent to

$$\mathscr{D}(\mathscr{G}v, w) < \max\{2\|v\|_\kappa, 2\|w\|_\kappa\} = 2\max\{\|v\|_\kappa, \|w\|_\kappa\}.$$

That is, $\frac{1}{2}\mathscr{D}(\mathscr{G}v, w) < \mathscr{D}(v, w)$.
Therefore, $\mathscr{G}$ is a triple F-Suzuki contraction mapping and its solution will represent the solution of the problem (11.79). □

11.5 Conclusion

In this article, we have generalized the F-Suzuki contraction mapping by using multiple F functions and certain conditions on the mapping. The conditions allow us to deal with the class of mappings whose iterates expand in general but may contract in some cases. Moreover, with the use of multiple F functions we have presented an idea that allows the use of weaker conditions for several types of differential equations. The new generalizations of F-Suzuki contraction mapping and their corresponding results will open up new avenues for researchers, allowing them to discover the existence of solutions to a wide range of differential equations.

11.6 Open Questions

- We have proved the above results with $p = 2$, but remained unable to further generalize F-contraction, F-Suzuki, and F-expanding mappings using $p > 2$.
- One may find the results regarding fixed points of some new generalizations of F-contraction, F-Suzuki, and F-expanding mappings using the mean of the iterations.
- There may be the possibility of finding fixed points of these generalized mappings for a complete G-metric space.

11.7 Declarations

Competing Interests: The author declares that they have no competing interests.

Funding: The author(s) received no financial support for the research, authorship, and/or publication of this article.

Author's Contribution: Naeem Saleem and Shahid Bashir conceived the presented idea. Shahid Bashir developed the theory and performed the computations. Syed Muhammad Husnine and Vahid Parvaneh verified the analytical methods. Naeem Saleem supervised the findings of this work. All the authors discussed the results and contributed to the final manuscript.

References

1. Alsulami, H.H., Karapınar, E., Piri, H.: Fixed points of generalized $F-$suzuki type contraction in complete $b-$metric spaces. Discrete Dyn. Nat. Soc. (2015), Article ID 969726. https://doi.org/10.1155/2015/969726
2. Altun, I., Minak, G., Dag, H.: Multivalued F-contractions on complete metric spaces. J. Nonlinear Convex Anal. **16**(4), 659–666 (2015)
3. Aydi, H., Mlaiki, N., Souayah, N., Abdeljawad, T.: Controlled metric type spaces and the related contraction principle. Mathematics **6**, 194 (2018)

4. Bashir, S., Saleem, N., Aydi, H., Husnine, S.M., Rwaily, A.A.: Developments of some new results that weaken certain conditions of fractional type differential equations. Adv. Differ. Equ. **2021**, 359 (2021). https://doi.org/10.1186/s13662-021-03519-1
5. Dung, N.V., Hang, V.L.: A fixed point theorem for generalized F-contractions on complete metric spaces. Vietnam J. Math. **43**, 743–753 (2015)
6. Gilić, E., Dolićanin-Dekić, D., Mitrović, Z., Pučić, D., Aydi, H.: On some recent results concerning $F-$Suzuki-contractions in $b-$metric spaces. Mathematics **8**, 940 (2020). https://doi.org/10.3390/math8060940
7. Gholidahaneh, A., Sedghi, S., Parvaneh, V.: Some fixed point results for Perov-Ćirić-Prešić type F-contractions with application. J. Funct. Spaces **2020**, Article ID 1464125, 9 pages
8. Goebel, K., Sims, B.: Mean Lipschitzian mappings. Contemp. Math. **513**, 157–167 (2010)
9. Gornicki, J.: Fixed point theorems for F-expanding mappings. Fixed Point Theory Appl. **9**, (2017). https://doi.org/10.1186/s13663-017-0602-3
10. Jain, R., Nashine, H.K., DelaSen, M.: An implicit relation, relational theoretic approach under w-distance and application to nonlinear matrix equations. J. Inequal. Appl. **169**(2021). https://doi.org/10.1186/s13660-021-02704-w
11. Jain, R., Nashine, H.K., Kumar, S.: An implicit relation approach in metric spaces underdistance and application to fractional differential equation. J. Funct. Spaces. **2021**, Article ID 9928881, 12 pages
12. Jain, R., Nashine, H.K., Parvaneh, V.: Extended Branciari quasi-b-distance spaces, implicit relations and application to nonlinear matrix equations. J. Inequal. Appl. **200**(2021). https://doi.org/10.1186/s13660-021-02736-2
13. Karapınar, E., Fulga, A., Agarwal, R.P.: A survey: $F-$contractions with related fixed point results. J. Fixed Point Theory Appl. **22**, 69 (2020). https://doi.org/10.1007/s11784-020-00803-7
14. Merryfield, J., Stein, Jr., J.D.: A generalization of the Banach contraction principle. J. Math. Anal. Appl. 112–120 (2002)
15. Mohammadi, B., Parvaneh, V., Aydi, H.: On extended interpolative Ciric-Reich-Rus type F-contractions and an application. J. Inequal. Appl. **2019** (2019)
16. Parvaneh, V., Hussain, N., Khorshidi, M., Mlaiki, N., Aydi, H.: Fixed point results for generalized F-contractions in modular b-metric spaces with applications. Mathematics **7**(10), 887 (2019)
17. Piri, H., Kumam, P.: Some fixed point theorems concerning F-contraction in complete metric spaces. Fixed Point Theory Appl. **210**, (2014). https://doi.org/10.1186/1687-1812-2014-210
18. Saleem, N., Iqbal, I., Iqbal, B., Radenović, S.: Coincidence and fixed points of multivalued F-contractions in generalized metric space with application. J. Fixed Point Theory Appl. **22**, 1–24 (2020)
19. Saleem, N., Vujaković, J., Baloch, W.U., Radenović, S.: Coincidence point results for multivalued Suzuki type mappings using θ-contraction in b-metric spaces. Mathematics **7**(11), 1017 (2019)
20. Saleem, N., Zhou, M., Bashir, S., Husnine, S.M.: Some new generalizations of F-contraction type mappings that weaken certain conditions on Caputo fractional type differential equations. Aims Math. **6**(11), 1–25 (2021)
21. Wardowski, D.: Fixed point theory of a new type of contractive mappings in complete metric spaces. Fixed Point Theory Appl. **94**, (2012). https://doi.org/10.1186/1687-1812-2013-277
22. Wardowski, D., Dung, N.V.: Fixed points of f-weak contractions on complete metric spaces. Demonstratio Mathematica **1**, 146–155 (2014)

Chapter 12
Fractional Elzaki Transformation with its Applications to Fractional Differential Equations and Special Functions

Shrinath D. Manjarekar

Abstract The aim of this chapter is to introduce some new properties and applications of the newly defined fractional Functional Operator called fractional Elzaki transformation by developing its relationship with fractional Laplace operator. We also apply it to find out the fractional functional transformation of some special functions under some conditions. As an application of fractional functional operator towards conformable fractional differential equations; we obtain the analytic solution of it and show the convergence of the solution graphically.

Keywords Fractional Elzaki transformation · Fractional differential equations · Hypergeometric functions

12.1 Introduction

The word "transform" [24] itself tells us the change of one form into another form. The majority of mathematical concepts relevant to the research of integral transformation originated from the work of A. Caley [1821–1895] in his book, "**Theory of Linear Transformations**", where he examines the composition of linear transformations.

The important and useful applications of most of the functional operators [12] were to solve the differential as well as partial differential equations with integer and fractional order also of Local order. It is the basic toolkit in the modern (1930's) variational theory of PDE.

Though the origin of fractional calculus were during 1695 [13] when L' Hospital asked the question of finding $\frac{d^n y}{dx^n}$ for $n = 0.5$ and then Leibnitz replied $d^{\frac{1}{2}}x = x\sqrt{dx : x}$. But the process of thinking for solution of fractional order starts from ancient Greek's.

Greek's have tried to solve the problem of solving Pythagorean triplets (x, y, z) satisfying $x^2 + y^2 = z^2$ and converting them it into the form $(\frac{x}{z})^2 + (\frac{y}{z})^2 = 1$ for

S. D. Manjarekar (✉)
Department of Mathematics, L.V.H. ASC College, Panchavati, Nashik 422003, Maharashtra, India
e-mail: shrimathematics@gmail.com

B. Hazarika et al. (eds.), *Advances in Functional Analysis and Fixed-Point Theory*, Industrial and Applied Mathematics,
https://doi.org/10.1007/978-981-99-9207-2_12

finding traiangles [23] with fractional/rational numbers which are fitted inside the unit radius circle.

Most of the applications [24] of functional analytic methods and functional analysis are towards the theory of ordinary differential equations and partial differential equations. The main important theorem in functional analytic techniques which plays an important role in differential equations was Picard–Lindelof Theorem in the existence and uniqueness towards the solutions of a large and important class of ordinary differential equations.

The aim of our chapter is to introduce the relationship between fractional operators and finding out there properties, fractional Elzaki transform of some special functions and gives applications towards the conformable fractional differential equation. So that, one might insight onto applications of fractional Elzaki transform fractional operator theory on Banach spaces, if possible.

The chapter has been organized as follows. In Sect. 12.2, Some preliminary definitions were given and proved some properties of fractional Elzaki operator. Section 12.3, consists of the relationship between fractional Finite Elzaki transform and Finite Laplace transform. In Sect. 12.4, we have found fractional Elzaki transform of some special functions. In Sect. 12.5, as an application, we have solved the conformable fractional differential equations by using fractional Elzaki transform along with the concluding remarks.

12.2 Definition and Properties

(1) **Fractional Laplace type Integral Transform**

Given a function f(t) which is piece wise continuous and of exponential order, then its fractional Laplace—type integral [21] transform is defined as follows:

$$\mathscr{L}_\alpha \{f(t); p\} = \int_0^\infty E_\alpha(-p, t) f(t) d_\alpha t \tag{12.1}$$

where $Re.(p) > 0$, $E_\alpha(c, t) = e^{c\frac{t^\alpha}{\alpha}}$, $c \in \Re$ and $d_\alpha t = t^{\alpha-1}dt$.

(2) **Elzaki–Tarig Transform**

If S $= \{f(t) : \exists k_1, k_2 > 0, |f(t)| < Me^{\frac{|t|}{k_j}}, t \in (-1)^j \times [0, \infty)\}$ where $M > 0$ then for the given function which satisfies the condition of the set S the Elzaki–Tarig transformation [9] of $f(t)$ is defined as,

$$T(f(t); p) = p\int_0^\infty f(t)e^{\frac{-t}{p}}dt, \quad p \neq 0. \tag{12.2}$$

(3) **Derivative of a Convolution**

Given any real-valued integrable function f and g, then convolution of f with g is denoted by $f \star g$ and is given by $(f \star g)(t) = \int\limits_0^\infty f(x-t)g(x)dx$ which have the property related to the derivative given by [25]

$$\frac{d}{dt}[f * g] = \frac{df}{dt} * g = f * \frac{dg}{dt}. \tag{12.3}$$

(4) **Fractional Elzaki Transform**

The fractional Elzaki transform of a function $f(t)$ of order α satisfying the conditions of the set S is given by [16]

$$E_\alpha^C[f(t)](p) = p \int_0^\infty K_\alpha(-p,t) f(t) d_\alpha t \tag{12.4}$$

where $K_\alpha(-p,t) = E_\alpha(-\frac{1}{p}, t),\ p > 0 \Rightarrow K_\alpha(-p,t) = e^{-\frac{t^\alpha}{p\alpha}}$.

(5) **Conformable fractional derivative**

If $f : [0, \infty) \to \Re$ is a real-valued function, then its conformable fractional derivative [21] of order $\alpha \in (0, 1]$ and $\forall t > 0$ is defined as

$$D^\alpha f(t) = T_\alpha f(t) = \lim_{\epsilon \to 0} \frac{f(t + \epsilon t^{1-\alpha}) - f(t)}{\epsilon} \tag{12.5}$$

where at $t = 0$ it is $f^\alpha(0) = \lim_{t \to 0^+} f^\alpha(t)$ and $f \in C^1$.

Before proving the properties and applications of fractional functional operator, one should talk about the meaning of solutions which were defined by the use of functional operator in terms of the problems where we stated some definition which comes after the time when modern theory of partial differential equations was developed later it extends to fractional differential equations and recently to conformable fractional differential equations [6, 16, 22].

12.2.1 Existence of Fractional Elzaki Transform

Using the definition (12.4) and the conditions of (12.2), the Fractional Elzaki transform $E_\alpha^C[f(t)](p)$ exists and finite provided the following conditions are satisfied:

[1] $\lim_{t \to \infty} \mid f(t) \mid$ exists and finite

[2] $Re.(p) > 0$ by using the relation between Laplace and Elzaki Transformation [20] while the inverse existence was by using [16].

From the above two statements, one can say that if $f(t)$ is a function which moves much faster towards ∞ than the kernel of the fractional Elzaki transform tends to 0. Then the fractional Elzaki transform of such function does not exist.

12.2.2 Existence of Inverse for Fractional Elzaki Transform

Given the function $f(t)$ which satisfies the conditions in (12.2), then its fractional Elzaki transform at a point 'p' is given by

$$E_\alpha^C[f(t)](p) = p\int_0^\infty K_\alpha(-p,t)f(t)d_\alpha t$$

where $K_\alpha(-p,t) = E_\alpha(-\frac{1}{p},t)$, $p > 0 \Rightarrow K_\alpha(-p,t) = e^{-\frac{t^\alpha}{p\alpha}}$.

Now using the relationship between fractional Elzaki and Laplace Transform and the existence of inverse for the Laplace transformation; the inverse fractional Elzaki transform of $f(t)$ is given by

$$f(t) = E_\alpha^{-1}(E_\alpha^C[f(t)](p)).$$

Thus, from the above definition and using its relationship with Laplace transformation, one can directly say that, to evaluate the inverse fractional Elzaki transform, one has to perform an integration in the complex region. Thus, for evaluating the Inverse fractional Elzaki transform with the help of Inverse Laplace transformation, it provides various exercises in contour integration. These properties were useful to solve the ordinary differential equations and hence for conformable fractional differential equations.

12.2.3 Linearity

If $f(t)$ and $g(t)$ are the two functions satisfying the conditions in (12.2), then for any constant α_1 and α_2, we have

$$E_\alpha^C[\alpha_1 f(t) + \alpha_2 g(t)](p) = \alpha_1 E_\alpha^C[f(t)](p) + \alpha_2 E_\alpha^C[g(t)](p).$$

Using the definition, it gives us

$$\begin{aligned}E_\alpha^C[\alpha_1 f(t) + \alpha_2 g(t)](p) &= p\int_0^\infty K_\alpha(-p,t)[\alpha_1 f(t) + \alpha_2 g(t)]d_\alpha t\\ &= p\alpha_1\int_0^\infty K_\alpha(-p,t)f(t)d_\alpha t + p\alpha_2\int_0^\infty K_\alpha(-p,t)g(t)d_\alpha t.\end{aligned}$$

Hence the property.

12.2.4 Relationship Between Fractional Elzaki and Fractional Laplace Transformation

Theorem 12.1 *Given the real-valued function $f(t)$ satisfying the conditions in (12.2) then the relationship between fractional Elzaki and fractional Laplace transformation [16] is given as: $E_\alpha^C[f(t)](p) = p\mathscr{L}_\alpha[f(t; \frac{1}{p})]$ with $p > 0$ and $\alpha \in (0, 1]$.*

Proof We know that the fractional Elzaki transform of $f(t)$ at a point ‘p’ is given by

$$E_\alpha^C[f(t)](p) = p\int_0^\infty K_\alpha(-p,t)f(t)d_\alpha t$$

where $K_\alpha(-p,t) = E_\alpha(-\frac{1}{p}, t)$, $p > 0 \Rightarrow K_\alpha(-p,t) = e^{-\frac{t^\alpha}{p\alpha}}$.

on the other hand, the fractional Laplace transformation of $f(t)$ at a point ‘p’ is given by

$$\mathscr{L}_\alpha\{f(t); p\} = \int_0^\infty E_\alpha(-p,t)f(t)d_\alpha t$$

where $Re.(p) > 0$, $E_\alpha(c,t) = e^{c\frac{t^\alpha}{\alpha}}$, $c \in \Re$ and $d_\alpha t = t^{\alpha-1}dt$.

Using the above two definitions gives us

$$E_\alpha^C[f(t)](p) = p\mathscr{L}_\alpha[f(t; \tfrac{1}{p})].$$

12.2.5 Relationship with Laplace Transformation

Theorem 12.2 *If $E_\alpha^C[f(t)](p)$ is a fractional Elzaki transformation of $f(t)$ at a point ‘p’ and $\mathscr{L}\left\{f\left((\alpha t)^{\frac{1}{\alpha}}\right); p\right\}$ were Laplace transformation of $\mathscr{L}\left\{f\left((\alpha t)^{\frac{1}{\alpha}}\right)\right\}$ at a point ‘p’, then the relationship between them is given by*

$$E_\alpha^C[f(t)](p) = p\mathscr{L}\left\{f\left((\alpha t)^{\frac{1}{\alpha}}\right)\right\}(\tfrac{1}{p})$$

with the condition that $p > 0$ and $\alpha \in (0, 1]$.

Proof By using the equation in the above definition (12.1, 12.2, 12.4) and relationship, we get

$$E_\alpha^C[f(t)](p) = p\mathscr{L}_\alpha[f(t)](\frac{1}{p}). \tag{12.6}$$

Now we know that

$$\mathscr{L}_\alpha[f(t)](\frac{1}{p}) = \mathscr{L}\left\{f\left((\alpha t)^{\frac{1}{\alpha}}\right)\right\}(\frac{1}{p}). \tag{12.7}$$

So, using Eqs. (12.6) and (12.7), we have the relation given by

$$E_\alpha^C[f(t)](p) = p\mathscr{L}\left\{f\left((\alpha t)^{\frac{1}{\alpha}}\right)\right\}(\tfrac{1}{p}).$$

From the above relation, one can say that the Parseval like theorem will not hold for fractional Elzaki transformation, this is because the Laplace transform does not have the sort of symmetry that is there for the Fourier transform. Thus, many fractional Elzaki transformable functions are not integrable because same is the true for Laplace. For example, if one takes any polynomial.

The way in which Laplace transform carries $L^1(0, \infty)$ into the Banach algebra $A_0(C^+)$ with dense image. Using the above relationship between fractional Elzaki transformation and Laplace transformation, one can say that fractional Elzaki transformation does the same thing formed by all continuous functions on $\bar{C}^+$ and analytic on C^+.

12.2.6 Shifting Property

Theorem 12.3 *If $g(t) = f(t+h)$, then its fractional Elzaki transformation is given as*

$$E_\alpha^C[g(t)](p) = p\int_h^\infty K_\alpha(-p, t-h)f(t)d_\alpha(t-h), \text{ where } h \geq 0$$

Proof $E_\alpha^C[g(t)](p) = \int_0^\infty pK_\alpha(-p,t)g(t)d_\alpha t$

$\Longrightarrow E_\alpha^C[g(t)](p) = \int_0^\infty pe^{-\frac{t^\alpha}{p^\alpha}} f(t+h)d_\alpha t$ where $p > 0$

substitute $t + h = x \Longrightarrow dt = dx$ gives

$E_\alpha^C[g(x)](p) = \int_h^\infty pe^{-\frac{(x-h)^\alpha}{p^\alpha}} f(x)d_\alpha(x-h)$

which can be rewritten as

$E_\alpha^C[g(t)](p) = p\int_h^\infty e^{-\frac{(t-h)^\alpha}{p^\alpha}} f(t)d_\alpha(t-h)$

$\Longrightarrow E_\alpha^C[g(t)](p) = p\int_h^\infty K_\alpha(-p, t-h)f(t)d_\alpha(t-h)$, where $h \geq 0$.
which gives the required relation and will be useful for solving differential equations.

12.2.7 Dilation Property

Theorem 12.4 *If $g(t) = f(th)$ $h > 0$, then its fractional Elzaki transformation is given as*

$$E_\alpha^C[g(t)](p) = \frac{p}{h^\alpha}\int_0^\infty K_\alpha(-ph^\alpha, t) f(t) d_\alpha t \quad \text{where } \alpha \in (0, 1]$$

Proof $E_\alpha^C[g(t)](p) = \int_0^\infty pK_\alpha(-p, t) g(t) d_\alpha t$

$$\Longrightarrow E_\alpha^C[g(t)](p) = \int_0^\infty pe^{-\frac{t^\alpha}{p^\alpha}} f(th) d_\alpha t$$

$$\Longrightarrow E_\alpha^C[g(t)](p) = \int_0^\infty pe^{-\frac{t^\alpha}{p^\alpha}} f(th) t^{\alpha-1} d_\alpha t$$

put $ht = x \Longrightarrow hdt = dx$

$$\Longrightarrow E_\alpha^C[f(x)](p) = p\int_0^\infty e^{-\frac{(\frac{x}{h})^\alpha}{p^\alpha}} f(x)\frac{x^{\alpha-1}}{h^\alpha} dx$$

$$\Longrightarrow E_\alpha^C[f(t)](p) = \frac{p}{h^\alpha}\int_0^\infty K_\alpha(-ph^\alpha, t) f(t) d_\alpha t \quad \text{where } \alpha \in (0, 1].$$

12.2.8 Boundedness of Fractional Elzaki Transformation

Theorem 12.5 *As we proved the relationship between fractional Elzaki Transformation and Laplace transformation which is given by*

$$E_\alpha^C[f(t)](p) = p\mathscr{L}_\alpha[f(t)](\tfrac{1}{p})$$

using the property of Laplace transformation, one can say that fractional Elzaki transform is bounded on $L^2(\Re_+)$.

Proof The Laplace transform of $f(t)$ at a point ‘p’ is defined to be

$$[\mathscr{L} f(t)](p) = \int_0^\infty f(t)e^{-pt} dt \text{ where }, \quad p > 0$$

where $f(t)$ is continuous, compactly supported and non-negative.

Now, for fractional Elzaki transformation

$$\| E_\alpha^C[f(t)](p) \|^2_{L^2(0,\infty)} = \| p\mathscr{L}\left\{f\left((\alpha x)^{\frac{1}{\alpha}}\right)\right\}(\tfrac{1}{p}) \|^2_{L^2(0,\infty)}.$$

Using the property of boundedness of Laplace transformation, one can say directly that the R. H. S. of the above equation is bounded, and hence the L. H. S. is also bounded. So that, the fractional Elzaki transformation is bounded in $L^2(0, \infty)$, with the relation

$$\| E_\alpha^C[f(t)](p) \|^2_{L^2(0,\infty)} \leq \pi \| f(t) \|^2_{L^2(0,\infty)}.$$

As we know that, the Laplace transform gives a bijection from $L^2(0, \infty)$ to Hardy space H^2 with the property that $\| F(s) \|_2 = \sqrt{2\pi} \| f(t) \|_2$.

Using the relationship between fractional Elzaki and Laplace transformation, one can say that, fractional Elzaki transform is also a bijection from $L^2(0, \infty)$ to Hardy space H^2 while the relation is slightly changed to $\| E_\alpha^C[f(t)](p) \|_2 = p\sqrt{2\pi} \| f(t) \|_2$.

12.3 Relationship Between Finite Fractional Elzaki Transform and Finite Laplace Transform

As of now, we have found the relationship between fractional Elzaki transformation and Laplace transformations directly. Now one can think about defining the finite fractional Elzaki transformation with the help of finite Laplace transformation [23], respectively.

One can define finite fractional Elzaki transformation as follows:

Let $\tau > 0, \omega \in \Re$. Then $F : (\omega, \infty) \to X$ is called a finite fractional Elzaki transform on $[0, M]$ if there exists an integrable function $f : [0, \tau] \to X$ with

$$F(p) = \int_0^\tau E_\alpha(-p, t) f(t) d_\alpha t \tag{12.8}$$

where $Re.(p) > 0$, $E_\alpha(c, t) = e^{c\frac{t^\alpha}{\alpha}}$, $c \in \Re$ and $d_\alpha t = t^{\alpha-1}dt$ and $p > \omega$.

Theorem 12.6 *The finite fractional Elzaki transformation is well defined and has a relation with the finite Laplace transform, which is given as*

$$F(p) = \frac{1}{\alpha}\left[\mathscr{L}(f(t^{\frac{1}{\alpha}}))\right](\tfrac{p}{\alpha}).$$

where L. H. S. is defined as $[0, \tau]$ *and R. H. S. as* $[0, \tau^{alpha}]$.

Proof By definition of finite fractional transformation, we have

$$F(p) = \int_0^\tau E_\alpha(-p, t) f(t) d_\alpha t$$

where $Re.(p) > 0$, $E_\alpha(c, t) = e^{c\frac{t^\alpha}{\alpha}}$, $c \in \Re$ and $d_\alpha t = t^{\alpha-1}dt$ and $p > \omega$.
Put, $t^\alpha = u \Rightarrow \alpha t^{\alpha-1}dt = du$ and using change of limits, we get

$$F(p) = \int_0^{\tau^\alpha} \frac{1}{\alpha} f(u^{\frac{1}{\alpha}}) e^{\frac{-pu}{\alpha}} du$$

Now using the definition of finite Laplace transform, it gives

$$F(p) = \frac{1}{\alpha}\left[\mathscr{L}(f(t^{\frac{1}{\alpha}}))\right](\tfrac{p}{\alpha}).$$

12.4 Fractional Elzaki Transformation of Some Special Functions

In this section, we will insight onto fractional Elzaki transform of some special functions which were mostly useful in solving differential equations by using the relationship between fractional Elzaki and Laplace operator.

1. If $f(t)$ is a constant function then its fractional Elzaki Transformation is given by

$$E_\alpha^C[f(t)](p) = Mp^2 \quad \text{where } f(t) = M \text{ is constant function}$$

Solution:

Given $f(t) = M$, the definition of fractional Elzaki transformation is given as

$$E_\alpha^C[f(t)](p) = p\int_0^\infty Me^{\frac{-t^\alpha}{p\alpha}}d_\alpha t$$
$$\Rightarrow E_\alpha^C[f(t)](p) = Mp\int_0^\infty Me^{\frac{-t^\alpha}{p\alpha}}t^{\alpha-1}dt$$

$$\text{put } \frac{t^\alpha}{p\alpha} = y \Rightarrow \alpha\frac{t^{\alpha-1}}{p\alpha}dt = dy$$

which gives us $t^{\alpha-1}dt = pdy \Rightarrow d_\alpha t = pdy$.

Hence, by the definition of fractional Elzaki Transformation, it resulted into

$$E_\alpha^C[f(t)](p) = Mp\int_0^\infty pe^{-y}dy$$

which finally resulted into

$$E_\alpha^C[f(t)](p) = Mp^2.$$

2. If $f(t) = t^q$ where $q \geq 0$ is a constant, then its fractional Elzaki Transformation at a point p is given by

$$E_\alpha^C[f(t)](p) = p^{2+\frac{q}{\alpha}}[\alpha^{\frac{q}{\alpha}}\Gamma(1+\tfrac{q}{\alpha})]$$

Solution:

Given $f(t) = t^q$ $q \geq 0$, using the relationship Laplace transformation and fractional Elzaki transformation. It gives us

$$E_\alpha^C[f(t)](p) = p\mathscr{L}f[(t\alpha)^{\frac{1}{\alpha}}](\tfrac{1}{p})$$

$$E_\alpha^C[f(t)](p) = p\int_0^\infty e^{\frac{-t}{p}}(t\alpha)^{\frac{q}{\alpha}}dt$$

Now by the substitution of $\frac{t}{p} = y \Rightarrow dt = pdy$ and $t = py$
It resulted into

$$E_\alpha^C[f(t)](p) = p\alpha^{\frac{q}{\alpha}}\int_0^\infty e^{-y}(py)^{\frac{q}{\alpha}}pdy$$
$$\Rightarrow E_\alpha^C[f(t)](p) = p^{2+\frac{q}{\alpha}}\int_0^\infty e^{-y}y^{\frac{q}{\alpha}}dy$$

Now using the definition of Gamma function, it resulted into

$$E_{\alpha}^{C}[f(t)](p) = p^{2+\frac{q}{\alpha}}\alpha^{\frac{q}{\alpha}}\Gamma(1+\tfrac{q}{\alpha})$$

which is the required solution.

3. If $f(t) = t^q K_{\alpha}(M, t)$, where $q \geq 0,\ M > 0$ is a constant, then its fractional Elzaki Transformation at a point p is given by

$$E_{\alpha}^{C}[f(t)](p) = p\alpha^{\frac{q}{\alpha}}(\tfrac{Mp}{(M-p)})^{1+\frac{q}{\alpha}}\Gamma(1+\tfrac{q}{\alpha})$$

Solution:

Given $f(t) = t^q t^q K_{\alpha}(M, t)\ q \geq 0$, the relationship with fractional Elzaki transformation and Laplace transformation is given as

$$E_{\alpha}^{C}[f(t)](p) = p\mathscr{L} f[(t\alpha)^{\frac{1}{\alpha}}](\tfrac{1}{p})dt$$

which gives us

$$E_{\alpha}^{C}[f(t)](p) = p\int_0^{\infty} e^{-\frac{t}{p}}(t\alpha)^{\frac{q}{\alpha}} e^{\frac{t^{\alpha}}{M\alpha}} dt$$

After re-arranging the terms, it turns out to be

$$E_{\alpha}^{C}[f(t)](p) = p\alpha^{\frac{q}{\alpha}} \int_0^{\infty} t^{\frac{q}{\alpha}} e^{-t(M-p)/Mp} dt$$

put $t\frac{(M-p)}{Mp} = y \Rightarrow dt = \frac{Mp}{(M-p)}dy$ with limit $y = 0\ \ to\ \ y = \infty$
Therefore, from above equation, it leads to the equation

$$E_{\alpha}^{C}[f(t)](p) = p\alpha^{\frac{q}{\alpha}}(\tfrac{Mp}{(M-p)})^{1+\frac{q}{\alpha}} \int_0^{\infty} y^{\frac{q}{\alpha}} e^{-y} dy$$

Now using the definition of Gamma function, it resulted into

$$E_{\alpha}^{C}[f(t)](p) = p\alpha^{\frac{q}{\alpha}}(\tfrac{Mp}{(M-p)})^{1+\frac{q}{\alpha}}\Gamma(1+\tfrac{q}{\alpha})$$

which is required solution.

4. If $g(t) = \int_0^t f(t)dt$, then its fractional Elzaki transformation at a point p is given by

$$E_{\alpha}^{C}[g(t)](p) = p\left(\mathscr{L}\left[\int_0^t f(t\alpha)^{\frac{1}{\alpha}}\right](\tfrac{1}{p})\right)$$

Solution:

Given that $g(t) = \int_0^t f(t)dt$, where $f(t)$ is the rapidly decreasing function or moreover $f(t)$ is in Lizorkin Space. Now, the relation between fractional Elzaki and Laplace transformation tells us that

$$E_{\alpha}^{C}[f(t)](p) = p\mathscr{L} f[(t\alpha)^{\frac{1}{\alpha}}](\tfrac{1}{p})dt$$

Also, we know that if $g(t) = \int_0^t f(t)dt$, then its Laplace transformation is given by

$$\mathscr{L}[g(t)](p) = \tfrac{1}{p}\mathscr{L}[f(t)] + \tfrac{f'(0)}{p}$$

Thus, using the above relationship, we have the fractional Elzaki Transformation of $g(t)$ given as follows:

$$E_\alpha^C[g(t)](p) = p\mathscr{L}[g(t\alpha)^{\frac{1}{\alpha}}](\tfrac{1}{p})$$
$$\Rightarrow E_\alpha^C[g(t)](p) = p\left(\mathscr{L}\left[\int_0^t f(t\alpha)^{\frac{1}{\alpha}}\right](\tfrac{1}{p})\right)$$

which is the required solution.

5. Now consider $f(t) = \sin(t)$ and $\alpha = 0.5$, then its fractional Elzaki Transformation of order α using the relation

$$E_\alpha^C[f(t)](p) = p\mathscr{L}\left\{f\left((\alpha t)^{\frac{1}{\alpha}}\right)\right\}(\tfrac{1}{p})$$

can be found as

$$E_{0.5}^C[\sin(t)](p) = p\mathscr{L}\left\{\sin\left((0.5t)^{\frac{1}{0.5}}\right)\right\}(\tfrac{1}{p})$$

Solution:

The Laplace transformation of sin function property gives us
$\mathscr{L}[\sin(0.5t)^{\frac{1}{0.5}}](p) = 2\sqrt{2\pi}\left[\frac{1}{2}\left(\frac{\sin(p^2)}{2} - \sqrt[4]{\frac{1}{p^4}}p\left(c(p\sqrt{\frac{2}{\pi}}\cos[p^2])+ p(p\sqrt{\frac{2}{\pi}}\sin[p^2])\right)\right) + \frac{1}{4}\cos(p^2)\right]$
Now using the relationship

$$E_\alpha^C[f(t)](p) = p\mathscr{L}\left\{f\left((\alpha t)^{\frac{1}{\alpha}}\right)\right\}(\tfrac{1}{p})$$

The fractional Elzaki Transformation of $f(t) = \sin(t)$ for order $\alpha = 0.5$ is given by

$$E_{0.5}^C[\sin(t)](p) = 2\sqrt{2\pi}\,p\left[\frac{1}{2}\left(\frac{\sin(\frac{1}{p^2})}{2} - \sqrt[4]{\frac{1}{\frac{1}{p^4}}}\frac{1}{p}\left(c(\tfrac{1}{p}\sqrt{\tfrac{2}{\pi}}\cos[\tfrac{1}{p^2}]) + \tfrac{1}{p}(\tfrac{1}{p}\sqrt{\tfrac{2}{\pi}}\sin[\tfrac{1}{p^2}])\right)\right) + \frac{1}{4}\cos(\tfrac{1}{p^2})\right]$$

which is the required fractional Elzaki transformation.

6. Now consider $f(t) = \cos(t)$ and $\alpha = 0.5$, then its fractional Elzaki Transformation of order α can be evaluated using the relation

$$E_\alpha^C[f(t)](p) = p\mathscr{L}\left\{f\left((\alpha t)^{\frac{1}{\alpha}}\right)\right\}(\tfrac{1}{p})$$

gives us

$$E_{0.5}^C[\cos(t)](p) = p\mathscr{L}\left\{\cos\left((0.5t)^{\frac{1}{0.5}}\right)\right\}(\tfrac{1}{p})$$

Solution:

The Laplace transformation of *cosine* function property we get,
$\mathscr{L}[\cos(0.5t)^{\frac{1}{0.5}}](p) = 4\sqrt{2\pi}\left[\frac{1}{4}\left(cp\sqrt{\frac{2}{\pi}}\sin(p^2) - \frac{1}{2}\sin(p^2)\right) - \frac{1}{4}p^2\sqrt{\frac{2}{\pi}}\cos (p^2) + \frac{1}{8}\cos(p^2)\right]$

Now using the relationship

$$E^C_{0.5}[\cos(t)](p) = p\mathcal{L}\left\{\cos\left((0.5t)^{\frac{1}{0.5}}\right)\right\}(\tfrac{1}{p})$$

$E^C_{0.5}[\cos(t)](p) = 4\sqrt{2\pi}\,p\left[\frac{1}{4}\left(\frac{c}{p}\sqrt{\frac{2}{\pi}}\sin(\frac{1}{p^2}) - \frac{1}{2}\sin(\frac{1}{p^2})\right) - \frac{1}{4}\frac{1}{p^2}\sqrt{\frac{2}{\pi}}\cos(\frac{1}{p^2}) + \frac{1}{8}\cos(\frac{1}{p^2})\right]$

which is the required fractional Elzaki transformation.

7. The fractional Elzaki transformation of conformable fractional derivative of $f(t)$ at a point 'p' which is denoted by $D^\alpha[f(t)](p)$ and its fractional Elzaki transformation were given by [10];

$$E^C_\alpha\left[D^\alpha\left(f(t)\right)\right](p) = \frac{1}{p}E^C_\alpha\left[f(t)\right](p) - pf(0)\quad \alpha \in (0, 1] \tag{12.9}$$

8. Now consider the Dirac Delta $f(t) = \delta(t - c)$, where $c > 0$. Then the fractional Elzaki transformation of Dirac Delta function can be find as follows:

Solution:

We know that the Laplace transformation of Dirac Delta function at a point 'p' is given by

$$\mathcal{L}[\delta(t - c)](p) = e^{-cp}$$

Now using the relationship between fractional Elzaki and Laplace transformation, the fractional Elzaki transformation of Dirac delta function is given as follows:

$$E^C_\alpha[f(t)](p) = p\mathcal{L}\left\{f\left((\alpha t)^{\frac{1}{\alpha}}\right)\right\}(\tfrac{1}{p}).$$

12.5 Applications to CFDE

Now as an applications towards conformable fractional differential equations and special functions; We consider the conformable fractional differential equation with initial condition and find out the functional analytic solution of it using fractional Elzaki transformation.

Consider the conformable fractional differential equation with given initial condition

$$D^\alpha_t\left(Z(t)\right) = E_\alpha(-1, t) + Z(t) \tag{12.10}$$

$Z(0) = 1$ and $\alpha \in (0, 1]$.

Solution:

The solution for $\alpha = 1$ by usual functional analytic method is given by

$$Z(t) = \frac{3}{2}e^{t} - \frac{1}{2}e^{-t}$$

Now Consider the equation given above for $\alpha \in (0, 1]$.

$$D_t^{\alpha}(Z(t)) = E_{\alpha}(-1, t) + Z(t)$$

Applying, fractional Elzaki Transformation on both sides of the above equation and by using the relationship

$$E_{\alpha}^{C}\left[D_t^{\alpha}(f(t))\right](p) = \frac{1}{p}E_{\alpha}^{C}[f(t)](p) - pf(0) \quad \alpha \in (0, 1].$$

we get

$$E_{\alpha}^{C}[D_t^{\alpha}(Z(t))] = E_{\alpha}^{C}[E_{\alpha}(-1, t) + Z(t)]$$

using the property of Linearity, it results into

$$\frac{1}{p}Z_{\alpha}(p) - pZ(0) = p\frac{1}{[1+\frac{1}{p}]} + Z_{\alpha}(p) \text{ with } \quad Z(0) = 1$$

$$\Rightarrow \frac{1}{p}Z_{\alpha}(p) - p = p\frac{1}{[1+\frac{1}{p}]} + Z_{\alpha}(p)$$

$$\Rightarrow \frac{1}{p}Z_{\alpha}(p) - p = \frac{p^2}{[1+p]} + Z_{\alpha}(p)$$

$$\Rightarrow \left(\frac{1-p}{p}\right)Z_{\alpha}(p) = \frac{p^2}{[1+p]} + p$$

$$\Rightarrow Z_{\alpha}(p) = \frac{p^3}{1-p^2} + \frac{p^2}{1-p}$$

which can be rewritten as

$$Z_{\alpha}(p) = p\left[\frac{-1}{1-\frac{1}{p^2}}\right] - p\left[\frac{1}{1-\frac{1}{p}}\right]$$

Applying inverse fractional Elzaki operator and relationship with inverse fractional Laplace transformation along the property of convolution, it leads to the solution as

$$Z(t) = \frac{-1}{2}e^{-\frac{t^{\alpha}}{\alpha}} + \frac{1}{2}e^{\frac{t^{\alpha}}{\alpha}} + e^{\frac{t^{\alpha}}{\alpha}}$$

which finally gives us

$$Z(t) = \frac{3}{2}e^{\frac{t^{\alpha}}{\alpha}} - \frac{1}{2}e^{-\frac{t^{\alpha}}{\alpha}}$$

For various values of α, the solutions were given in tabular form as

Value of α	Solution	Color in image
$\alpha = 0.5$	$Z(t) = \frac{3}{2}e^{\frac{t^{0.5}}{0.5}} - \frac{1}{2}e^{-\frac{t^{0.5}}{0.5}}$	Red
$\alpha = 0.6$	$Z(t) = \frac{3}{2}e^{\frac{t^{0.6}}{0.6}} - \frac{1}{2}e^{-\frac{t^{0.6}}{0.6}}$	Blue
$\alpha = 0.9$	$Z(t) = \frac{3}{2}e^{\frac{t^{0.9}}{0.9}} - \frac{1}{2}e^{-\frac{t^{0.9}}{0.9}}$	Green
$\alpha = 1$	$Z(t) = \frac{3}{2}e^{t} - \frac{1}{2}e^{-t}$	Violet

The solution for various values of α is shown graphically as follows:

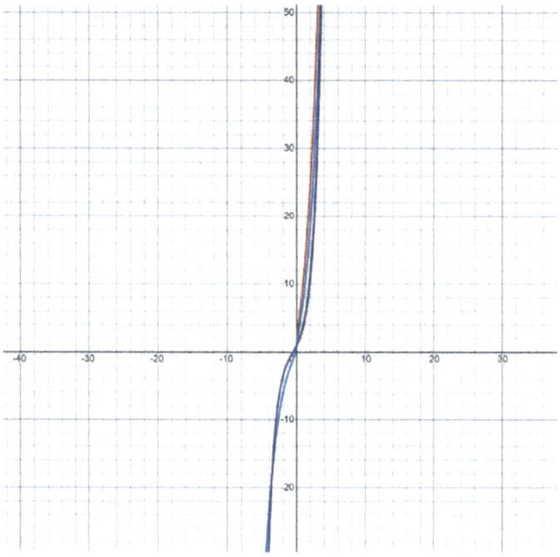

12.6 Conclusion

The newly defined fractional Elzaki transformation, i.e. fractional functional operator, was useful to solve fractional differential equations. As this newly defined fractional functional operator has its relationship with Laplace as well as fractional Laplace transformations,

it will be helpful towards finding the theoretical approach in the area of fractional functional analysis which is the topic of the current research. As an application of the fractional functional operator, we have also proved some properties of fractional Elzaki transformation and found out the fractional Elzaki transformation of some special functions.

The newly defined fractional functional operator is used to find out the analytic solution of conformable fractional differential equation compare the solution for various values of α and proved that as value of α tends towards 1, the solution also tends towards the actual one.

References

1. Ansari, A.: The Generalized Laplace Transform and fractional Differential Equations of distributed orders. Differ. Equ. Control Process. **N3**, 128–137 (2012)
2. Atangana, A., Baleanu, D.: New Fractional Derivatives with Non Local and Non Singular Kernel: Theory and Applications to Heat Transfer Model. Thermal Science–online first (2016)
3. Atangana, A., Baleanu, D.: New fractional derivative with non-singular kernel. New Trends Math. Sci. **20**(2), 757–763 (2016)
4. Baleanu, D., Purohit, S.D., Agarwal, P.: On fractional integral inequalities involving hyper geometric operators. Chin. J. Math. **2014**, 609476 (2014)
5. Baleanu, D., Agarwal, P.: Certain inequalities involving the fractional q-integral operators. Abstract Appl. Anal. **2014**, 371274 (2014)
6. Douglas, A., Camrud, E., Ulness, D.J.: On the nature of the conformable derivative and its applications to Physics. JFCA **10**(2), 92–135 (2019)
7. Debnath, P., Srivastava, H.M., Kumam, P., Hazarika, B.: Fixed Point Theory and Fractional Calculus: Recent Advances and Applications. Springer, Singapore (2022)
8. Debnath, P., Konwar, N., Radenovic, S.N.: Metric Fixed Point Theory: Applications in Science. Engineering and Behavioral Sciences. Springer, Singapore (2021)
9. Elzaki, T.: The new integral transform-Elzaki Transform. Global J. Pure Appl. Math. **1**, 57–64 (2011)
10. Elzaki, T., Elzaki, S.M.: The new integral transform Tarig Transform properties and applications to differential equations. Elixir Appl. Math. **38**, 4239–4242 (2011)
11. Elzaki, T., Elzaki, S.M.: On connections between Laplace and Elzaki Tarig Transform **6**, 1–11 (2011)
12. John, S.: Mathematics and its History. Springer Publication (2010)
13. Khalil, R., Al Horani, M., Sababheh, Y.A.: A new definition of fractional derivative. J. Comput. Appl. Math. **264**, 65–70 (2014)
14. Liang, S., Wu, R., Chen, L.: Laplace transform of fractional order differential equations. Electron. J. Differ. Equ. **139**, 1–15 (2015)
15. Manjarekar, S., Bhadane, A.: Generalized Elzaki-Tarig Transformation and its applications to new fractional derivative with non-singular kernel. Progr. Fract. Differ. Appl. **3**(3), 1–6 (2017)
16. Manjarekar, S., Bhadane, A.: New conformable fractional ELZAKI transformation: Theory and applications. Malaya J. Matematik **S1**, 619–625 (2019)
17. Manjarekar, S., Bhadane, A.: Applications of tarig transformation to new fractional derivatives with non-singular kernel. JFCA **9**(1), 160–166 (2018)
18. Ozakan, O., Kurt, A.: The analytical solutions for conformable integral equations and integro–differential equations by conformable Laplace transform. Opt. Quant. Electron. **50**(81) (2018)
19. Podlubny, I.: Fractional Differential Equation. Academic Press (1999)
20. Pospisil, P., Shripkova, L.P.: Sturm's theorems for conformable fractional differential equations. Math. Commun. **21**, 273–281 (2016)
21. Silva, S., Moreira, D.M., Moret, M.A.: Conformable laplace transform of fractional differential equations. Axioms **7**(55) (2018)
22. Thabet, A.: On conformable fractional Calculus. J. Comput. Appl. Math. **279**, 57–60 (2015)
23. Valentin, K., Muler, C., Vieten, P.: Finite and local laplace transforms in banach spaces. Proceed. Edinburgh Math. Soc. **46**, 357–372 (2003)
24. Xiao-Jun, Y.: Local Fractional Functional Analysis and its Applications. Asian Academic Publisher (2011)
25. Zhao, D., Luo, M.: General conformable fractional derivative and its physical Interpretation. Calcolo **54**, 903–917 (2017)
26. Zhang, X., Agarwal, P., Liu, Z., Peng, H.: The general solution for impulsive differential equations with Riemann—Liouville fractional—order $q \in (1, 2)$. **13**(1) (2015)

Chapter 13
Existence, Uniqueness, and Stability Results for Tempered Fractional Integro-Differential Equations via Fixed Point Techniques

Pratibha Verma, Sushmita Anand, Parul Saini, and Amar Deep

Abstract In this chapter, we consider the most important and problems used frequently in random walks and tempered fractional equation that describes the particle plume shape. This work investigates the basic results of the tempered fractional integro-differential equation (TF-IDE) via the concept of Caputo fractional derivative. The existence and uniqueness of a solution for BVP-FDEs are discussed by utilizing the Banach fixed point theorem and Schaefer's fixed point theorem. Then, we provide the Ulam–Hyers stability results under the Lipschitz condition. These theoretical results are significant before solving the problems and help in finding the solutions to the problems.

Keywords Caputo fractional derivative · Tempered fractional operator · Ulam–Hyers stability · Fixed point theorem

Mathematics Subject Classification (2010) 47H10 · 35R11

P. Verma
Faculty of Engineering Technology (ITER), Siksha Ö Anusandhan (Deemed to be University), Bhubaneswar 751030, Odisha, India

S. Anand
Department of Mathematics, Motilal Nehru National Institute of Technology Allahabad, Prayagraj 211004, UP, India
e-mail: sushmitaanand@mnnit.ac.in

P. Saini (✉)
Mathematics Discipline, PDPM-IIITDM, Jabalpur 482005, MP, India
e-mail: iniaslurap1212@gmail.com

A. Deep
Department of Applied Sciences (Math), IIMT Engineering College, Meerut 250001, UP, India

B. Hazarika et al. (eds.), *Advances in Functional Analysis and Fixed-Point Theory*, Industrial and Applied Mathematics,
https://doi.org/10.1007/978-981-99-9207-2_13

13.1 Introduction

Fractional calculus (FC) becomes a very attractive and important topic to discuss in the science and engineering fields. Because of its advantages, it provides improved results as compared to standard calculus. The FC gives the extended vision to view the model's behavior [19, 20]. Many researchers focused on this direction and try to develop new theories and definitions for fractional operators. Several fractional operators are discussed in the previous works. Here, we have adopted the famous and frequently used fractional operator called the Caputo derivative [10, 11].

Also, there are some special fractional operators that exist in this field such as Riesz, tempered, and Mitter–Laffer fractional operators. The tempered fractional (TF) derivative is one of the generalized forms of fractional derivatives. Multiplying classical fractional derivative exponential factors leads to the TFD. The TFD depends on a ρ called the parameter. In special cases, $\rho = 0$, then the TFD becomes a Riemann–Liouville and Caputo fractional derivative [13, 18]. Tempered fractional Brownian motion can show semi-long range dependency, with correlations falling off like a power law at moderate time scales but becoming short-range dependent at long time scales. The Kolmogorov model for turbulence is now extended to encompass low frequencies. The renowned Davenport spectrum of wind speed is used to construct electric power production facilities, and tempered fractional Brownian motion gives a time-domain stochastic process model for it. For various diffusion processes, the mean squared displacement (MSD) varies as a power law of a unique exponent, which can be described by a constant fractional diffusion equation (FDE) [14].

In [5], the existence of the solution of fractional hybrid differential equations is discussed using the infinite system with given initial conditions by converting them to integral equations in the space $C(I, m(\theta)^{\beta})$ which is provided based on Darbo fixed point theorem. Zaky [22] studied the results of the existence and uniqueness solutions for a class of nonlinear tempered fractional boundary value problems. Also, the author discussed the sensitivity of the solution established using perturbations in the given parameters of the problem.

From the above work, we motivate and establish important and valuable theoretical works to solve the considered problem. This study provides the existence and uniqueness results with new assumptions, and stability is a new study of this type of equation. In previous work, there are many fixed point theorems for the existence and uniqueness conditions developed for the several differential equations on the Banach space. A complete normed vector space called a Banach space in functional analysis enables the calculation of vector length. Each vector other than the zero vector has a length higher than zero when the vector space is normed. Thus, it is possible to determine the length and separation between two vectors.

Currently, many researchers focus on theoretical analysis of fractional differential equations (FDEs) and provide the existence and uniqueness of the solution for different problems. The stability analysis of the solution has also become a very hot topic in this field. For more details, see the references in [1, 2, 4, 7].

Now, we consider the tempered fractional integro-differential equations (TF-IDEs) [13], that is,

$$\begin{aligned} D_{0,\theta}^{\sigma,\rho} y(\theta) &= \mathscr{F}(\theta) + \int_0^\theta \mathscr{K}(\theta,\mu) y(\mu) d\mu, 0 < \theta, \mu \leq 1, 0 < \sigma < 1, \\ y(0) &= \ell, \end{aligned} \tag{13.1}$$

where $D_{0,\theta}^{\sigma,\rho}$ denotes the $\sigma-$th-order tempered derivative of $y(\theta)$, and $\mathscr{K}(\theta,\mu)$, $\mathscr{F}(\theta)$ are known functions.

In this work, we use a standard fixed point theorems and obtained some new results about the existence and uniqueness of solutions. We consider problem (13.1) tried to use fixed point theorems proved for tempered operators to establish new results with a wide range of applications for existence and uniqueness. Furthermore, we give the stability analysis of the solution for the TF-IDEs via Ulam–Hyers stability.

The chapter is organized in the following way: in the next section, some preliminary discussions of tempered fractional integrals and derivatives in Caputo sense are reported. In Sect. 13.3, the existence results provide for the TF-IDEs using Schaefer and Arzela–Ascoli theorems. In Sect. 13.4, we devote for the uniqueness results via the Banach fixed point theorem for the TF-IDEs. Finally, the results are concluded in Sect. 13.5.

13.2 Preliminaries

In this section, we provide some basic definitions of tempered fractional operators and fundamental theorems which are needed in this study.

Definition 13.1 ([13, 18, 21]) For any fixed parameter $\rho \geq 0$ and order $\sigma > 0$, the Riemann–Liouville tempered fractional integral of the function $y(\theta)$ is defined as

$$D_{0,\theta}^{\sigma,\rho} y(\theta) = e^{-\rho\theta} I_{0,\theta}^{\sigma}(e^{\rho\theta} y(\theta)) = \frac{1}{\Gamma(\sigma)} \int_0^\theta \frac{1}{(\theta-\mu)^{\sigma-1}} e^{-\rho(\theta-\mu)} y(\mu) d\mu, \theta > 0. \tag{13.2}$$

If $\rho = 0$, then the Riemann–Liouville tempered fractional integral reduces to the Riemann–Liouville fractional integral.

Definition 13.2 ([13, 18, 21]) For any fixed parameter $\rho \geq 0$ and order $\sigma > 0$, the Riemann–Liouville tempered fractional derivative of the function $y(\theta)$ is defined as

$$D_{0,\theta}^{\sigma,\rho} y(\theta) = e^{-\rho\theta} D_{0,\theta}^{\sigma}(e^{\rho\theta} y(\theta)) = \frac{e^{\rho\theta}}{\Gamma(1-\sigma)} \left(\frac{d}{d\theta}\right) \int_0^\theta \frac{1}{(\theta-\mu)^{\sigma}} e^{-\rho\theta} y(\mu) d\mu, \theta > 0. \tag{13.3}$$

If $\rho = 0$, then the Riemann–Liouville tempered fractional derivative reduces to the Riemann–Liouville fractional derivative.

Definition 13.3 ([13, 18, 21]) For any fixed parameter $\rho \geq 0$ and order $\sigma > 0$, the Caputo tempered fractional derivative of the function $y(\theta)$ is defined as

$$D_{0,\theta}^{\sigma,\rho} y(\theta) = e^{-\rho\theta} D_{0,\theta}^{\sigma}(e^{\rho\theta} y(\theta)) = \frac{e^{\rho\theta}}{\Gamma(1-\sigma)} \int_0^{\theta} \frac{1}{(\theta-\mu)^{\sigma}} \left(\frac{d}{d\theta} e^{-\rho\theta} y(\mu)\right) d\mu, \theta > 0. \tag{13.4}$$

If $\rho = 0$, then the Caputo tempered fractional derivative reduces to the Caputo fractional derivative.

Definition 13.4 An operator is called completely continuous if it is continuous and maps bounded sets into precompact sets.

Definition 13.5 Let $(Y, \|\cdot\|)$ be a normed space. A contraction of Y is mapping $P : Y \to Y$ that satisfies, for every $\nu_1, \nu_2 \in Y$,

$$\|P(Y_1) - P(\nu_2)\| \leq \delta \|\nu_1 - \nu_2\|, \tag{13.5}$$

for some real values $0 \leq \delta < 1$.

Lemma 13.1 ([15–17]) *Let $\sigma > 0$, then a general solution to the homogeneous equation*

$$D_{0+}^{\sigma} \psi(\nu) = 0 \tag{13.6}$$

is given by

$$\psi(\nu) = a_0 + a_1\nu + a_2\nu^2 + a_3\nu^3 + \cdots + a_{m-1}\nu^{m-1}, a_j \in \mathbb{R}, j = 1, 2, \ldots, m-1 (m = [\sigma] + 1). \tag{13.7}$$

Lemma 13.2 ([15–17]) *Let $\sigma > 0$, we have*

$$\begin{aligned} J_{0+}^{\sigma} D_{0+}^{\sigma} \psi(\nu) = \psi(\nu) + a_0 + a_1\nu + a_2\nu^2 + a_3\nu^3 + \cdots + a_{m-1}\nu^{m-1}, a_j \in \mathbb{R}, \\ j = 1, 2, \ldots, m-1 (m = [\sigma] + 1). \end{aligned} \tag{13.8}$$

Theorem 13.1 (Banach fixed point theorem, [15–17]) *Every contraction mapping on a complete metric space has a unique fixed point.*

Theorem 13.2 (Schaefer fixed point theorems, [15–17]) *If Y is a closed bounded convex subset of a Banach space X and $P : Y \to Y$ is completely continuous, then P has a fixed point in P.*

Theorem 13.3 (Arzelà–Ascoli Theorem, [15–17]) *Let Y be a compact metric space. Let $C(Y, \mathbb{R})$ be given the sup norm metric. Then a set $P \subset C(Y)$ is compact iff P is bonded, closed, and equicontinuous.*

We denote $C(I, \mathbb{R})$ as the Banach space of all continuous functions from $I = (0, 1]$ into $\mathbb{R}$ with the norm $\|\cdot\|_{infty}$ defined by $\|y\|_{\infty} := \sup\{|y(\theta) : \theta \in I|\}$ (see [8, 12]).

13.3 Main Results

Before establishing and proving the main results, we have to introduce some following assumptions for the problem (13.1):
$[P_1]$: There exists a function $\mathscr{K} \in C(I, \mathbb{R})$, the set of all positive continuous function on $\mathscr{E} = \{(\theta, \mu) \in I \times I\}$ such that

$$\mathscr{K}^* = \sup_{\theta \in I} \int_0^\theta \mathscr{K}(\theta, \mu) d\mu < \infty.$$

$[P_2]$: Assume that the function $\mathscr{F} : I \rightarrowtail \mathbb{R}$ is continuous.
$[P_3]$: There exists constant $\mathscr{L} > 0$, such that, for any $\theta \in I$

$$|\mathscr{F}| \leq \mathscr{L}.$$

13.3.1 Uniqueness Result of the Solution for TF-IDEs

In this section, we investigate the uniqueness result of the solution for the problem given by problem (13.1) by using the Banach fixed point theorem.

Theorem 13.4 *Assume that the hypotheses* $[P_1] - [P_2]$ *hold if*

$$\left[\frac{\mathscr{K}^*}{\Gamma(\sigma + 1)} \right] < 1 \tag{13.9}$$

is satisfied, then problem (13.1) *has unique solution in* $C(I, \mathbb{R})$.

Proof We will transform problem (13.1) into a fixed point problem. Consider the operator $\Delta : C(I, \mathbb{R}) \to C(I, \mathbb{R})$, defined as

$$\Delta y(\theta) = \ell + \frac{1}{\Gamma(\sigma)} \int_0^\theta \frac{1}{(\theta - a)^{\sigma - 1}} e^{\rho(\theta - a)} \left(\int_0^\mu \mathscr{K}(\theta, \mu) y(\mu) d\mu \right) da. \tag{13.10}$$

We can see that if $y \in C(I, \mathbb{R})$ is a fixed point of Δ, then y is the solution of problem (13.1).
Now, we shall prove Δ has a fixed point of y in $C(I, \mathbb{R})$. Let $y_1, y_2 \in C(I, \mathbb{R})$ and $\forall$ $\theta \in I$ such that

Now, we shall prove Δ has a fixed point of y in $C(I, \mathbb{R})$. For that $y_1, y_2 \in C(I, \mathbb{R})$ and $\forall$ $\theta \in I$ such that

$$y_1(\theta) = \ell + \frac{1}{\Gamma(\sigma)} \int_0^\theta \frac{1}{(\theta - a)^{\sigma - 1}} e^{\rho(\theta - a)} \left(\int_0^\mu \mathscr{K}(\theta, \mu) y_1(\mu) d\mu \right) da \tag{13.11}$$

and

$$y_2(\theta) = \ell + \frac{1}{\Gamma(\sigma)} \int_0^\theta \frac{1}{(\theta - a)^{\sigma-1}} e^{\rho(\theta-a)} \Big(int_0^\mu \mathscr{K}(\theta, \mu) y_2(\mu) d\mu \Big) da. \quad (13.12)$$

Consequently, we get

$$\begin{aligned} |\Delta y_1(\theta) - \Delta y_1(\theta)| &\leq \frac{1}{\Gamma(\sigma)} \int_0^\theta \frac{1}{(\theta - a)^{\sigma-1}} e^{\rho(\theta-a)} \Big(\int_0^\mu |\mathscr{K}(\theta, \mu)||y_1(\mu) - y_2(\mu)| d\mu \Big) da \\ &\leq \frac{\mathscr{K}^*}{\Gamma(1+\sigma)} |y_1(\theta) - y_2(\theta)|. \end{aligned}$$

Finally, we obtain

$$\|\Delta y_1(\theta) - \Delta y_1(\theta)\|_\infty \leq \Sigma(\sigma, \mathscr{K}^*) ||\psi_1(y) - \psi_2(y)||_\infty, \quad (13.13)$$

where $\Sigma(\sigma, \mathscr{K}^*) = \frac{\mathscr{K}^*}{\Gamma(1+\sigma)}$.
Consequently, by using the condition (13.13), Δ is a contraction with the help of the Banach fixed point theorem, we conclude that the operator Δ has a fixed point that is a solution of problem (13.1). Thus, problem (13.1) attains unique solution on I. □

Now, we are going to prove the second result for the existence of problem (13.1), which is based on Schaefer's fixed point theorem.

13.3.2 Existence Result of the Solution for TF-IDEs

In this section, we present the existence of the solution for the problem given by (13.1) by using Schaefer's fixed point theorem in the Banach space.

Theorem 13.5 *Assuming that the conditions* $[P_1] - [P_3]$ *are satisfied. Then problem* (13.1) *has at least one solution in* I.

Proof The proof consists of four steps. We shall prove that the operator Δ is continuous.
Let $(y_n)_{n \in \mathbb{N}}$ be a sequence such that $y_n \to y$ in $C(I, \mathbb{R})$. Then, for every $\theta \in I$, we have

$$\begin{aligned} |\Delta y_n(\theta) - \Delta y(\theta)| &\leq \frac{1}{\Gamma(\sigma)} \int_0^\theta \frac{1}{(\theta - a)^{\sigma-1}} e^{\rho(\theta-a)} \Big(\int_0^\mu |\mathscr{K}(\theta, \mu)||y_n(\mu) - y(\mu)| d\mu \Big) da \\ &\leq \frac{\mathscr{K}^*}{\Gamma(1+\sigma)} |y_n(\theta) - y(\theta)|. \end{aligned}$$

Using the assumption (B_2), we have

$$\|\Delta y_n(\theta) - \Delta y(\theta)\|_\infty \to 0, n \to +\infty.$$

Thus, the operator Δ is continuous on $C(I, \mathbb{R})$.
Step 2: The map Δ maps bounded sets into bounded sets in $C(I, \mathbb{R})$.
Let $\mathbb{B}_\epsilon = \{y \in C(I, \mathbb{R}) : \|y\|_\infty \leq \epsilon, \epsilon > 0\}$, $\|y\|_\infty = \{\sup |y|, \theta \in I\}$. We have, for all $y \in \mathbb{B}_\epsilon$

$$
\begin{aligned}
|(\Delta y)(\theta)| &= |\ell| + \frac{1}{\Gamma(\sigma)} \int_0^\theta \frac{1}{(\theta - a)^{\sigma-1}} |e^{\rho(\theta-a)}| \left(\int_0^\mu |\mathscr{K}(\theta, \mu)||y(\mu)| d\mu \right) da \\
&\leq |\ell| + \frac{\mathscr{K}^* \mathscr{L} y^\sigma}{\Gamma(1+\sigma)} \\
&\leq |\ell| + \frac{\mathscr{K}^* \mathscr{L}}{\Gamma(1+\sigma)}.
\end{aligned}
$$

Finally, we have

$$
\|(\Delta y)(\theta)\|_\infty \leq \ell| + \frac{\mathscr{K}^* \mathscr{L}}{\Gamma(1+\sigma)}. \tag{13.14}
$$

Thus, $\|(\Delta y)(\theta)\|_\infty \leq \ell| + \frac{\mathscr{K}^* \mathscr{L}}{\Gamma(1+\sigma)} := C$, for every $\theta \in I$.
Hence, we have $\|(\Delta y)(\theta)\|_\infty < \infty$, for every $\theta \in I$.
Therefore, $\|(\Delta y)(\theta)\|_\infty \leq C$ for every $y \in \mathbb{B}_\epsilon$, which implies that $T\mathbb{B}_\epsilon \subset \mathbb{B}_C$.
Thus, we conclude that the operator Δ maps bounded sets into bounded sets in $C(I, \mathbb{R})$.
Step 3: We show that Δ is equicontinuous on $C(I, \mathbb{R})$.
Let us consider $y \in \mathbb{B}_\epsilon$, $\theta_1, \theta_2 \in I$, where $\theta_1 < \theta_2$. We have

$$
\begin{aligned}
|(\Delta y)(\theta_1) - (\Delta y)(\theta_2)| \leq \Bigg| & \frac{1}{\Gamma(\sigma)} \int_0^{\theta_1} \frac{1}{(\theta_1 - a)^{\sigma-1}} e^{\rho(\theta_1-a)} \left(\int_0^\mu \mathscr{K}(\theta_1, \mu) y(\mu) d\mu \right) da \\
& - \left(\frac{1}{\Gamma(\sigma)} \int_0^{\theta_2} \frac{1}{(\theta_2 - a)^{\sigma-1}} e^{\rho(\theta_2-a)} \left(\int_0^\mu \mathscr{K}(\theta_2, \mu) y(\mu) d\mu \right) da \right) \Bigg|.
\end{aligned}
$$

Consequently,

$$
\begin{aligned}
|(\Delta y)(\theta_1) - (\Delta y)(\theta_2)| &\leq \left(\int_{\theta_2}^{\theta_1} (\theta_2 - a)^{\sigma-1} \left(\int_0^a |K(a, \mu)||\psi(\mu)| d\mu \right) ds \right. \\
&\quad \left. + \int_0^{\theta_1} (\theta_1 - a)^{\sigma-1} - (\theta_2 - a)^{\sigma-1} \left(\int_0^a |K(a, \mu)||y(\mu)| d\mu \right) da \right) \\
&\leq \left(\int_{\theta_2}^{\theta_1} (\theta_2 - a)^{\sigma-1} \left(\int_0^a |K(a, \mu)||y(\mu)| d\mu \right) da \right. \\
&\quad \left. + \int_0^{\theta_1} (\theta_1 - a)^{\sigma-1} - (\theta_2 - a)^{\sigma-1} \left(\int_0^a |K(a, \mu)||y(\mu)| d\mu \right) da \right) \\
&\leq \frac{\mathscr{K}^* \mathscr{L}}{\Gamma(\sigma+1)} \left(2(\theta_2 - \theta_1)^\sigma + (\theta_1^\sigma - \theta_2^\sigma) \right) \\
&\leq \frac{\mathscr{K}^* \mathscr{L}}{\Gamma(\sigma+1)} 2(\theta_2 - \theta_1)^\sigma.
\end{aligned}
$$

Finally, we obtain

$$\|(\Delta y)(\theta_1) - (\Delta y)(\theta_2)\|_\infty \leq \frac{\mathscr{K}^*\mathscr{L}}{\Gamma(\sigma+1)}\Big(2(\theta_2-\theta_1)^\sigma\Big).$$

This implies, $\|(\Delta y)(\theta_1) - (\Delta y)(\theta_2)\|_\infty \to 0$ as $\theta_1 \to \theta_2$. That is the set $\{T\mathbb{B}_\epsilon\}$ is equicontinuous. By the Arzelà–Ascoli theorem [3], we deduce that Δ is a completely continuous operator.
Step 4: Let $V = \{y \in C(I,\mathbb{R});\ y = \varepsilon\Delta(z), 0 < \varepsilon < 1\}$.
We show that the set V is bounded. Let $y \in V$, then $z = \varepsilon\Delta(z)$, for some $0 < \varepsilon < 1$, for each $\theta \in I$, we have

$$|y(\theta)| = |\varepsilon\Delta(z)| \tag{13.15}$$

$$\leq \varepsilon \times \left(|\ell| + \frac{1}{\Gamma(\sigma)}\int_0^\theta (\theta-a)^{\sigma-1}\left(\int_0^a |K(a,\mu)||y(\mu)|d\mu\right)da\right). \tag{13.16}$$

From Equation (13.15), we obtain the following expression:

$$|y(\theta)| \leq \varepsilon \times \left(|\ell| + \frac{\mathscr{K}^*\mathscr{L}}{\Gamma(1+\sigma)}\right).$$

Hence $\|y\|_\infty \leq \infty$.
The above proof shows that V is bounded, as consequence of Schaefer's fixed point theorem, problem (13.1) has at least one solution on $C(I,\mathbb{R})$. □

13.4 Stability Analysis

In the present section, we discuss the stability of the solution for problem (13.1) using Ulam–Hyers (U-H) stabilities [6, 9].

Definition 13.6 The adopted problem (13.1) is U-H stable if for $\delta > 0$ and $y \in C(I,\mathbb{R})$ be any solution of the inequality

$$\|D_{0,\theta}^{\sigma,\rho} y(\theta) - \mathscr{F}(\theta) + \int_0^\theta \mathscr{K}(\theta,\mu)y(\mu)d\mu\| \leq \delta, \theta \in I, \tag{13.17}$$

there exists unique solution y' of the TF-IDEs (13.1) with $\mathscr{C}_k > 0$ such that

$$\|y(\theta) - y'(\theta)\| \leq \mathscr{C}_k\delta, \theta \in I. \tag{13.18}$$

Further, the TF-IDEs (13.1) will be generalized U-H (G-U-H) stable if there exists $\mathscr{O} \in C(\mathbb{R}^+, \mathbb{R}^+)$ with $\mathscr{O}(0) = 0$, for any solution y of Equation (13.17) and y' be unique solution of problem (13.1) such that

$$\|y(\theta) - y'(\theta)\| \leq \mathscr{O}(\delta), \theta \in I, \tag{13.19}$$

then the TF-IDEs (13.1) is G-U-H stable.

Remark 1: If there exists $\mathscr{Y} \in C(I, \mathbb{R})$, the $y \in C(I, \mathbb{R})$ satisfies inequality (13.17) if (i) $|\mathscr{Y}(\theta)| \leq \delta, \forall \theta \in I$,
(ii) $D_{0,\theta}^{\sigma,\rho} y(\theta) = \mathscr{F}(\theta) + \int_0^\theta \mathscr{K}(\theta, \mu) y(\mu) d\mu, \theta \in I$.
Consider the corresponding TF-IDEs (13.1) as

$$D_{0,\theta}^{\sigma,\rho} y(\theta) = \mathscr{F}(\theta) + \int_0^\theta \mathscr{K}(\theta, \mu) y(\mu) d\mu + \mathscr{Y}(\theta), \tag{13.20}$$

with the given condition

$$y(0) = \ell. \tag{13.21}$$

Lemma 13.3 *The following inequality holds to the TF-IDEs* (13.1).

$$\left(\frac{\delta}{\Gamma(\sigma + 1)} \right).$$

Proof With the help of Lemma (13.1), the solution of TF-IDEs (13.1) is given by

$$y(\theta) = \ell + I_{0,\theta}^{\sigma,\rho} \left(\mathscr{F}(\theta) + \int_0^\theta \mathscr{K}(\theta, \mu) y(\mu) d\mu + \mathscr{Y}(\theta) \right). \tag{13.22}$$

We have

$$\begin{aligned} |y(\theta) - \Delta y(\theta)| &\leq I_{0,\theta}^{\sigma,\rho} \Big(\mathscr{Y}(\theta) \Big) \\ &\leq \left(\frac{\delta}{\Gamma(\sigma + 1)} \right). \end{aligned} \tag{13.23}$$

Theorem 13.6 *Under the hypotheses of Lemma (13.3), the solution of the tempered problem* (13.1) *is U-H stable as well as G-U-H stable if* $\Sigma(\sigma, \mathscr{K}^*) \leq 1$.

Proof Let $y \in C(I, \mathbb{R})$ be any solution and $y' \in C(I, \mathbb{R})$ be unique solution of the tempered problem (13.1), then we have

$$\begin{aligned}|y(\theta) - y'(\theta)| &= |y(\theta) - \Delta y(\theta)| \\ &\quad + |\Delta y(\theta) - y'(\theta)| \\ &\leq \left(\frac{\delta}{\Gamma(\sigma+1)}\right) + \Sigma(\sigma, \mathscr{K}^*)|y(\theta) - y'(\theta)| \\ &\leq \frac{\left(\frac{\delta}{\Gamma(\sigma+1)}\right)}{1 - \Sigma(\sigma, \mathscr{K}^*)}.\end{aligned}$$

Thus, the proposed problem (13.1) is U-H stable as well as G-U-H stable. □

13.5 Conclusion

This research adopted the tempered fractional integro-differential equations using the Caputo with a parameter λ. Under some fixed point theorems on the Banach space, we obtained new results for the existence and uniqueness of solutions to the problem involving the initial condition. Additionally, we discuss the Ulam–Hyer stability for the TF-IDEs using the tempered fractional operators in the Caputo sense. These obtained aspects become important results in the analysis of the system.

References

1. Aslam, M., Gomez-Aguilar, J.F., Ur-Rahman, G., Murtaza, R.: Existence, uniqueness, and Hyers-Ulam stability of solutions to nonlinear p-Laplacian singular delay fractional boundary value problems. Math. Methods Appl. Sci. (2021). https://doi.org/10.1002/mma.7608
2. Ali, A., Khalid, S., Rahmat Kamran, G., Ali, G., Sooppy Nisar, K., Alshahrani, B.: Controllability and Ulam-Hyers stability of fractional order linear systems with variable coefficients. Alexandria Eng. J. **61**(8), 6071–6076 (2021)
3. Bachir, M., Nazaret, B.: Metrization of probabilistic metric spaces. Applications to fixed point theory and Arzela-Ascoli type theorem. Topol. Appl. **289**, 107549 (2021)
4. Deng, J., Ma, L.: Existence and uniqueness of solutions of initial value problems for nonlinear fractional differential equations. Appl. Math. Lett. **23**(6), 676–680 (2010)
5. Das, A., Hazarika, B., Deuri, B.C.: Existence of an infinite system of fractional hybrid differential equations in a tempered sequence space. Fract. Cal. Appl. Anal. **25**, 2113–2125 (2022)
6. Haq, F., Shah, K., Ur Rahman, G., Shahzad, M.: Hyers-Ulam stability to a class of fractional differential equations with boundary conditions. Int. J. Appl. Comput. Math. **3**, 1135–1147 (2017)
7. Ibrahim, R.W., Momani, S.: On the existence and uniqueness of solutions of a class of fractional differential equations. J. Math. Anal. Appl. **334**(1), 1–10 (2007)
8. Ziqiang, L., Zhu, Y.: Nonlinear impulsive problems for uncertain fractional differential equations. Chaos Solitons Fract. **157**, 111958 (2022)
9. Li, S., Shu, L., Shu, X.-B., Fei, X.: Existence and Hyers-Ulam stability of random impulsive stochastic functional differential equations with finite delays. Stochastics **91**(6), 857–872 (2019)

10. Logeswari, K., Ravichandran, C., Nisar, K.S.: Mathematical model for spreading of COVID-19 virus with the Mittag-Leffler kernel. Numer. Methods Partial Diff. Equ. (2020). https://doi.org/10.1002/num.22652
11. Nisar, K.S., Logeswari, K., Vijayaraj, V., Baskonus, H.M., Ravichandran, C.: Fractional order modeling the gemini virus in capsicum annuum with optimal control. Fract. Fract. (2022). https://www.mdpi.com/2504-3110/6/2/61
12. Saadi, C., Lakhal, H., Slimani, K., Dob, S.: Existence and uniqueness of distributional solution for semilinear fractional elliptic equation involving new operator and some numerical results. Math. Methods Appl. Sci. **45**(7), 3843–3854 (2022)
13. Sultana, F., Singh, D., Pandey, R.K., Zeidan, D.: Numerical schemes for a class of tempered fractional integro-differential equations. Appl. Numer. Math. **157**, 110–134 (2020)
14. Shiria, B., Wua, G.-C., Baleanu, D.: Collocation methods for terminal value problems of tempered fractional differential equations. Appl. Numer. Math. **156**, 385–395 (2020)
15. Verma, P., Kumar, M.: Analytical solution with existence and uniqueness conditions of nonlinear initial value multi-order fractional differential equations using Caputo derivative. Eng. Comput. (2020). https://doi.org/10.1007/s00366-020-01061-4
16. Verma, P., Kumar, M.: An analytical solution with existence and uniqueness conditions for fractional integro differential equations. Int. J. Model. Simul. Sci. Comput. (2020). https://doi.org/10.1142/S1793962320500452
17. Verma, P., Kumar, M.: New existence, uniqueness results for multi-dimensional multi-term Caputo time-fractional mixed sub-diffusion and diffusion-wave equation on convex domains. J. Appl. Anal. Comput. **11**(3), 1–26 (2021)
18. Verma, P., Kumar, M.: Exact solution with existence and uniqueness conditions for multi-dimensional time-space tempered fractional diffusion-wave equation. Eng. Comput. **38**, 271–281 (2022)
19. Wang, Z.-B., Liu, D.-Y., Boutat, D.: Algebraic estimation for fractional integrals of noisy acceleration based on the behaviour of fractional derivatives at zero. Appl. Math. Comput. **430**, 127254 (2022)
20. Karatas, E., AliAkgul, A., Yavuz, M.: New illustrative applications of integral transforms to financial models with different fractional derivatives. Chaos Solitons Fract. **146**, 110877 (2021)
21. Yuan, L., Zheng, S., Wei, Z.: Comparison theorems of tempered fractional differential equations. European Phys. J. Spec. Top. (2022). https://doi.org/10.1140/epjs/s11734-022-00486-w
22. Zaky, M.A.: Existence, uniqueness and numerical analysis of solutions of tempered fractional boundary value problems. Appl. Numer. Math. **145**, 429–457 (2019)

Chapter 14
Existence and Stability of Best Proximity Point Sets for a New Type of Multivalued Generalized F-Contraction Mappings in Metric Spaces

Raju Roy and Nilakshi Goswami

Abstract Let A and B be two nonempty subsets of a metric space (X, d) and $CB(X)$ denote the set of all nonempty closed and bounded subsets of X. Considering the multivalued mappings $S : A \longrightarrow CB(B)$ and $\phi : CB(B) \longrightarrow CB(A)$, we introduce multivalued generalized ϕ-F-contraction mappings. We establish the existence of the best proximity point for such types of mappings in complete metric space. Moreover, we define multivalued generalized ϕ-F-contraction pair of mappings and obtain best proximity point results for these. A stability result is obtained for best proximity point sets, and an analogous study is done for fixed point sets via fixed point formulation for such mappings. Suitable examples are provided in support of defined concepts.

Keywords Best proximity point · Contraction mapping

14.1 Introduction

The study of multivalued mappings has attracted many authors for its significant applications in various fields such as optimal control problems, differential inclusion, game theory, and real and complex analysis. Amongst the plenty of theories concerning multivalued mappings, proximity point theory is an important one in set-valued analysis. In 1969, Nadler [21] first introduced the concept of the Banach contraction principle based on multivalued mappings. Following Nadler's work on fixed point theory for multivalued mappings, many authors carried out work on fixed point as well as best proximity point theory for multivalued mappings (refer to [1, 4, 11, 18], etc.).

Let (X, d) be a metric space and A, B be two subsets of X. Let $CB(B)$ denote the set of all nonempty closed and bounded subsets of B. A point u is said to be a fixed point of a multivalued mapping $T : A \longrightarrow CB(B)$, if $u \in Tu$. If $A \cap B = \emptyset$,

R. Roy · N. Goswami (✉)
Department of Mathematics, Gauhati University, Guwahati 781014, Assam, India
e-mail: nila_g2003@yahoo.co.in

B. Hazarika et al. (eds.), *Advances in Functional Analysis and Fixed-Point Theory*, Industrial and Applied Mathematics,
https://doi.org/10.1007/978-981-99-9207-2_14

then the mapping T does not have any fixed point. In this case, one can find out an element u so that the distance between u and Tu will be optimal. The concept of best proximity point comes from this optimality. This idea was carried forward to study proximity points, and the subsequent introduction of the approximation theorem was done by Fan in [14]. Since then, many researchers have extended his work in many directions (refer to [3, 6, 15, 16, 26], etc.).

Motivated by this, in this work, we define multivalued generalized ϕ-F-contraction mapping and derive some best proximity point results in complete metric spaces. Moreover, the stability of best proximity point sets is discussed for the defined type of mappings.

14.1.1 Preliminaries

In 2012, Wardowski [29] defined F-contraction which generalizes the Banach contraction principle [8]. Following that, several researchers have explored the generalization of F-contraction mappings under various circumstances (refer to [17, 22, 23]).

Definition 14.1 [29] Let $F : \mathbb{R}^+ \longrightarrow \mathbb{R}$ be a mapping satisfying the following conditions:

(F1) F is strictly increasing,

(F2) for each sequence of positive numbers $\{u_n\}$,
$\lim_{n\to\infty} u_n = 0$ if and only if $\lim_{n\to\infty} F(u_n) = -\infty$,

(F3) there exists a number $k \in (0, 1)$ such that $\lim_{y\to 0^+} y^k F(y) = 0$.

Let $\mathscr{F}$ denote the set of all functions F satisfying (F1), (F2) and (F3). Some examples of F are

$$F(u) = \ln u, \ \ F(u) = \ln u + u.$$

For a metric space (X, d), a self-mapping $T : X \longrightarrow X$ is called F-contraction if there exists $\lambda > 0$ and $F \in \mathscr{F}$ such that

$$d(Tu, Tv) > 0 \text{ implies } \lambda + F(d(Tu, Tv)) \leq F(d(u, v)) \text{ for all } u, v \in X.$$

For a metric space (X, d), a mapping $H : CB(X) \times CB(X) \longrightarrow [0, \infty)$ defined by

$$H(A, B) = \max\{\sup_{u\in A} d(u, B), \sup_{v\in B} d(v, A)\} \text{ for all } A, B \in CB(X)$$

is known as Pompeiu-Hausdorff metric (refer to [21]).

Following the work of Wardowski [29] and Nadler [21], Altun et al. in [7] defined multivalued F-contraction considering Hausdorff metric.

Definition 14.2 [7] For a metric space (X, d), a mapping $T : X \longrightarrow CB(X)$ is called multivalued F-contraction if there exists $F \in \mathscr{F}$ and $\lambda > 0$ such that

$$H(Tu, Tv) > 0 \text{ implies } \lambda + F(H(Tu, Tv)) \leq F(d(u, v)) \text{ for all } u, v \in X.$$

Recently, Ali et al. [5] introduced the notion of α-F-contraction of Hardy-Roger-type multivalued mappings combining α-ψ-contraction [24] and F-contraction.

Definition 14.3 [5] For a metric space (X, d), a mapping $T : X \longrightarrow CB(X)$ is called α-F-contraction of Hardy-Roger-type if there exists $F \in \mathscr{F}$, $\lambda > 0$ and a mapping $\alpha : X \times X \longrightarrow [0, \infty)$ such that

$$\min\{\alpha(u, v)H(Tu, Tv), N(u, v)\} > 0 \text{ implies } \lambda + F(\alpha(u, v)H(Tu, Tv)) \leq F(N(u, v))$$

for all $u, v \in X$, where

$$N(u, v) = b_1 d(u, v) + b_2 d(u, Tu) + b_3 d(v, Tv) + b_4 d(u, Tv) + Ld(v, Tu),$$

with $b_1, b_2, b_3, b_4, L \geq 0$ satisfying $b_1 + b_2 + b_3 + 2b_4 = 1$ and $b_3 \neq 1$.

For two nonempty subsets A and B of a metric space (X, d), the following notations will be useful:

$$A_0 = \{u \in A : d(u, v) = d(A, B) \text{ for some } v \in B\},$$

$$B_0 = \{v \in B : d(u, v) = d(A, B) \text{ for some } u \in A\},$$

where $d(A, B) = \inf\{d(u, v) : u \in A, v \in B\}$ (refer to [6]).

Definition 14.4 [12] For two nonempty subsets A and B of a metric space (X, d), a point $u \in A$ is called a best proximity point of a multivalued mapping $T : A \longrightarrow CB(B)$ if $d(u, Tu) = d(A, B)$.

A point $u \in A$ is called a fixed point of a multivalued mapping $T : A \longrightarrow CB(A)$ if $u \in Tu$.

In 2011, Shankar Raj investigated fixed point results for weakly contractive mappings using P-property.

Definition 14.5 [25] Let A and B be two nonempty subsets of a metric space (X, d) with $A_0 \neq \emptyset$. Then the pair (A, B) is said to satisfy P-property if and only if

$$d(u_1, v_1) = d(A, B) = d(u_2, v_2) \text{ implies } d(u_1, u_2) = d(v_1, v_2),$$

where $u_1, u_2 \in A_0$ and $v_1, v_2 \in B_0$.

For some examples of P-property, one can follow [2].

To obtain an improvement of the best proximity point result of Theorem 3.1 of [10] in a simple way, Zhang et al. [30] introduced the notion of weak P-property.

Definition 14.6 [30] The pair (A, B) is said to satisfy the weak P-property if and only if

$$d(u_1, v_1) = d(A, B) = d(u_2, v_2) \text{ implies } d(u_1, u_2) \leq d(v_1, v_2),$$

where $u_1, u_2 \in A_0$ and $v_1, v_2 \in B_0$.

It is obvious that P-property of (A, B) implies the weak P-property, but the converse is not true.

To extend some results of Eldred et al. [13] in a metric space, Suzuki et al. [27] introduced the concept of UC property.

Definition 14.7 [27] For two nonempty subsets A and B of a metric space (X, d), the pair (A, B) is said to satisfy UC property if for sequences $\{x_n\}$, $\{x'_n\}$ in A and $\{y_n\}$ in B,

$$\lim_{n\to\infty} d(x_n, y_n) = d(A, B) = \lim_{n\to\infty} d(x'_n, y_n) \text{ implies that } \lim_{n\to\infty} d(x_n, x'_n) = 0.$$

For two nonempty subsets A and B of (X, d) if $d(A, B) = 0$, it is evident that the pair (A, B) satisfies the UC property.

The following lemmas are useful to obtain the best proximity point results and stability of best proximity point sets.

Lemma 14.1 [9] *Let A and B be two nonempty subsets of $CB(X)$. Then*

(i) $d(u, B) \leq d(u, v)$ for all $u \in A$ and $v \in B$,
(i) $d(u, B) \leq H(A, B)$ for all $u \in A$.

Lemma 14.2 [11] *Let A and B be two nonempty compact subsets of a metric space (X, d). Let $q \geq 1$. Then for every $u \in A$ there exists $v \in B$ such that*

$$d(u, v) \leq qH(A, B).$$

Lemma 14.3 [11] *Let A and B be two nonempty compact subsets of a metric space (X, d) and $T : A \longrightarrow CB(B)$ be a multivalued mapping. Let $q \geq 1$. Then for $u, v \in A$ and $x \in Tu$, there exists $y \in Tv$ such that*

$$d(x, y) \leq qH(Tu, Tv).$$

14.2 Main Results

We define the following new type of multivalued generalized F-contraction mappings.

Definition 14.8 Let A and B be two nonempty subsets of a metric space (X, d) and $S : A \longrightarrow CB(B)$ and $\phi : CB(B) \longrightarrow CB(A)$ be two multivalued mappings. The mapping S is said to be multivalued generalized ϕ-F-contraction mapping if there exists $\lambda > 0$ such that

$$H(Su, Sv) > 0 \text{ implies } \lambda + F(H(Su, Sv)) \le F(M(u, v)) \text{ for all } u, v \in A, \tag{14.1}$$

where $M(u, v) = \max\{d^*(u, v), d^*(u, Su), d^*(v, Sv), \frac{d^*(u,Sv)+d^*(v,Su)}{2},$
$\frac{d^*(\phi Su,u)+d^*(\phi Su,v)}{2}, \frac{d^*(\phi Su,Su)+d^*(\phi Su,Sv)}{2}\}$
with $d^*(x, y) = d(x, y)$ if $x, y \in A$ or $x, y \in B$; $d^*(x, y) = d(x, y) - d(A, B)$ if $x \in A$, $y \in B$; $d^*(P, Q) = d(P, Q)$ if $P, Q \subseteq A$ or $P, Q \subseteq B$ and $d^*(P, Q) = d(P, Q) - d(A, B)$ if $P \subseteq A$ and $Q \subseteq B$.

The notations $d^*(x, P)$, $d^*(x, Q)$, $d^*(P, y)$ and $d^*(Q, y)$ where $x \in A$, $y \in B$ and $P \subseteq A$, $Q \subseteq B$ can be defined analogously.

Example 14.1 Consider $X = \mathbf{R}^2$ with metric $d((u_1, u_2), (v_1, v_2)) = |u_1 - v_1| + |u_2 - v_2|$ for all $(u_1, u_2), (v_1, v_2) \in X$. Let $A = \{(u, 1) : u \in \mathbf{R}\}$ and $B = \{(u, 2) : u \in \mathbf{R}\}$. Then $d(A, B) = 1$. Define the mappings $S : A \longrightarrow CB(B)$ and $\phi : CB(B) \longrightarrow CB(A)$ by

$$S(u, 1) = \begin{cases} \{(0, 2)\} & \text{if } u \le 0, \\ \{(\frac{1}{2}, 2), (0, 2)\} & \text{if } 0 < u \le 10, \\ \{(1, 2)\} & \text{if } u > 10 \end{cases}$$

and

$$\phi(Y) = \{(5, 1)\} \text{ for all } Y \subseteq B.$$

Now, for $u \le 0$ and $v \le 0$ $(u \ne v)$,

$$H(S(u, 1), S(v, 1)) = 0.$$

For $u \le 0$ and $0 < v \le 10$,

$$H(S(u, 1), S(v, 1)) = H(\{(0, 2)\}, \{(\tfrac{1}{2}, 2), (0, 2)\}) = \tfrac{1}{2}$$

and

$$\begin{aligned} &\tfrac{1}{2}[d^*(\phi S(u, 1), S(u, 1)) + d^*(\phi S(u, 1), S(v, 1))] \\ &= \tfrac{1}{2}[d^*(\{(5, 1)\}, \{(0, 2)\}) + d^*(\{(5, 1)\}, \{(\tfrac{1}{2}, 2), (0, 2)\})] \\ &= \tfrac{19}{4}. \end{aligned}$$

For $u \le 0$ and $v > 10$,

$$H(S(u, 1), S(v, 1)) = H(\{(0, 2)\}, \{(1, 2)\}) = 1$$

and

$$\begin{aligned} &\tfrac{1}{2}[d^*(\phi S(u, 1), S(u, 1)) + d^*(\phi S(u, 1), S(v, 1))] \\ &= \tfrac{1}{2}[d^*(\{(5, 1)\}, \{(0, 2)\}) + d^*(\{(5, 1)\}, \{(1, 2)\})] \\ &= \tfrac{9}{2}. \end{aligned}$$

In all the cases, it is found that $H(S(u, 1), S(v, 1)) < M(u, v)$. Thus S is generalized multivalued ϕ-F-contraction for $F(u) = \log u$ and $\lambda = \log \frac{6}{5}$.

Now we define multivalued generalized ϕ-F-contraction pair of mappings.

Definition 14.9 Let A and B be two nonempty subsets of metric space (X, d) and $S : A \longrightarrow CB(B)$, $T : B \longrightarrow CB(A)$ and $\phi : CB(B) \longrightarrow CB(A)$ be multivalued mappings. The pair of mappings (S, T) is said to be multivalued generalized ϕ-F-contraction pair if there exists $\lambda > 0$ such that

$$H(Su, Tv) > 0 \text{ implies } \lambda + F(H(Su, Tv)) \leq F(M(u, v)) \text{ for all } u \in A \text{ and } v \in B, \tag{14.2}$$

where $M(u, v) = \max\{d^*(u, v), d^*(u, Su), d^*(v, Tv), \frac{d^*(u,Tv)+d^*(v,Su)}{2},$

$$\frac{d^*(\phi Su,u)+d^*(\phi Su,v)}{2}, \frac{d^*(\phi Su,Su)+d^*(\phi Su,Tv)}{2}\}.$$

Example 14.2 Consider $X = \mathbb{R}$ with usual metric $d(u, v) = |u - v|$ for all $u, v \in X$. Let A and B be two nonempty subsets of X such that $A = [1, 2]$, $B = [-2, -1] \cup \{0, \frac{1}{2}\}$. Then $d(A, B) = \frac{1}{2}$. Define the mappings $S : A \longrightarrow CB(B)$, $T : B \longrightarrow CB(A)$ and $\phi : CB(B) \longrightarrow CB(A)$ by

$$Su = \begin{cases} \{\frac{1}{2}\} & \text{if } u \in [1, 2), \\ \{0\} & \text{if } u = 2, \end{cases}$$

$$Tv = \begin{cases} \{1\} & \text{if } v \in (-2, -1) \cup \{0, \frac{1}{2}\}, \\ \{1, \frac{6}{5}\} & \text{if } v = -1, \\ \{\frac{5}{4}\} & \text{if } v = -2, \end{cases}$$

and

$$\phi(Y) = \{2\} \text{ for all } Y \subseteq B.$$

Now, for $u \in [1, 2)$ and $v \in (-2, -1) \cup \{0, \frac{1}{2}\}$,

$$H(Su, Tv) = H(\{\frac{1}{2}\}, \{1\}) = \frac{1}{2}$$

and

$$\frac{1}{2}[d^*(\phi Su, Su) + d^*(\phi Su, Tv)] = \frac{1}{2}[d^*(\{2\}, \frac{1}{2}) + d^*(\{2\}, \{1\})] \\ = \frac{3}{4}.$$

For $u \in [1, 2)$ and $v = -2$,

$$H(Su, Tv) = H(\{\frac{1}{2}\}, \{\frac{5}{4}\}) = \frac{3}{4}$$

and

$$d^*(-2, T(-2)) = d^*(-2, \{\frac{5}{4}\}) = \frac{11}{4}.$$

For $u \in [1, 2)$ and $v = -1$,

$$H(Su, Tv) = H(\{\frac{1}{2}\}, \{1, \frac{6}{5}\}) = \frac{7}{10}$$

and

$$d^*(-1, T(-1)) = d^*(-1, \{1, \frac{6}{5}\}) = \frac{3}{2}.$$

In all three cases, it is found that $H(Su, Tv) < M(u, v)$. Similarly, $H(Su, Tv) < M(u, v)$ holds for other cases. Thus, the pair (S, T) is a multivalued generalized ϕ-F-contraction pair for $F(u) = \log u$ and $\lambda = \log \frac{11}{10}$.

In the next result, we show the existence of the best proximity point for multivalued generalized ϕ-F-contraction mapping under certain conditions.

Theorem 14.1 *Let A and B be two nonempty closed subsets of a complete metric space (X, d) and S be a multivalued generalized ϕ-F-contraction mapping. If the following conditions are satisfied:*
(i) (A, B) satisfies weak P-property,
(ii) S is continuous and $Su_0 \subseteq B_0$ for all $u_0 \in A_0$,
(iii) $d(\phi Su, v) \leq d(Su, v)$ for all $u, v \in A$,
then the mapping S has a best proximity point in A.

Proof Let $u_0 \in A_0$. Since $Su_0 \subseteq B_0$, there exists $v_0 \in Su_0 \subseteq B_0$. By definition of B_0, we can choose $u_1 \in A_0$ such that

$$d(u_1, v_0) = d(A, B). \tag{14.3}$$

If $v_0 \in Su_1 \subseteq B_0$, then

$$d(A, B) \leq d(u_1, Su_1) \leq d(u_1, v_0) = d(A, B),$$

which implies that u_1 is a best proximity of S. Let $v_0 \notin Su_1$. Then there exists $v_1 \in Su_1$ such that

$$d(v_0, v_1) \leq H(Su_0, Su_1).$$

By (F1),

$$\begin{aligned} F(d(v_0, v_1)) &\leq F(H(Su_0, Su_1)) \\ &< \lambda + F(H(Su_0, Su_1)). \end{aligned}$$

Since S is a multivalued generalized ϕ-F-contraction, so,

$$\begin{aligned} F(d(v_0, v_1)) &\leq F(H(Su_0, Su_1)) \\ &\leq F(M(u_0, u_1)) - \lambda, \end{aligned} \tag{14.4}$$

where

$$
\begin{aligned}
M(u_0, u_1) = \max\{&d^*(u_0, u_1), d^*(u_0, Su_0), d^*(u_1, Su_1), \frac{d^*(u_0, Su_1) + d^*(u_1, Su_0)}{2},\\
&\frac{d^*(\phi Su_0, u_0) + d^*(\phi Su_0, u_1)}{2}, \frac{d^*(\phi Su_0, Su_0) + d^*(\phi Su_0, Su_1)}{2}\}\\
\leq \max\{&d^*(u_0, u_1), d^*(u_0, v_0), d^*(u_1, v_1), \frac{d^*(u_0, v_1) + d^*(u_1, v_0)}{2},\\
&\frac{d^*(Su_0, u_0) + d^*(Su_0, u_1)}{2}, \frac{d^*(Su_0, v_0) + d^*(Su_0, v_1)}{2}\}\\
\leq \max\{&d^*(u_0, u_1), d^*(u_0, v_0), d^*(u_1, v_1), \frac{d^*(u_0, v_1) + d^*(u_1, v_0)}{2},\\
&\frac{d^*(v_0, u_0) + d^*(v_0, u_1)}{2}, \frac{d^*(v_0, v_0) + d^*(v_0, v_1)}{2}\}\\
\leq \max\{&d^*(u_0, u_1), d^*(u_0, v_0), d^*(u_1, v_1), \frac{d^*(u_0, v_1)}{2}, d^*(v_0, v_1)\}\\
\leq \max\{&d^*(u_0, u_1), d^*(u_0, u_1) + d^*(u_1, v_0), d^*(u_1, v_0) + d^*(v_0, v_1),\\
&\frac{d^*(u_0, u_1) + d^*(u_1, v_1)}{2}, d^*(v_0, v_1)\}\\
\leq \max\{&d^*(u_0, u_1), d^*(v_0, v_1)\}.
\end{aligned}
$$

If $\max\{d^*(u_0, u_1), d^*(v_0, v_1)\} = d^*(v_0, v_1)$, then from (14.4),

$$F(d(v_0, v_1)) < F(d^*(v_0, v_1)),$$

which is a contradiction.
Therefore, $\max\{d^*(u_0, u_1), d^*(v_0, v_1)\} = d^*(u_0, u_1)$.
Thus,

$$F(d(v_0, v_1)) \leq F(d^*(u_0, u_1)) - \lambda. \tag{14.5}$$

Since $v_1 \in Su_1 \subseteq B_0$, there exists $u_2 \in A_0$ such that

$$d(u_2, v_1) = d(A, B). \tag{14.6}$$

If $v_1 \in Su_2$, then u_2 will be a best proximity point of S. Let $v_1 \notin Su_2$. Then there exists $v_2 \in Su_2$ such that

$$d(v_1, v_2) \leq H(Su_1, Su_2).$$

Following the previous step, we get

$$
\begin{aligned}
F(d(v_1, v_2)) &\leq F(d^*(u_1, u_2)) - \lambda\\
&= F(d(u_1, u_2)) - \lambda.
\end{aligned}\tag{14.7}
$$

Since (A, B) satisfies weak P-property,

$$d(u_2, v_1) = d(A, B) = d(u_1, v_0)$$

$$\text{implies } d(u_1, u_2) \leq d(v_0, v_1).$$

So using (14.7),

$$\begin{aligned} F(d(v_1, v_2)) &\leq F(d^*(u_1, u_2)) - \lambda \\ &= F(d(u_1, u_2)) - \lambda \\ &\leq F(d(u_0, u_1)) - 2\lambda, \text{ (using 14.5)}. \end{aligned}$$

Continuing the same process, we get two sequences $\{u_n\}$ and $\{v_n\}$ such that

$$v_n \in Su_n,$$

$$d(u_{n+1}, v_n) = d(A, B)$$

and

$$F(d(u_n, u_{n+1})) \leq F(d(v_{n-1}, v_n)) \leq F(d(u_0, u_1)) - n\lambda. \tag{14.8}$$

Taking limit as $n \to \infty$ on both sides of (14.8), we have

$$\lim_{n\to\infty} F(d(u_n, u_{n+1})) = -\infty$$

$$\Rightarrow \lim_{n\to\infty} d(u_n, u_{n+1}) = 0.$$

Let $\beta_n = d(u_n, u_{n+1})$ for each $n \in \mathbb{N}$. Then by (F_3), we have $k \in (0, 1)$ such that

$$\lim_{n\to\infty} \beta_n^k F(\beta_n) = 0.$$

From (14.8), we have

$$\beta_n^k F(d(u_n, u_{n+1})) - \beta_n^k F(d(u_0, u_1)) \leq -\beta_n^k n\lambda \leq 0 \text{ for each } n \in \mathbb{N}. \tag{14.9}$$

Taking limit as $n \to \infty$ on both sides of (14.9), we have

$$\lim_{n\to\infty} \beta_n^k n = 0.$$

Therefore, there exists $k_1 \in \mathbb{N}$ such that

$$\beta_n^k n \leq 1 \text{ for each } n \geq k_1,$$

i.e.,

$$\beta_n \leq \frac{1}{n^k} \text{ for each } n \geq k_1.$$

To show that $\{u_n\}$ is a Cauchy sequence, let $m, n \in \mathbb{N}$ with $m > n > k_1$. Now,

$$\begin{aligned} d(u_n, u_m) &\leq d(u_n, u_{n+1}) + d(u_{n+1}, u_{n+2}) + \cdots + d(u_{m-1}, u_m) \\ &= \sum_{i=n}^{m-1} \beta_i \\ &\leq \sum_{i=n}^{\infty} \beta_i \\ &\leq \sum_{i=n}^{\infty} \frac{1}{i^k}. \end{aligned}$$

Thus, $d(u_n, u_m) \to 0$ as $n \to \infty$. This shows that $\{u_n\}$ is a Cauchy sequence in A. Since A is closed, there exists an element $u \in A$ such that

$$u_n \to u \text{ as } n \to \infty.$$

Now,

$$d(u_{n+1}, Su_n) \leq d(u_{n+1}, v_n) = d(A, B). \tag{14.10}$$

Taking limit as $n \to \infty$ on both sides of (14.10) and using continuity of S, we get

$$d(u, Su) = d(A, B).$$

Hence, u is a best proximity point of S. □

It is seen that analogous result holds in case of multivalued generalized ϕ-F-contraction pair of mappings using UC property of (A, B).

Theorem 14.2 *Let A and B be two nonempty closed subsets of a complete metric space (X, d) and (S, T) be a multivalued generalized ϕ-F-contraction pair of mappings. If the following conditions are satisfied:*

(i) (A, B) and (B, A) satisfy UC property,

(ii) S and T are continuous,

(iii) $d(\phi Sx, x) \leq d(Sx, x)$ for all $x \in A$,

then there exists $u \in A$ and $v \in B$ such that $d(u, v) = d(A, B)$ and $d(u, Su) = d(A, B) = d(v, Tv)$.

Proof Let $u_0 \in A$. Since $Su_0 \subseteq B$, we can choose $u_1 \in Su_0$. Since $Su_0 \neq Tu_1$ then there exists $u_2 \in Tu_1$ such that

$$d^*(u_1, u_2) \leq H(Su_0, Tu_1).$$

Using (F1),

$$\begin{aligned} F(d^*(u_1, u_2)) &\leq F(H(Su_0, Tu_1)) \\ &< \lambda + F(H(Su_0, Tu_1)). \end{aligned} \tag{14.11}$$

From the definition of multivalued generalized ϕ-F-contraction pair, we get

$$\lambda + F(H(Su_0, Tu_1)) < F(M(u_0, u_1)), \tag{14.12}$$

where

$$\begin{aligned} M(u_0, u_1) &= \max\{d^*(u_0, u_1), d^*(u_0, Su_0), d^*(u_1, Tu_1), \frac{d^*(u_1, Su_0) + d^*(u_0, Tu_1)}{2}, \\ &\quad \frac{d^*(\phi Su_0, u_0) + d^*(\phi Su_0, u_1)}{2}, \frac{d^*(\phi Su_0, Su_0) + d^*(\phi Su_0, Tu_1)}{2}\} \\ &\leq \max\{d^*(u_0, u_1), d^*(u_1, u_2), \frac{d^*(u_0, u_2)}{2}, \frac{d^*(u_1, u_0)}{2}, \frac{d^*(u_1, u_2)}{2}\} \\ &\leq \max\{d^*(u_0, u_1), d^*(u_1, u_2), \frac{d^*(u_0, u_2)}{2}\} \\ &= \max\{d^*(u_0, u_1), d^*(u_1, u_2)\}. \end{aligned}$$

If $\max\{d^*(u_0, u_1), d^*(u_1, u_2)\} = d^*(u_1, u_2)$, then using (14.11) and (14.12),

$$F(d^*(u_1, u_2)) < F(M(u_0, u_1)) \leq F(d^*(u_1, u_2)),$$

which is a contradiction.
Therefore, $\max\{d^*(u_0, u_1), d^*(u_1, u_2)\} = d^*(u_0, u_1)$.
From (14.12), we have

$$\lambda + F(H(Su_0, Tu_1)) \leq F(M(u_0, u_1)) \leq F(d^*(u_0, u_1))$$

$$\Rightarrow F(H(Su_0, Tu_1)) \leq F(d^*(u_0, u_1)) - \lambda.$$

Thus,

$$F(d^*(u_1, u_2)) \le F(d^*(u_0, u_1)) - \lambda. \tag{14.13}$$

Again, since $Su_2 \neq Tu_1$, there exists $u_3 \in Su_2$ such that

$$\begin{aligned} F(d^*(u_2, u_3)) &\le F(d^*(u_1, u_2)) - \lambda \\ &\le F(d^*(u_0, u_1)) - 2\lambda. \end{aligned} \tag{14.14}$$

Continuing in the same way, we will get two sequences $\{u_{2n}\}$ and $\{u_{2n+1}\}$ such that

$$u_{2n+1} \in Su_{2n}$$

and

$$F(d^*(u_{2n}, u_{2n+1})) < F(d^*(u_0, u_1)) - 2n\lambda. \tag{14.15}$$

Taking limit on both sides of (14.15), we get

$$\lim_{n\to\infty} F(d^*(u_{2n}, u_{2n+1})) = -\infty$$

$$\Rightarrow \lim_{n\to\infty} d^*(u_{2n}, u_{2n+1}) = 0$$

$$\Rightarrow \lim_{n\to\infty} d(u_{2n}, u_{2n+1}) = d(A, B).$$

Since (A, B) and (B, A) satisfy UC property,

$$\lim_{n\to\infty} d(u_{2n}, u_{2n+2}) = 0 = \lim_{n\to\infty} d(u_{2n-1}, u_{2n+1}). \tag{14.16}$$

To show the sequence $\{u_{2n}\}$ is a Cauchy sequence in A, let $m, n \in \mathbb{N}$ with $m > n \geq K_1$.
Now,

$$d(u_{2n}, u_{2m}) \le d(u_{2n}, u_{2n+2}) + d(u_{2n+2}, u_{2n+4}) + \cdots + d(u_{2m-2}, u_{2m}). \tag{14.17}$$

Taking limit as $n \to \infty$ on both sides of (14.17), we get

$$\lim_{n\to\infty} d(u_{2n}, u_{2m}) = 0.$$

Therefore, the sequence $\{u_{2n}\}$ is a Cauchy sequence in A. Similarly, it can be shown that $\{u_{2n+1}\}$ is a Cauchy sequence in B. Since A and B are closed, there exist $u \in A$ and $v \in B$ such that

$$u_{2n} \longrightarrow u \text{ and } u_{2n+1} \longrightarrow v.$$

Therefore, $\lim_{n\to\infty} d(u_{2n}, u_{2n+1}) = d(u, v) = d(A, B)$.
Now, using continuity of S,

$$\begin{aligned} d(A, B) \le d(u, Su) &= \lim_{n\to\infty} d(u_{2n}, Su_{2n}) \\ &\le \lim_{n\to\infty} d(u_{2n}, u_{2n+1}) \\ &= d(A, B). \end{aligned}$$

Hence, $d(u, Su) = d(A, B)$, i.e., u is a best proximity point of S.
Similarly, $d(v, Tv) = d(A, B)$. □

In the following theorem, we show the existence of best proximity point of S and T without assuming continuity property of S and T.

Theorem 14.3 *Let A and B be two nonempty closed subsets of a complete metric space (X, d) with $d(A, B) > 0$ and (S, T) be a multivalued generalized ϕ-F-contraction pair of mappings. If the following conditions are satisfied:*

(i) (A, B) and (B, A) satisfy UC property,

(ii) $\max\{d(\phi Sx, x), d(\phi Sx, Sx)\} \le d(Sx, x)$ and

$d(\phi Sx, Ty) \le d(Sx, y)$ for all $x \in A, y \in B$,

then there exists $u \in A$ and $v \in B$ such that $d(u, v) = d(A, B)$ and $d(u, Su) = d(A, B) = d(v, Tv)$.

Proof As in the proof of Theorem 14.2, we get two Cauchy sequences $\{u_{2n}\}$ and $\{u_{2n+1}\}$ in A and B respectively with $u \in A$, $v \in B$ such that

$$u_{2n} \longrightarrow u,\ u_{2n+1} \longrightarrow v \text{ and } d(u, v) = d(A, B).$$

Furthermore,

$$d(u_{2n+1}, Tv) \le H(Su_{2n}, Tv). \tag{14.18}$$

Using (F1), we get

$$F(d(u_{2n+1}, Tv)) \le F(H(Su_{2n}, Tv)),$$

and so,

$$F(d(u_{2n+1}, Tv)) < \lambda + F(H(Su_{2n}, Tv)).$$

Now, from the definition of multivalued generalized ϕ-F-contraction pair of mappings, we have

$$\lambda + F(H(Su_{2n}, Tv)) \leq F(M(u_{2n}, v)).$$

Hence,

$$F(d(u_{2n+1}, Tv)) < F(M(u_{2n}, v)), \tag{14.19}$$

$$i.e.,\ d(u_{2n+1}, Tv) < M(u_{2n}, v), \tag{14.20}$$

where

$$\begin{aligned} M(u_{2n}, v) &= \max\{d^*(u_{2n}, v), d^*(u_{2n}, Su_{2n}), \\ &\quad d^*(v, Tv), \frac{d^*(u_{2n}, Tv) + d^*(v, Su_{2n})}{2}, \\ &\quad \frac{d^*(\phi Su_{2n}, u_{2n}) + d^*(\phi Su_{2n}, v)}{2}, \\ &\quad \frac{d^*(\phi Su_{2n}, Su_{2n}) + d^*(\phi Su_{2n}, Tv)}{2}\} \\ &\leq \max\{d^*(u_{2n}, v), d^*(u_{2n}, u_{2n+1}), d^*(v, Tv), \\ &\quad \frac{d^*(u_{2n}, Tv) + d^*(v, u_{2n+1})}{2}, \\ &\quad \frac{d^*(u_{2n+1}, u_{2n}) + d^*(u_{2n+1}, v)}{2}, \\ &\quad \frac{d^*(u_{2n+1}, u_{2n+1}) + d^*(u_{2n+1}, v)}{2}\}. \end{aligned}$$

$$\begin{aligned} \text{So,} \quad \lim_{n\to\infty} M(u_{2n}, v) &\leq \max\{d^*(u, v), d^*(v, Tv), \frac{d^*(u, Tv) + d^*(v, v)}{2}, d^*(v, v)\} \\ &\leq \max\{d^*(u, v), d^*(v, Tv)\}. \end{aligned}$$

If $\max\{d^*(u, v), d^*(v, Tv)\} = d^*(v, Tv)$,
then taking limit on both sides of (14.20), we get

$$d(v, Tv) \leq d^*(v, Tv) < d(v, Tv),$$

which is a contradiction (since $d(A, B) > 0$). So,

$$\max\{d^*(u, v), d^*(v, Tv)\} = d^*(u, v) = 0.$$

Therefore, $d^*(v, Tv) = 0$, i.e., $d(v, Tv) = d(A, B)$. Thus, v is a best proximity point of T. Similarly, it can be shown that u is a best proximity point of S. □

Now we derive the fixed point formulation for multivalued generalized ϕ-F-contraction mapping. For this, we take $A = B = X$ and $S = T$ in Definition (14.9).

Theorem 14.4 *Let (X, d) be a complete metric space and $\phi : CB(X) \longrightarrow CB(X)$, $S : X \longrightarrow CB(X)$ be multivalued mappings such that S is a multivalued generalized*

ϕ-F-contraction mapping satisfying the following conditions:

(i) S is continuous,

(ii) $d(\phi Su, v) \leq d(Su, v)$ for all $u, v \in X$.

Then the mapping S has a fixed point in X.

Proof Suppose $u_0 \in X$. Since Su_0 is nonempty, choose $u_1 \in Su_0$. If $u_1 \in Su_1$, then u_1 is a fixed point of S. Let $u_1 \notin Su_1$. Then there exists $u_2 \in Su_1$ such that

$$d(u_1, u_2) \leq H(Su_0, Su_1). \tag{14.21}$$

Using $(F1)$, we get

$$\begin{aligned} F(d(u_1, u_2)) &\leq F(H(Su_0, Su_1)) \\ &< \lambda + F(H(Su_0, Su_1)). \end{aligned} \tag{14.22}$$

Since S is multivalued generalized ϕ-F-contraction, we get

$$\lambda + F(H(Su_0, Su_1)) \leq F(M(u_0, u_1)), \tag{14.23}$$

where

$$\begin{aligned} M(u_0, u_1) &= \max\{d(u_0, u_1), d(u_0, Su_0), d(u_1, Su_1), \frac{d(u_0, Su_1) + d(u_1, Su_0)}{2}, \\ &\quad \frac{d(\phi Su_0, u_0) + d(\phi Su_0, u_1)}{2}, \frac{d(\phi Su_0, Su_0) + d(\phi Su_0, Su_1)}{2}\} \\ &\leq \max\{d(u_0, u_1), d(u_0, u_1), d(u_1, u_2), \frac{d(u_0, u_2) + d(u_1, u_1)}{2}, \\ &\quad \frac{d(u_1, u_0) + d(u_1, u_1)}{2}, \frac{d(u_1, u_1) + d(u_1, u_2)}{2}\} \\ &\leq \max\{d(u_0, u_1), d(u_1, u_2), \frac{d(u_0, u_1) + d(u_1, u_2)}{2}\} \\ &= \max\{d(u_0, u_1), d(u_1, u_2)\}. \end{aligned}$$

Now if $\max\{d(u_0, u_1), d(u_1, u_2)\} = d(u_1, u_2)$, then

$$F(d(u_1, u_2)) < F(M(u_0, u_1)) = F(d(u_1, u_2)),$$

which is contradiction. Therefore, $\max\{d(u_0, u_1), d(u_1, u_2)\} = d(u_0, u_1)$.

From (14.23), we have

$$\lambda + F(H(Su_0, Su_1)) \le F(M(u_0, u_1)) \tag{14.24}$$
$$\Rightarrow F(d(u_1, u_2)) \le F(d(u_0, u_1)) - \lambda. \tag{14.25}$$

Again, if $u_2 \notin Su_2$, then there exists $u_3 \in Su_2$ such that

$$F(d(u_2, u_3)) \le F(d(u_1, u_2)) - \lambda$$
$$\le F(d(u_0, u_1)) - 2\lambda.$$

Continuing this process, we will get a sequence $\{u_n\}$ such that

$$F(d(u_n, u_{n+1})) \le F(d(u_0, u_1)) - n\lambda. \tag{14.26}$$

Taking limit on both sides of (14.26), we get

$$\lim_{n\to\infty} F(d(u_n, u_{n+1})) = -\infty$$

$$\Rightarrow \lim_{n\to\infty} d(u_n, u_{n+1}) = 0. \tag{14.27}$$

As in the proof of Theorem 14.1, it follows that $\{u_n\}$ is a Cauchy sequence. X being complete, there exists $u \in X$ such that $u_n \longrightarrow u$ as $n \to \infty$. Now,

$$d(u_n, Su_n) \le d(u_n, u_{n+1}). \tag{14.28}$$

Taking limit on both sides of (14.28), we get

$$\lim_{n\to\infty} d(u_n, Su_n) = d(u, Su) = 0.$$

Hence, $u \in Su$ and so u is a fixed point of S. □

14.3 Stability of Best Proximity Point Sets

Nadler [21] in 1969 and Markin [20] in 1976 investigated the stability of fixed point sets for multivalued mappings assuming some stronger conditions. Lim [19] established the stability of fixed point sets for multivalued mappings removing two strong conditions of [20]. Extending these concepts, Abkar et al. [1] introduced the stability of proximity point sets for multivalued mappings. In this section, we discuss the stability of proximity point sets for the multivalued generalized ϕ-F-contraction mappings considering compact subsets.

Theorem 14.5 *Let A and B be two nonempty compact subsets of a complete metric space (X, d), $\phi : C(B) \longrightarrow C(A)$ be a mapping and $S_1, S_2 : A \longrightarrow C(B)$ be continuous multivalued generalized ϕ-F-contraction mappings satisfying the following conditions:*

(i) for each $u \in A_0$, $S_i u \subseteq B_0$, $(i = 1, 2)$,

(ii) (A, B) satisfies weak P-property,

(iii) $d(\phi S_i u, v) \leq d(S_i u, v)$ for all $u \in A, v \in B$, $(i = 1, 2)$,

(iv) for each closed ball $B(b_i, r_i)$ in A, there exists $p \in (0, 1)$ such that

$$d(b_i, S_i b_i) + d(A, B) \leq (1 - \sqrt{p})\frac{r_i}{2}, \ (i = 1, 2). \tag{14.29}$$

Then

$$H(B_{S_1}, B_{S_2}) \leq P + 2d(A, B) + R, \tag{14.30}$$

where $P = \sup_{x \in A} H(S_1 x, S_2 x)$, $R > 0$ is a constant and B_{S_i} denotes the set of all best proximity points of S_i, $(i = 1, 2)$.

Proof By Theorem 14.1, B_{S_1} and B_{S_2} are nonempty. Let $u_0 \in B_{S_1}$. Since $S_2 u_0 \neq \emptyset$, we have $v_0 \in S_2 u_0$. Now,

$$\begin{aligned} d(u_0, v_0) &\leq d(u_0, S_1 u_0) + d(S_1 u_0, v_0) \\ &\leq d(A, B) + H(S_1 u_0, S_2 u_0). \end{aligned}$$

Since $v_0 \in S_2 u_0 \subseteq B_0$, we have $u_1 \in A_0$ such that

$$d(u_1, v_0) = d(A, B).$$

Without loss of generality, we take $b_2 = u_0$ and $r_2 = r$ (say) in (14.29) to get

$$d(u_0, S_2 u_0) + d(A, B) \leq (1 - \sqrt{p})\frac{r}{2}.$$

Now,

$$\begin{aligned}d(u_0, u_1) &\leq d(u_0, S_2u_0) + d(S_2u_0, u_1)\\ &\leq d(u_0, S_2u_0) + d(v_0, u_1)\\ &\leq d(u_0, S_2u_0) + d(A, B)\\ &\leq (1-\sqrt{p})\frac{r}{2}.\end{aligned}$$

Hence, $u_1 \in B(u_0, r) \subseteq A$. From Lemma 14.3, for $q = 1$, there exists $v_1 \in S_2u_1$ such that

$$d(v_0, v_1) \leq H(S_2u_0, S_2u_1). \tag{14.31}$$

Now, by (F1) and the definition of S_2,

$$\begin{aligned}F(d(v_0, v_1)) &\leq F(H(S_2u_0, S_2u_1))\\ &\leq F(M(u_0, u_1)) - \lambda\\ &< F(d^*(u_0, u_1)) - \lambda \text{ (as in the proof of Theorem 14.1)}.\end{aligned} \tag{14.32}$$

Therefore, from $(F1)$ and (14.32), we have

$$d(v_0, v_1) < d(u_1, u_0).$$

Since $v_1 \in S_2u_1 \subseteq B_0$, there exists $u_2 \in A_0$ such that

$$d(u_2, v_1) = d(A, B).$$

Now,

$$\begin{aligned}d(u_2, u_0) &\leq d(u_2, u_1) + d(u_1, u_0)\\ &\leq d(v_1, v_0) + d(u_1, u_0), \text{ (using weak P-property of } (A, B))\\ &< d(u_1, u_0) + d(u_1, u_0)\\ &\leq 2d(u_1, u_0)\\ &\leq (1-\sqrt{p})r.\end{aligned}$$

This implies $u_2 \in B(u_0, r) \subseteq A$.
Continuing this process, we get two sequences $\{u_n\} \subseteq A_0$ with $u_n \in B(u_0, r) \subseteq A$ and $\{v_n\} \subseteq B_0$ with $v_n \in S_2u_n$ such that

$$d(u_n, v_{n-1}) = d(A, B). \tag{14.33}$$

Moreover,

$$\begin{aligned} d(u_n, u_{n+1}) &\leq d(v_{n-1}, v_n) \\ &< d(u_{n-1}, u_n) \\ &\leq d(u_1, u_0) \text{ for all } n \in \mathbb{N}. \end{aligned}$$

Let $\beta_n = d(u_n, u_{n+1})$, $n \in \mathbb{N}$. For $n > m$, we have

$$\begin{aligned} d(u_n, u_m) &\leq \sum_{i=m}^{n-1} d(u_i, u_{i+1}) \\ &= \sum_{i=m}^{n-1} \beta_i \\ &\leq \sum_{i=m}^{\infty} \beta_i < \infty. \end{aligned}$$

As in Theorem 14.1, we can show that $\{u_n\}$ is a Cauchy sequence in $B(u_0, r)$. In a similar way, $\{v_n\}$ is a Cauchy sequence in B. Since $B(u_0, r)$ and B are closed, there exist $u \in B(u_0, r)$ and $v \in B$ such that

$$u_n \longrightarrow u \text{ and } v_n \longrightarrow v.$$

Taking limit as $n \to \infty$ in (14.33), we get

$$d(u, v) = d(A, B).$$

Clearly, $v \in S_2 u$. Therefore,

$$d(A, B) \leq d(u, S_2 u) \leq d(u, v) = d(A, B),$$

i.e., $d(u, S_2 u) = d(A, B)$. Hence, u is a best proximity point of S_2.
Now,

$$\begin{aligned} d(u_0, u) &\leq d(u_0, v_0) + d(v_0, u_1) + d(u_1, u) \\ &\leq d(u_0, v_0) + d(v_0, u_1) + \sum_{i=1}^{n} d(u_i, u_{i+1}) + d(u_{n+1}, u). \end{aligned}$$

So, as $n \to \infty$,

$$d(u_0, u) \leq d(u_0, v_0) + d(v_0, u_1) + \sum_{i=1}^{\infty} d(u_i, u_{i+1})$$

$$\leq H(S_1u_0, S_2u_0) + 2d(A, B) + \sum_{i=1}^{\infty} \beta_i$$

$$= H(S_1u_0, S_2u_0) + 2d(A, B) + R, \text{ where } R = \sum_{i=1}^{\infty} \beta_i < \infty.$$

Thus,

$$H(B_{S_1}, B_{S_2}) \leq \sup_{x \in A} H(S_1x, S_2x) + 2d(A, B) + R$$

$$= P + 2d(A, B) + R, \text{ where } P = \sup_{x \in A} H(S_1x, S_2x).$$ □

Following is the stability result for fixed point sets.

Theorem 14.6 *Let (X, d) be a complete metric space and $\phi : C(X) \longrightarrow C(X)$, $S_i : X \longrightarrow C(X)$, $(i = 1, 2)$ be multivalued mappings such that S_i is a continuous multivalued generalized ϕ-F-contraction mapping satisfying condition (ii) of Theorem 14.4. Then*

$$H(F(S_1), F(S_2)) \leq m + R,$$

where $m = \sup_{x \in X} H(S_1x, S_2x)$, $R = \sum_i \beta_i < \infty$ and $F(S_i)$ denotes the set of all fixed points of S_i, $(i = 1, 2)$.

Proof From Theorem 14.4, we have $F(S_i)$ is nonempty $(i = 1, 2)$.
Let $u_0 \in F(S_1)$, i.e., $u_0 \in S_1u_0$. Then by Lemma 14.3 for $q = 1$, there exists $u_1 \in S_2u_0$ such that

$$d(u_0, u_1) \leq H(S_1u_0, S_2u_0). \tag{14.34}$$

From Lemma 14.3 taking $q = 1$, for $u_1 \in S_2u_0$ there exists $u_2 \in S_2u_1$ such that

$$d(u_1, u_2) \leq H(S_2u_0, S_2u_1).$$

Since each S_i, $(i = 1, 2)$ is a multivalued generalized ϕ-F-contraction, as in the proof of Theorem 14.4, we get a sequence $\{u_n\}$ such that

$$u_{n+1} \in S_2u_n,$$

$$d(u_{n+1}, u_n) \leq d(u_n, u_{n-1}),$$

and $\{u_n\}$ is a Cauchy sequence in X. Hence, there exists $u \in X$ such that

$$u_n \longrightarrow u \text{ as } n \rightarrow \infty.$$

Moreover, u is a fixed point of S_2, i.e., $u \in S_2 u$.
Now,

$$
\begin{aligned}
d(u_0, u) &\leq \sum_{i=0}^{n} d(u_i, u_{i+1}) + d(u_{n+1}, u) \\
&\leq d(u_0, u_1) + \sum_{i=0}^{n} d(u_i, u_{i+1}) + d(u_{n+1}, u).
\end{aligned}
$$

Taking limit as $n \to \infty$ in the above inequality and using (14.34), we get

$$
\begin{aligned}
d(u_0, u) &\leq d(u_0, u_1) + \sum_{i=0}^{\infty} d(u_i, u_{i+1}) \\
&= d(u_0, u_1) + \sum_{i=0}^{\infty} \beta_i, \text{ (where } \beta_i = d(u_i, u_{i+1})) \\
&\leq H(S_1 u_0, S_2 u_0) + R, \text{ (where } R = \sum_i \beta_i < \infty).
\end{aligned}
$$

Thus,

$$
\begin{aligned}
H(F(S_1), F(S_2)) &\leq \sup_{x \in X} H(S_1 x, S_2 x) + R \\
&= m + R, \text{(where } m = \sup_{x \in X} H(S_1 x, S_2 x)).
\end{aligned}
$$

Conclusion

We have shown the existence of the best proximity point for multivalued generalized ϕ-F-contraction mappings in complete metric spaces with stability analysis of best proximity point sets. In 2016, Uddin et al. [28] introduced the notion of one-step iteration scheme involving multivalued mappings in CAT(0) spaces. Concerning this, the study of some new type of iteration involving multivalued generalized ϕ-F-contraction in CAT(0) spaces is a scope of further study. Moreover, the formulation of the existence of the best proximity point for multivalued generalized ϕ-F-contraction mappings considering noncomplete metric spaces is another interesting topic for future discussion.

References

1. Abkar, A., Gabeleh, M.: The existence of best proximity points for multivalued non-self mappings. RACSAM **107**, 319–325 (2013). https://doi.org/10.1007/s13398-012-0074-6
2. Ali, A., Moosa, G.: Global optimal solutions of noncyclic mappings in metric spaces. J. Optim. Theory Appl. **153**(2), (2012)
3. Abkar, A., Gabeleh, M.: A note on some best proximity point theorems proved under P-property. Abst. Appl. Anal. **2013**, (2013)
4. Ali, M.U., Kiran, Q., Shahzad, N.: Fixed point theorems for multivalued mappings involving α-function. Abstr. Appl. Anal. **2014**, (2014)
5. Ali, M.U., Kamran, T.: Multivalued F-contractions and related fixed point theorems with an application. Filomat **30**(14), 3779–3793 (2016)
6. Al-Thagafi, M.A., Shahzad, N.: Convergence and existence results for best proximity points. Nonlinear Anal. **70**(10), 3665–3671 (2009). https://doi.org/10.1016/j.na.2008.07.022
7. Altun, I., Mınak, G., Dag, H.: Multivalued F-contractions on complete metric space. J. Nonlinear Convex Anal. **16**(4), 659–666 (2015)
8. Banach, S.: Sur les operations dans les ensembles abstrits et leur applications aux equations integrals. Fund. Math. **3**, 133–181 (1992)
9. Boriceanu, M., Petrusel, A., Rus, I.: Fixed point theorems for some multivalued generalized contractions in b-metric spaces. Internat. J. Math. Stat. **6**, 65–76 (2010)
10. Caballero, J., Harjani, J., Sadarangani, K.: A best proximity point theorem for Geraghty-contractions. Fixed Point Theory Appl. **2012**(1), 1–9 (2012)
11. Choudhuri, B.S., Metiya, N., Bandyopadhyay, C.: Fixed points of multivalued α-admissible mappings and stability of fixed point sets in metric spaces. Rend. Circ. Mat. Palermo **64**, 43–55 (2015)
12. Debnath, P.: Optimization through best proximity points for multivalued F-contractions. Miskolc Math. Notes **22**(1), 143–151 (2021)
13. Eldred, A.A., Veeramani, P.: Existence and convergence of best proximity points. J. Math. Anal. Appl. **323**(2), 51001–1006 (2006)
14. Fan, K.: Extensions of two fixed point theorems of F.E. Browder. Mathematische Zeitschrift **112**(3), 234–240 (1969)
15. Goswami, N., Roy, R., Mishra, V.N., Sánchez Ruiz, L.M.: Common best proximity point results for T-GKT cyclic ϕ-contraction mappings in partial metric spaces with some applications. Symmetry **13**(6), 1–13 (2021)
16. Goswami, N., Roy, R.: Some coupled best proximity point results for weak GKT cyclic ø-contraction mappings on metric spaces. Proc. Jangjeon Math. Soc. **23**(4), 483–500 (2020)
17. Khammahawong, K., Kumam, P.: A best proximity point theorem for Rogar-Hardy type generalized F-contractive mappings in complete metric spaces with some examples. RACSAM **112**, 1503–1519 (2018)
18. Khammahawong, K., Sa Ngiamsunthorn, P., Kumam, P.: On best proximity points for multivalued cyclic F-contraction mappings. Int. J. Nonlinear Anal. Appl. **7**(1), 363–374 (2016)
19. Lim, T.C.: On fixed point stability for set-valued contractive mappings with application to generalized differential equations. J. Math. Appl. **110**, 436–441 (1985)
20. Markin, T.: A fixed point stability theorem for nonexpansive set-valued mappings. J. Math. Anal. **54**, 289–291 (1976)
21. Nadler, S.B.: Multi-valued contraction mappings. Pac. J. Math. **30**(2), 475–488 (1969)
22. Omidvari, M., Vaezpour, S.M., Saadati, R.: Best proximity point theorems for F-contractive non-self mappings. Miskolc Math. Notes **15**, 615–623 (2014)
23. Salamatbakhsh, M., Haghi, R.H.: On best proximity points of generalized almost-F-contraction mappings. J. Indones. Math. Soc. **1**, 16–23 (2019)
24. Samet, B.: Some results on best proximity points. J. Optim. Theory Appl. **159**(1), 281–291 (2013)
25. Shankar Raj, V.: A best proximity point theorem for weakly contractive non-self mappings. Nonlinear Anal. **74**(14), 4804–4808 (2011)

26. Suzuki, T.: The existence of best proximity points with the weak P-property. Fixed Point Theory Appl. **2013**(1), 1–6 (2013)
27. Suzuki, T., Kikkawa, M., Vetro, C.: The existence of the best proximity points in metric spaces with property UC. Nonlinear Anal. Theory Methods Appl. **71**, 2918–2926 (2009)
28. Uddin, I., Nieto, J.J., Ali, J.: One-step iteration scheme for multivalued nonexpansive mappings in CAT(0) spaces. Mediterr. J. Math. **13**, 1211–1225 (2016)
29. Wardowski, D.: Fixed points of a new type of contractive mappings in complete metric spaces. Fixed Point Theory Appl. **94**, (2012)
30. Zhang, J., Su, Y., Cheng, Q.: A note on best proximity point theorem for Geraghty contractions. Fixed Point Theory Appl. **2013**(99), 1–4 (2013)

Chapter 15
Some New Fixed-Point Results in Extended Branciari b-distance Spaces and Applications

Reena Jain and Hemant Kumar Nashine

Abstract We introduce $\alpha - \lambda -$ rational contractive conditions and derive fixed points and periodic point results in the framework of an extended Branciari b-distance spaces. We offer a few illustrations to support our work. As an application of the main finding, we finally obtain sufficient conditions for the existence of solutions for the two-point boundary-value differential equation.

Keywords Fixed point · Extended branciari b-distance spaces · Periodic point

15.1 Introductory Notes

Denote $\mathbb{R} :=$ the set of real numbers, $\mathbb{R}_+ := [0, +\infty)$, $\mathbb{N} :=$ the set of natural numbers, and $\mathbb{N}^* := \mathbb{N} \cup \{0\}$.

Numerous authors [8, 10–20, 23–26] introduce and generalize the concept of distance within the metric fixed point theory in a variety of ways. Bakhtin [2] defines the concept of b-metric space, which Czerwik employs in [3, 4]. Kamran et al.[23] introduce the concept of extended b-metric space, whereas Branciari [5] extends the metric space and introduces the concept of Branciari distance by replacing the triangle inequality property with a quadrilateral inequality.

Definition 15.1 [23] Let $\Omega \neq \emptyset$ be a set and $w : \Omega^2 \to \mathbb{R}_+ \backslash (0, 1)$. A function $\sigma_e : \Omega^2 \to \mathbb{R}_+$ is said to be an extended b-metric (σ_e-metric, for short) if the following conditions are met:

(eb1) $\sigma_e(\xi, \zeta) = 0 \iff \xi = \zeta$;

R. Jain · H. K. Nashine
Mathematics Division, School of Advanced Sciences and Languages, VIT Bhopal University, Bhopal-Indore Highway, Kothrikalan, Sehore, Madhya Pradesh 466114, India
e-mail: reena.jain@vitbhopal.ac.in

H. K. Nashine (✉)
Department of Mathematics and Applied Mathematics, University of Johannesburg, Kingsway Campus, Auckland Park 2006, South Africa
e-mail: hemantkumar.nashine@vitbhopal.ac.in

B. Hazarika et al. (eds.), *Advances in Functional Analysis and Fixed-Point Theory*, Industrial and Applied Mathematics,
https://doi.org/10.1007/978-981-99-9207-2_15

(eb2) $\sigma_e(\xi, \zeta) = \sigma_e(\zeta, \xi)$;
(eb3) $\sigma_e(\xi, \zeta) \leq w(\xi, \zeta)[\sigma_e(\xi, \upsilon) + \sigma_e(\upsilon, \zeta)]$,

for all $\xi, \zeta, \upsilon \in \Omega$. The symbol (Ω, σ_e) denotes a σ_e-metric space.

Definition 15.2 [5] Let $\Omega \neq \emptyset$ be a set and let $\sigma_b : \Omega^2 \to \mathbb{R}_+$ such that, for all $\xi, \zeta \in \Omega$ and all distinct $u, v \in \Omega \backslash \{\xi, \zeta\}$,

(bd1) $\sigma_b(\xi, \zeta) = 0$ if and only if $\xi = \zeta$;
(bd2) $\sigma_b(\xi, \zeta) = \sigma_b(\zeta, \xi)$;
(bd3) $\sigma_b(\xi, \zeta) \leq \sigma_b(\xi, u) + \sigma_b(u, v) + \sigma_b(v, \zeta)$.

The symbol (Ω, σ_b) denotes Branciari distance space and is abbreviated as "BDS". Abdeljawad et al. [1] extended the above notions as follows:

Definition 15.3 [1] Let $\Omega \neq \emptyset$ be a set and $w : \Omega^2 \to \mathbb{R}_+ \backslash (0, 1)$. We say that a function $b_e : \Omega^2 \to \mathbb{R}_+$ is an extended Branciari b-metric (b_e-metric, in short) if it satisfies:

(B1) $b_e(\xi, \zeta) = 0$ if and only if $\xi = \zeta$,
(B2) $b_e(\xi, \zeta) = b_e(\zeta, \xi)$,
(B3) $b_e(\xi, \zeta) \leq w(\xi, \zeta)[b_e(\xi, \nu) + b_e(\nu, \varrho) + b_e(\varrho, \zeta)]$

for all $\xi, \zeta \in \Omega$ all distinct $\nu, \varrho \in \Omega \backslash \{\xi, \zeta\}$. The symbol (Ω, b_e) denotes an extended Branciari b-distance space (EBbDS, in short). For $w(\xi, \zeta) = 1$, (Ω, b_e) will be called a Branciari b-distance space (BbDS, in short).

Example 15.1 [1] Let $\Omega = C([0, 1], \mathbb{R})$ and define $b_e : \Omega^2 \to \mathbb{R}_+$ by $b_e(J, K) = \int_0^1 (J(\varsigma) - K(\varsigma))^2 d\varsigma$ with $w(J, K) = |J(\varsigma)| + |K(\varsigma)| + 2$. Note that $b_e(J, K) \geq 0$ for all $J, K \in \Omega$, and $b_e(J, K) = 0$ if and only if $J = K$. Also $b_e(J, K) = b_e(K, J)$. As a result, it is clear that (Ω, b_e) is an EBbDS, but not a BDS or a metric space.

Definition 15.4 [1] Let $(\Omega \neq \emptyset, b_e)$ be an extended Branciari b-distance space and $w \colon \Omega^2 \to \mathbb{R}_+$.

(a) A sequence $\{\xi_n\}$ in Ω converges to ξ if for every $\epsilon > 0$ there exists $N = N(\epsilon) \in \mathbb{N}$ such that $b_e(\xi_n, \xi) < \epsilon$ for all $n \geq N$. In this particular instance, we write $\lim_{n\to\infty} \xi_n = \xi$.
(b) A sequence $\{\xi_n\}$ in Ω is called Cauchy if for every $\epsilon > 0$ there exists $N = N(\epsilon) \in \mathbb{N}$ such that $b_e(\xi_m, \xi_n) < \epsilon$ for all $m, n \geq N$.
(c) A b_e-metric space (Ω, b_e) is complete if every Cauchy sequence in Ω is convergent.

Definition 15.5 For a $\Omega \neq \emptyset$ set, let $\alpha : \Omega \times \Omega \to [0, \infty)$ and $\Im \colon \Omega \to \Omega$ be two mappings. Then $\Im$ is called :

1. [27] α-admissible if:

$$\xi, \zeta \in \Omega \text{ with } \alpha(\xi, \zeta) \geq 1 \Rightarrow \alpha(\Im\xi, \Im\zeta) \geq 1.$$

2. [28] weakly α-admissible if:

$$\xi \in \Omega \text{ with } \alpha(\xi, \Im\xi) \geq 1 \Rightarrow \alpha(\Im\xi, \Im\Im\xi) \geq 1.$$

For a $\Omega \neq \emptyset$ set and a mapping $\alpha : \Omega \times \Omega \to \mathbb{R}_+$, we use

$$\mathfrak{A}(\Omega, \alpha) := \{\text{All } \alpha - \text{admissible mappings on } \Omega\}$$

and

$$\mathfrak{W}(\Omega, \alpha) := \{ \text{ All weakly } \alpha - \text{admissible mappings on } \Omega\}.$$

It is noted that

$$\mathfrak{A}(\Omega, \alpha) \subset \mathfrak{W}(\Omega, \alpha).$$

15.2 Main Results

In this section, the concept of an $\alpha - \lambda-$ rational contraction in an EBbDS is introduced.

We begin with the following concepts.

Denote $\Theta = \{\theta : (0, \infty) \to [1, \infty)\}$ satisfying the conditions [22].

(θ_1) θ is strictly increasing;
(θ_2) θ is continuous;
(θ_3) for each sequence $\{\tau_n\} \subseteq (0, \infty)$, $\lim_{n\to\infty} \theta(\tau_n) = 1$ if and only if $\lim_{n\to\infty} \tau_n = 0$.

Definition 15.6 Let (Ω, b_e) be an EBbDS and $\alpha\colon \Omega^2 \to \mathbb{R}_+$ and $\lambda_i : \Omega \to [0, 1)$ $(i = 1, 2, 3, 4, 5)$. A mapping $\Im : \Omega \to \Omega$ is said to be $\alpha - \lambda-$ rational contraction, if there exist $\gamma \in [0, 1)$ such that

$$\begin{aligned} &\zeta, \xi \in \Omega \text{ with } \alpha(\zeta, \xi) \geq 1 \text{ and } b_e(\zeta, \xi) > 0,\, b_e(\Im\zeta, \Im\xi) > 0 \text{ implies} \\ &\theta(w(\zeta, \xi) b_e(\Im\zeta, \Im\xi)) \\ &\leq \left[\theta\left(\begin{array}{l} \lambda_1(\zeta)\, b_e(\zeta, \xi) + \lambda_2(\zeta)\, b_e(\zeta, \Im\zeta) + \lambda_3(\zeta)\, b_e(\xi, \Im\xi) \\ +\lambda_4(\zeta)\, \frac{b_e(\xi,\Im\xi)[1+b_e(\zeta,\Im\zeta)]}{w(\zeta,\xi)[1+b_e(\zeta,\xi)]} + \lambda_5(\zeta)\, \frac{b_e(\zeta,\Im\zeta).b_e(\xi,\Im\xi)}{w(\zeta,\xi).b_e(\zeta,\xi)} \end{array}\right)\right]^{\gamma} \end{aligned} \tag{15.1}$$

where $\lambda = \sum_{i=1}^{5} \lambda_i < 1$.

The entire set of all rational-$\alpha - \lambda - JS$-contractive mappings on (Ω, b_e) is denoted by $\mathfrak{J}(\Omega, \alpha, \lambda)$.

Denote $Fix(\Im) = \{\zeta : \Im : \Omega \to \Omega \text{ such that } \Im\zeta = \zeta\}$.

Theorem 15.1 *Let (Ω, b_e) be a complete EBbDS and $\alpha\colon \Omega^2 \to \mathbb{R}_+$. Suppose $\Im : \Omega \to \Omega$ be a mapping satisfying*

($\mathscr{B}1$) *there exists $\zeta_0 \in \Omega$ such that $\alpha(\zeta_0, \Im\zeta_0) \geq 1$;*
($\mathscr{B}2$) *$\Im \in \mathfrak{J}(\Omega, \alpha, \lambda) \cap \mathfrak{W}(\Omega, \alpha)$;*
($\mathscr{B}3$) *$\Im$ is continuous.*

Then $Fix(\Im) \neq \emptyset$.

Proof Using the condition ($\mathscr{B}_1$) and any starting point in Ω, we can consider the sequence $\{\zeta_n\}$ by $\zeta_{n+1} = \Im\zeta_n$. If there exists $n_0 \in \mathbb{N}$ such that $\zeta_{n_0} = \zeta_{n_0+1}$, then $\zeta_{n_0} \in Fix(\Im)$ and we have completed. Consequently, we conclude that $\zeta_n \neq \zeta_{n+1}$ for all $n \in \mathbb{N}$.
It follows that

$$b_e(\zeta_n, \zeta_{n+1}) > 0, \ \forall\, n \in \mathbb{N}.$$

Using ($\mathscr{B}_1$)-($\mathscr{B}_2$),

$$\alpha(\zeta_1, \zeta_2) = \alpha(\Im\zeta_0, \Im\Im\zeta_0) \geq 1.$$

By carrying on in this manner, we have

$$\alpha(\zeta_n, \zeta_{n+1}) \geq 1 \quad \forall n \in \mathbb{N}.$$

• To begin, let's establish

$$\lim_{n\to\infty} b_e(\zeta_n, \zeta_{n+1}) = 0. \tag{15.2}$$

Using ($\mathscr{B}_2$)

$$\begin{aligned}
&\theta(w(\zeta_{n-1}, \zeta_n)b_e(\Im\zeta_{n-1}, \Im\zeta_n)) \\
&\leq \left[\theta\left(\begin{array}{c} \lambda_1(\zeta_{n-1})b_e(\zeta_{n-1}, \zeta_n) + \lambda_2(\zeta_{n-1})b_e(\zeta_{n-1}, \Im\zeta_{n-1}) \\ +\lambda_3(\zeta_{n-1})b_e(\zeta_n, \Im\zeta_n) + \lambda_4(\zeta_{n-1})\frac{b_e(\zeta_n,\Im\zeta_n)[1+b_e(\zeta_{n-1},\Im\zeta_{n-1})]}{w(\zeta_{n-1},\zeta_n)[1+b_e(\zeta_{n-1},\zeta_n)]} \\ +\lambda_5(\zeta_{n-1})\frac{b_e(\zeta_{n-1},\Im\zeta_{n-1}).b_e(\zeta_n,\Im\zeta_n)}{w(\zeta_{n-1},\zeta_n).b_e(\zeta_{n-1},\zeta_n)} \end{array}\right)\right]^{\gamma} \\
&= \left[\theta\left(\begin{array}{c} \lambda_1(\zeta_{n-1})b_e(\zeta_{n-1}, \zeta_n) + \lambda_2(\zeta_{n-1})b_e(\zeta_{n-1}, \zeta_n) \\ +\lambda_3(\zeta_{n-1})b_e(\zeta_n, \zeta_{n+1}) + \lambda_4(\zeta_{n-1})\frac{b_e(\zeta_n,\zeta_{n+1})[1+b_e(\zeta_{n-1},\zeta_n)]}{w(\zeta_{n-1},\zeta_n)[1+b_e(\zeta_{n-1},\zeta_n)]} \\ +\lambda_5(\zeta_{n-1})\frac{b_e(\zeta_{n-1},\zeta_n).b_e(\zeta_n,\zeta_{n+1})}{w(\zeta_{n-1},\zeta_n).b_e(\zeta_{n-1},\zeta_n)} \end{array}\right)\right]^{\gamma} \\
&\leq \left[\theta\left(\begin{array}{c} \lambda_1(\zeta_{n-1})b_e(\zeta_{n-1}, \zeta_n) + \lambda_2(\zeta_{n-1})b_e(\zeta_{n-1}, \zeta_n) \\ +\lambda_3(\zeta_{n-1})b_e(\zeta_n, \zeta_{n+1}) + \lambda_4(\zeta_{n-1})b_e(\zeta_n, \zeta_{n+1}) \\ +\lambda_5(\zeta_{n-1})b_e(\zeta_n, \zeta_{n+1}) \end{array}\right)\right]^{\gamma}.
\end{aligned} \tag{15.3}$$

Since θ is strictly rising and $\gamma < 1$, we conclude that

$$\begin{aligned} w(\zeta_{n-1}, \zeta_n)b_e(\Im\zeta_{n-1}, \Im\zeta_n) \le \lambda_1(\zeta_{n-1})b_e(\zeta_{n-1}, \zeta_n) + \lambda_2(\zeta_{n-1})b_e(\zeta_{n-1}, \zeta_n) \\ + \lambda_3(\zeta_{n-1})b_e(\zeta_n, \zeta_{n+1}) + \lambda_4(\zeta_{n-1})b_e(\zeta_n, \zeta_{n+1}) \\ + \lambda_5(\zeta_{n-1})b_e(\zeta_n, \zeta_{n+1}). \end{aligned} \tag{15.4}$$

If $b_e(\zeta_{n-1}, \zeta_n) \le b_e(\zeta_n, \zeta_{n+1})$ for some $n \in \mathbb{N}$, then, from (15.4), we have $w(\zeta_n, \zeta_{n+1}) b_e(\zeta_n, \zeta_{n+1}) \le \lambda(\zeta_{n-1})b_e(\zeta_n, \zeta_{n+1})$, which is contradictory because $w \ge 1$ and $\lambda < 1$. Thus, $b_e(\zeta_n, \zeta_{n+1}) \le b_e(\zeta_{n-1}, \zeta_n)$ for all $n \in \mathbb{N}$ and the sequence $\{b_e(\zeta_n, \zeta_{n+1})\}$ is a deceasing sequence of real numbers. As a result, there is ζ such that

$$\lim_{n\to\infty} b_e(\zeta_n, \zeta_{n+1}) = \zeta.$$

From (15.3), we have

$$\theta(w(\zeta_{n-1}, \zeta_n)b_e(\zeta_n, \zeta_{n+1})) \le \theta(b_e(\zeta_{n-1}, \zeta_n))^{\gamma} \le \theta(b_e(\zeta_{n-2}, \zeta_{n-1}))^{\gamma^2}.$$

Thus,

$$1 \le \theta(b_e(\zeta_{n-1}, \zeta_n))^{\gamma} \le \theta(b_e(\zeta_{n-2}, \zeta_{n-1}))^{\gamma^2} \le .. \le \theta(b_e(\zeta_0, \zeta_1))^{\gamma^n}.$$

Applying the limit in the above relation, we get

$$\lim_{n\to\infty} \theta(b_e(\zeta_{n-1}, \zeta_n)) = 1$$

and by the property (θ_3) of θ,

$$\lim_{n\to\infty} b_e(\zeta_{n-1}, \zeta_n) = 0.$$

- To show that $\{\zeta_n\}$ is a Cauchy sequence, we need only show that it exists for $m > n$.

$$\lim_{n,m\to\infty} b_e(\zeta_n, \zeta_m) = 0.$$

Using (ebb3), we have

$$\begin{aligned} & b_e(\zeta_n, \zeta_m) \\ \le\ & w(\zeta_n, \zeta_m)[b_e(\zeta_n, \zeta_{n+1}) + b_e(\zeta_{n+1}, \zeta_{n+2}) + b_e(\zeta_{n+2}, \zeta_m)] \\ \le\ & w(\zeta_n, \zeta_m)b_e(\zeta_n, \zeta_{n+1}) + w(\zeta_n, \zeta_m)b_e(\zeta_n, \zeta_{n+1}) + \\ & w(\zeta_n, \zeta_m)b_e(\zeta_{n+2}, \zeta_m) \\ \le\ & w(\zeta_n, \zeta_m)b_e(\zeta_n, \zeta_{n+1}) + w(\zeta_n, \zeta_m)b_e(\zeta_n, \zeta_{n+1}) + \\ & w(\zeta_n, \zeta_m)w(\zeta_{n+2}, \zeta_m)[b_e(\zeta_{n+2}, \zeta_{n+3}) + b_e(\zeta_{n+3}, \zeta_{n+4}) \\ & +b_e(\zeta_{n+4}, \zeta_m)] \end{aligned}$$

$$\begin{aligned}
&\leq w(\zeta_n, \zeta_m)b_e(\zeta_n, \zeta_{n+1}) + w(\zeta_n, \zeta_m)b_e(\zeta_n, \zeta_{n+1}) + \\
&\quad w(\zeta_n, \zeta_m)w(\zeta_{n+2}, \zeta_m)b_e(\zeta_n, \zeta_{n+1}) + w(\zeta_n, \zeta_m) \\
&\quad w(\zeta_{n+2}, \zeta_m)b_e(\zeta_n, \zeta_{n+1}) + w(\zeta_n, \zeta_m)w(\zeta_{n+2}, \zeta_m)b_e(\zeta_{n+4}, \zeta_m) \\
&\vdots \\
&\leq w(\zeta_n, \zeta_m)b_e(\zeta_n, \zeta_{n+1}) + w(\zeta_n, \zeta_m)b_e(\zeta_n, \zeta_{n+1}) + \\
&\quad w(\zeta_n, \zeta_m)w(\zeta_{n+2}, \zeta_m)b_e(\zeta_n, \zeta_{n+1}) + w(\zeta_n, \zeta_m) \\
&\quad w(\zeta_{n+2}, \zeta_m)b_e(\zeta_n, \zeta_{n+1}) + \ldots + \\
&\quad w(\zeta_n, \zeta_m)w(\zeta_{n+2}, \zeta_m)\ldots w(\zeta_{m-2}, \zeta_m)b_e(\zeta_n, \zeta_{n+1}) + \\
&\quad w(\zeta_n, \zeta_m)w(\zeta_{n+2}, \zeta_m)\ldots w(\zeta_{m-2}, \zeta_m)b_e(\zeta_n, \zeta_{n+1}) \\
&\leq w(\zeta_n, \zeta_m)b_e(\zeta_n, \zeta_{n+1}) + w(\zeta_n, \zeta_m)w(\zeta_{n+1}, \zeta_m)b_e(\zeta_n, \zeta_{n+1}) + \\
&\quad w(\zeta_n, \zeta_m)w(\zeta_{n+1}, \zeta_m)w(\zeta_{n+2}, \zeta_m)b_e(\zeta_n, \zeta_{n+1}) + w(\zeta_n, \zeta_m) \\
&\quad w(\zeta_{n+1}, \zeta_m)w(\zeta_{n+2}, \zeta_m)w(\zeta_{n+3}, \zeta_m)b_e(\zeta_n, \zeta_{n+1}) + \ldots + \\
&\quad w(\zeta_n, \zeta_m)w(\zeta_{n+1}, \zeta_m)w(\zeta_{n+2}, \zeta_m)\ldots w(\zeta_{m-2}, \zeta_m)b_e(\zeta_n, \zeta_{n+1}) + \\
&\quad w(\zeta_n, \zeta_m)w(\zeta_{n+1}, \zeta_m)w(\zeta_{n+2}, \zeta_m)\ldots w(\zeta_{m-2}, \zeta_m)w(\zeta_{m-1}, \zeta_m) \\
&\quad b_e(\zeta_n, \zeta_{n+1}).
\end{aligned}$$

Passing $n, m \to \infty$ and using (15.2),

$$\lim_{n,m\to\infty} b_e(\zeta_n, \zeta_m) = 0.$$

Hence, $\{\zeta_n\}$ is a Cauchy sequence. Since (Ω, b_e) is a complete $EBbDS$, then $\exists$ a point $\zeta^* \in \Omega$ such that $\zeta_n \to \zeta^*$ as $n \to +\infty$,

$$\lim_{n\to+\infty} b_e(\zeta_n, \zeta^*) = 0.$$

Next, we prove that $\zeta^* \in Fix(\Im)$. Indeed, we write

$$b_e(\zeta^*, \Im\zeta^*) \leq w(\zeta^*, \Im\zeta^*)[b_e(\zeta^*, \zeta_n) + b_e(\zeta_n, \zeta_{n+1}) + b_e(\zeta_{n+1}, \Im\zeta^*)].$$

Since $\Im$ is continuous, on letting $n \to +\infty$, we obtain $b_e(\zeta^*, \Im\zeta^*) = 0$, $\Im\zeta^* = \zeta^*$ and hence $\zeta^* \in Fix(\Im)$. □

In order to demonstrate the $\zeta^* \in Fix(\Im)$ is singleton, we impose a further condition.

$(A4)$ For every pair $\zeta^*, \xi^* \in Fix(\Im)$, $\alpha(\zeta^*, \xi^*) \geq 1$.

Theorem 15.2 *Inclusion of a condition* $(A4)$ *in Theorem 15.1,* $Fix(\Im)$ *contains only one unique element.*

Proof As a result of the Theorem 15.1, $\zeta^* \in Fix(\Im)$. To prove $Fix(\Im)$ is a singleton set, assume that there exist $\zeta^*, \xi^* \in Fix(\Im)$ with $\zeta^* \neq \xi^*$, and by $(A4)$, we have $\alpha(\zeta^*, \xi^*) \geq 1$. It follows from $\Im \in \Lambda(\Omega, \alpha)$ that

$$\begin{aligned}
&\theta(w(\zeta^*, \xi^*)b_e(\Im\zeta^*, \Im\xi^*)) \\
&\leq \left[\theta\left(\begin{array}{c}\lambda_1(\zeta^*)b_e(\zeta^*, \xi^*) + \lambda_2(\zeta^*)b_e(\zeta^*, \Im\zeta^*) + \lambda_3(\zeta^*)b_e(\xi^*, \Im\xi^*) \\ +\lambda_4(\zeta^*)\frac{b_e(\xi^*,\Im\xi^*)[1+b_e(\zeta^*,\Im\xi^*)]}{w(\zeta^*,\xi^*)[1+b_e(\zeta^*,\xi^*)]} + \lambda_5(\zeta^*)\frac{b_e(\zeta^*,\Im\zeta^*).b_e(\xi^*,\Im\xi^*)}{w(\zeta^*,\xi^*)).b_e(\zeta^*,\xi^*)}\end{array}\right)\right]^\gamma \\
&\leq \left[\theta(\lambda_1(\zeta^*)b_e(\zeta^*, \xi^*))\right]^\gamma .
\end{aligned}$$

Since θ is strictly rising and $\gamma < 1$, we conclude that

$$w(\zeta^*, \xi^*)b_e(\zeta^*, \xi^*) \leq \lambda_1(\zeta^*)b_e(\zeta^*, \xi^*),$$

a contradiction and hence $\zeta^* = \xi^*$. □

15.3 A Periodic Point Result

Definition 15.7 [21] A mapping $\Im : \Omega \to \Omega$ is said to have the property (P) if it has no periodic points, i.e., if $Fix(\Im^n) = Fix(\Im)$ for every $n \in \mathbb{N}$, where $Fix(\Im) := \{\zeta \in \Omega : \Im\zeta = \zeta\}$.

Theorem 15.3 *Let (Ω, b_e) be a complete EBbDS, and $\alpha : \Omega^2 \to \mathbb{R}_+$ and $\lambda_i : \Omega \to [0, 1)$ $(i = 1, 2, 3, 4, 5)$. A mapping $\Im : \Omega \to \Omega$ be a given mapping satisfying*

$\zeta \in \Omega$ with $\alpha(\zeta, \Im\zeta) \geq 1$ and $b_e(\zeta, \Im\zeta) > 0$, $b_e(\Im\zeta, \Im^2\zeta) > 0$ implies

$$\begin{aligned}
&\theta(w(\zeta, \Im\zeta)b_e(\Im\zeta, \Im^2\zeta)) \\
&\leq \left[\theta\left(\begin{array}{c}\lambda_1(\zeta)\, b_e(\zeta, \Im\zeta) + \lambda_2(\zeta)\, b_e(\zeta, \Im\zeta) + \lambda_3(\zeta)\, b_e(\Im\zeta, \Im^2\zeta) \\ +\lambda_4(\zeta)\, \frac{b_e(\Im\zeta,\Im^2\zeta)[1+b_e(\zeta,\Im\zeta)]}{w(\zeta,\Im\zeta)[1+b_e(\zeta,\Im\zeta)]} + \lambda_5(\zeta)\, \frac{b_e(\zeta,\Im\zeta).b_e(\Im\zeta,\Im^2\zeta)}{w(\zeta,\Im\zeta).b_e(\zeta,\Im\zeta)}\end{array}\right)\right]^\gamma \qquad (15.5)
\end{aligned}$$

where $\gamma \in [0, 1)$ and $\lambda = \sum_{i=1}^{5} \lambda_i < 1$. Then $\Im$ has the property (P).

Proof Beginning with the given $\zeta_0 \in \Omega$, define the sequence $\{\zeta_n\}$ by $\zeta_n = \Im^n\zeta_0 = \Im\zeta_{n-1}$, $n \in \mathbb{N}$.

If there exists $n_0 \in \mathbb{N}$ such that $\zeta_{n_0} = \zeta_{n_0+1} = \Im\zeta_{n_0}$, then $\zeta_{n_0} \in Fix(\Im)$. Hence, we assume $\zeta_n \neq \zeta_{n+1}$, i.e., $b_e(\Im\zeta_{n-1}, \Im^2\zeta_{n-1}) > 0$ for all $n \in \mathbb{N}$.

From (15.5), we have

$$\theta(w(\zeta_{n-1}, \Im\zeta_{n-1})b_e(\Im\zeta_{n-1}, \Im^2\zeta_{n-1}))$$

$$\leq \left[\theta\begin{pmatrix}\lambda_1(\zeta_{n-1})\, b_e(\zeta_{n-1},\Im\zeta_{n-1}) + \lambda_2(\zeta_{n-1})\, b_e(\zeta_{n-1},\Im\zeta_{n-1})\\ +\lambda_3(\zeta_{n-1})\, b_e(\Im\zeta_{n-1},\Im^2\zeta_{n-1})\\ +\lambda_4(\zeta_{n-1})\, \frac{b_e(\Im\zeta_{n-1},\Im^2\zeta_{n-1})[1+b_e(\zeta_{n-1},\Im\zeta_{n-1})]}{w(\zeta_{n-1},\Im\zeta_{n-1})[1+b_e(\zeta_{n-1},\Im\zeta_{n-1})]}\\ +\lambda_5(\zeta_{n-1})\, \frac{b_e(\zeta_{n-1},\Im\zeta_{n-1}).b_e(\Im\zeta_{n-1},\Im^2\zeta_{n-1})}{w(\zeta_{n-1},\Im\zeta_{n-1}).b_e(\zeta_{n-1},\Im\zeta_{n-1})}\end{pmatrix}\right]^{\gamma}$$

$$\leq \left[\theta\begin{pmatrix}\lambda_1(\zeta_{n-1})\, b_e(\zeta_{n-1},\zeta_n) + \lambda_2(\zeta_{n-1})\, b_e(\zeta_{n-1},\zeta_n)\\ +\lambda_3(\zeta_{n-1})\, b_e(\zeta_n,\zeta_{n+1})\\ +\lambda_4(\zeta_{n-1})\, \frac{b_e(\zeta_n,\zeta_{n+1})[1+b_e(\zeta_{n-1},\zeta_n)]}{w(\zeta_{n-1},\zeta_n)[1+b_e(\zeta_{n-1},\zeta_n)]}\\ +\lambda_5(\zeta_{n-1})\, \frac{b_e(\zeta_{n-1},\zeta_n).b_e(\zeta_n,\zeta_{n+1})}{w(\zeta_{n-1},\zeta_n).b_e(\zeta_{n-1},\zeta_n)}\end{pmatrix}\right]^{\gamma}$$

$$\leq \left[\theta\begin{pmatrix}(\lambda_1(\zeta_{n-1}) + \lambda_2(\zeta_{n-1}))\, b_e(\zeta_{n-1},\zeta_n)\\ +(\lambda_3(\zeta_{n-1}) + \lambda_4(\zeta_{n-1}) + \lambda_5(\zeta_{n-1}))b_e(\zeta_n,\zeta_{n+1})\end{pmatrix}\right]^{\gamma}.$$

Since θ is strictly rising and $\gamma < 1$, we conclude that

$$\begin{aligned} w(\zeta_{n-1},\Im\zeta_{n-1})b_e(\zeta_n,\zeta_{n+1})) &\leq [\lambda_1(\zeta_{n-1}) + \lambda_2(\zeta_{n-1})]\, b_e(\zeta_{n-1},\zeta_n)\\ &+ [\lambda_3(\zeta_{n-1}) + \lambda_4(\zeta_{n-1}) + \lambda_5(\zeta_{n-1})]b_e(\zeta_n,\zeta_{n+1}). \end{aligned} \tag{15.6}$$

If there exists $n \in \mathbb{N}$ such that $b_e(\zeta_{n-1},\zeta_n) \leq b_e(\zeta_n,\zeta_{n+1})$, then using property of $w(\zeta,\xi) \geq 1$,(15.6) becomes

$$b_e(\zeta_n,\zeta_{n+1})) \leq \lambda(\zeta_{n-1})b_e(\zeta_n,\zeta_{n+1}),$$

a contradiction. As a result, we conclude that

$$b_e(\zeta_n,\zeta_{n+1}) \leq b_e(\zeta_{n-1},\zeta_n).$$

Using a similar strategy to that used to prove the Theorem 15.1, we get $\Im\eta = \eta$. Thus, $Fix(\Im^n) = Fix(\Im)$ is true for $n = 1$. It is best to prove that it is true for $n > 1$. Suppose it is not true, that is, $\varrho \in Fix(\Im^n)$ and $\varrho \notin Fix(\Im)$; then $d_e(\varrho,\Im\varrho) > 0$. Applying (15.5), we get

$$\theta(w(\Im^{n-1}\varrho,\varrho)b_e(\varrho,\Im\varrho))$$

$$= \theta(w(\Im^{n-1}\varrho,\Im(\Im^{n-1}\varrho))b_e(\Im(\Im^{n-1}\varrho),\Im^2(\Im^{n-1}\varrho))$$

$$\leq \left[\theta \begin{pmatrix} \lambda_1(\Im^{n-1}\varrho)\, b_e(\Im^{n-1}\varrho, \Im(\Im^{n-1}\varrho)) + \lambda_2(\Im^{n-1}\varrho)\, b_e(\Im^{n-1}\varrho, \Im(\Im^{n-1}\varrho)) \\ +\lambda_3(\Im^{n-1}\varrho)\, b_e(\Im(\Im^{n-1}\varrho), \Im^2(\Im^{n-1}\varrho)) \\ +\lambda_4(\Im^{n-1}\varrho)\, \frac{b_e(\Im(\Im^{n-1}\varrho),\Im^2(\Im^{n-1}\varrho))[1+b_e(\Im^{n-1}\varrho,\Im(\Im^{n-1}\varrho)]}{w(\Im^{n-1}\varrho,\Im(\Im^{n-1}\varrho))[1+b_e(\Im^{n-1}\varrho,\Im(\Im^{n-1}\varrho)]} \\ +\lambda_5(\Im^{n-1}\varrho)\, \frac{b_e(\Im^{n-1}\varrho,\Im(\Im^{n-1}\varrho)).b_e(\Im(\Im^{n-1}\varrho),\Im^2(\Im^{n-1}\varrho))}{w(\Im^{n-1}\varrho,\Im(\Im^{n-1}\varrho)).b_e(\Im^{n-1}\varrho,\Im(\Im^{n-1}\varrho))} \end{pmatrix} \right]^{\gamma}$$

$$= \left[\theta \begin{pmatrix} \lambda_1(\Im^{n-1}\varrho)\, b_e(\Im^{n-1}\varrho, \varrho) + \lambda_2(\Im^{n-1}\varrho)\, b_e(\Im^{n-1}\varrho, \varrho) \\ +\lambda_3(\Im^{n-1}\varrho)\, b_e(\varrho, \Im\varrho) \\ +\lambda_4(\Im^{n-1}\varrho)\, \frac{b_e(\varrho,\Im\varrho)[1+b_e(\Im^{n-1}\varrho,\varrho)]}{w(\Im^{n-1}\varrho,\varrho)[1+b_e(\Im^{n-1}\varrho,\varrho)]} \\ +\lambda_5(\Im^{n-1}\varrho)\, \frac{b_e(\Im^{n-1}\varrho,\varrho).b_e(\varrho,\Im\varrho)}{w(\Im^{n-1}\varrho,\varrho).b_e(\Im^{n-1}\varrho,\varrho)} \end{pmatrix} \right]^{\gamma}.$$

Since θ is strictly rising and $\gamma < 1$, we conclude that

$$\begin{aligned} w(\Im^{n-1}\varrho, \varrho)b_e(\varrho, \Im\varrho) &\leq \lambda_1(\Im^{n-1}\varrho)\, b_e(\Im^{n-1}\varrho, \varrho) \\ &\quad + \lambda_2(\Im^{n-1}\varrho)\, b_e(\Im^{n-1}\varrho, \varrho) + \lambda_3(\Im^{n-1}\varrho)\, b_e(\varrho, \Im\varrho) \\ &\quad + \lambda_4(\Im^{n-1}\varrho)\, \frac{b_e(\varrho, \Im\varrho)[1 + b_e(\Im^{n-1}\varrho, \varrho)]}{w(\Im^{n-1}\varrho, \varrho)[1 + b_e(\Im^{n-1}\varrho, \varrho)]} \\ &\quad + \lambda_5(\Im^{n-1}\varrho)\, \frac{b_e(\Im^{n-1}\varrho, \varrho).b_e(\varrho, \Im\varrho)}{w(\Im^{n-1}\varrho, \varrho).b_e(\Im^{n-1}\varrho, \varrho)} \\ &\leq [\lambda_1(\Im^{n-1}\varrho) + \lambda_2(\Im^{n-1}\varrho)]\, b_e(\Im^{n-1}\varrho, \varrho) \\ &\quad + [\lambda_3(\Im^{n-1}\varrho) + \lambda_4(\Im^{n-1}\varrho) + \lambda_5(\Im^{n-1}\varrho)]b_e(\varrho, \Im\varrho). \end{aligned}$$

Using property of $w(\zeta, \xi) \geq 1$, we have

$$b_e(\varrho, \Im\varrho) \leq \lambda(\Im^{n-1}\varrho)\, b_e(\Im^{n-1}\varrho, \varrho).$$

Repeating this process, we get

$$b_e(\varrho, \Im\varrho) \leq \Pi_{i=1}^{n}[\lambda(\Im^{i-1}\varrho)\, b_e(\Im^{i-1}\varrho, \varrho)].$$

Taking $n \to \infty$ in the above inequality and using properties of $\lambda < 1$, we encounter a contradiction, leading us to the conclusion that $b_e(\varrho, \Im\varrho) = 0$. Therefore, $Fix(\Im^n) = Fix(\Im)$ for every $n \in \mathbb{N}$, □

15.4 Illustrations

Example 15.2 Let $\Omega = \{1, \frac{1}{2}, \frac{1}{3}, \frac{1}{4}, \frac{1}{5}\}$. Define $b_e : \Omega^2 \to \mathbb{R}_+$ so that $b_e(\zeta, \xi) = b_e(\xi, \zeta)$ for all $\zeta, \xi \in \Omega$, and

$$b_e(\frac{1}{2},\frac{1}{3}) = 0.07, b_e(\frac{1}{2},\frac{1}{4}) = 0.015, b_e(\frac{1}{4},\frac{1}{5}) = 0.02,$$

$$b_e(\frac{1}{3},\frac{1}{4}) = 0.02, b_e(\frac{1}{3},\frac{1}{5}) = 0.02,$$

$b_e(\zeta,\xi) = (\zeta - \xi)^2$ otherwise. Then (Ω, b_e) is an EBbDS with $w(\zeta,\xi) = \zeta + \xi + 2$ but neither a BDS (Ω, b) nor a metric space (Ω, d). For instances,

$$b_e(\frac{1}{2},\frac{1}{3}) = 0.07 \nleq 0.035 = b_e(\frac{1}{2},\frac{1}{4}) + b_e(\frac{1}{4},\frac{1}{3})$$

and

$$b_e(\frac{1}{2},\frac{1}{3}) = 0.07 \nleq 0.055 = b_e(\frac{1}{2},\frac{1}{4}) + b_e(\frac{1}{4},\frac{1}{5}) + b_e(\frac{1}{5},\frac{1}{3})$$

but

$$b_e(\frac{1}{2},\frac{1}{3}) = 0.07 \leq 0.154 = w(\zeta,\xi)[b_e(\frac{1}{2},\frac{1}{4}) + b_e(\frac{1}{4},\frac{1}{5}) + b_e(\frac{1}{5},\frac{1}{3})].$$

Define self-mapping $\Im$ on Ω, $\alpha\colon \Omega^2 \to \mathbb{R}_+$ as

$$\Im : \begin{pmatrix} \frac{1}{5} & \frac{1}{4} & \frac{1}{3} & \frac{1}{2} & 1 \\ \\ \frac{1}{3} & \frac{1}{2} & \frac{1}{5} & \frac{1}{2} & \frac{1}{4} \end{pmatrix},$$

$$\alpha(\zeta,\xi) = \begin{cases} 1, & (\zeta,\xi) \in (1,\frac{1}{5}) \cup (1,\frac{1}{2}) \cup (\frac{1}{5},1) \cup (\frac{1}{2},1) \\ 0, & \text{otherwise,} \end{cases}$$

and $\lambda_i : \Omega \to [0,1)$ by $\lambda_i(\zeta) = \lambda(\zeta) = \frac{2\zeta}{3}$ for all $\zeta \in \Omega$, $i \in \{1,2,3,4,5\}$. Take $\theta(t) = e^{t.e^t}$ and $\gamma = \frac{1}{2}$.

It is easy to see that $\Im \in \mathfrak{W}(\Omega,\alpha)$. We will verify that $\Im$ satisfy (15.1) for $\zeta \neq \xi$ with $\Im(\zeta) \neq \Im(\xi)$ and $\alpha(\zeta,\xi) > 1$. We demonstrate by three non-trivial possible cases. Here $w(\zeta,\xi) \in [2.4, 6]$.

Case 1. $\zeta = \frac{1}{2}, \xi = 1$ (or vice versa if ζ, ξ swap the values). Then $b_e(\Im\zeta, \Im\xi) = 0.015$, $w(\zeta,\xi) = 3.5$, $\lambda(\zeta) = 0.333$. Then

$$\theta(w(\zeta,\xi) b_e(\Im\zeta,\Im\xi)) = 1.056889$$

and

$$\left[\theta\begin{pmatrix} \lambda_1(\zeta)\, b_e(\zeta,\xi) + \lambda_2(\zeta)\, b_e(\zeta,\Im\zeta) + \lambda_3(\zeta)\, b_e(\xi,\Im\xi) \\ +\lambda_4(\zeta)\, \frac{b_e(\xi,\Im\xi)[1+b_e(\zeta,\Im\zeta)]}{w(\zeta,\xi)[1+b_e(\zeta,\xi)]} + \lambda_5(\zeta)\, \frac{b_e(\zeta,\Im\zeta).b_e(\xi,\Im\xi)}{w(\zeta,\xi).b_e(\zeta,\xi)} \end{pmatrix}\right]^{\gamma}$$

$$= 1.239063137.$$

Therefore, (15.1) implies that $1.056889 < 1.239063137$, and (15.1) is holds true.

Case 2. $\zeta = \frac{1}{5}, \xi = 1$ (or vice versa if ζ, ξ swap the values). Then $b_e(\Im\zeta, \Im\xi) = 0.02$, $w(\zeta, \xi) = 3.2$, $\lambda(\zeta) = 0.1333$. Then

$$\theta(w(\zeta, \xi)b_e(\Im\zeta, \Im\xi)) = 1.070611428$$

and

$$\left[\theta\begin{pmatrix}\lambda_1(\zeta)\, b_e(\zeta, \xi) + \lambda_2(\zeta)\, b_e(\zeta, \Im\zeta) + \lambda_3(\zeta)\, b_e(\xi, \Im\xi) \\ +\lambda_4(\zeta)\, \frac{b_e(\xi,\Im\xi)[1+b_e(\zeta,\Im\zeta)]}{w(\zeta,\xi)[1+b_e(\zeta,\xi)]} + \lambda_5(\zeta)\, \frac{b_e(\zeta,\Im\zeta).b_e(\xi,\Im\xi)}{w(\zeta,\xi).b_e(\zeta,\xi)}\end{pmatrix}\right]^{\gamma}$$
$$= 1.112406667.$$

Therefore, (15.1) implies that $1.070611428 < 1.112406667$, and (15.1) is holds true.

As a result, all of the theorem's conditions are met, and the $\Im$ has a unique fixed point. (which is $\zeta^* = \frac{1}{2}$).

It is worth noting that in this example, the use of weakly α-admissibility and $\lambda(\zeta)$ were crucial because, e.g., if we take $\zeta = \frac{1}{5}, \xi = \frac{1}{2}$, we get $b_e(\Im\zeta, \Im\xi) = 0.07$, $w(\zeta, \xi) = 2.7$, $\lambda(\zeta) = 0.1333$. Then

$$\theta(w(\zeta, \xi)b_e(\Im\zeta, \Im\xi)) = 1.25648701$$

and

$$\left[\theta\begin{pmatrix}\lambda_1(\zeta)\, b_e(\zeta, \xi) + \lambda_2(\zeta)\, b_e(\zeta, \Im\zeta) + \lambda_3(\zeta)\, b_e(\xi, \Im\xi) \\ +\lambda_4(\zeta)\, \frac{b_e(\xi,\Im\xi)[1+b_e(\zeta,\Im\zeta)]}{w(\zeta,\xi)[1+b_e(\zeta,\xi)]} + \lambda_5(\zeta)\, \frac{b_e(\zeta,\Im\zeta).b_e(\xi,\Im\xi)}{w(\zeta,\xi).b_e(\zeta,\xi)}\end{pmatrix}\right]^{\gamma}$$
$$= 1.00746961.$$

Therefore (15.1) implies that $1.25648701 > 1.00746961$, and (15.1) is does not hold true. Even no contractive condition for any $\lambda(\zeta) < 1$ can be chosen which would hold for these points.

Example 15.3 Consider $\Omega = [0, 1]$ and define $b_e : \Omega^2 \to \mathbb{R}_+$ by $b_e(\zeta, \xi) = |\zeta - \xi|^2$. Then (Ω, b_e) is an EBbDS with $w(\zeta, \xi) = \zeta + \xi + 2.5$ but neither a BDS (Ω, b) nor a metric space (Ω, d). For instances,

$$b_e(0, 1) = 1 \nleq 0.5 = b_e(0, 0.5) + b_e(0.5, 1)$$

and

$$b_e(0, 1) = 1 \nleq 0.4902 = b_e(0, 0.5) + b_e(0.5, 0.99) + b_e(0.99, 1)$$

but

$$\begin{aligned} b_e(\zeta,\xi) &= |\zeta-\xi|^2 \\ &= |\zeta-\mu+\mu-\upsilon+\upsilon-\xi|^2 \\ &\le |\zeta-\mu|^2+|\mu-\upsilon|^2+|\upsilon-\xi|^2+2|\zeta-\mu||\mu-\upsilon| \\ &\quad +2|\mu-\upsilon||\upsilon-\xi|+2|\upsilon-\xi||\zeta-\mu| \\ &\le \left(\zeta+\xi+\frac{5}{2}\right)[|\zeta-\mu|^2+|\mu-\upsilon|^2+|\upsilon-\xi|^2] \\ &= w(\zeta,\xi)[b_e(\zeta,\mu)+b_e(\mu,\ \upsilon)+b_e(\upsilon,\xi)] \end{aligned}$$

for all $\zeta, \xi, \mu, \upsilon \in \Omega$.

Define self-mapping $\Im$ on Ω given by $\Im(\zeta) = \frac{\zeta^2}{2}$. Taking $\lambda_i : \Omega \to [0, 1)$ by $\lambda_i(\zeta) = \lambda(\zeta) = \frac{8+\zeta}{10}$ for all $\zeta \in \Omega$, $i \in \{1, 2, 3, 4, 5\}$ and $\alpha : \Omega^2 \to [0, \infty)$ as $\alpha(\zeta, \xi) = 1$ for $\zeta, \xi \in \Omega$. Take $\theta(t) = exp(texp(t))$ and $\gamma = \frac{9}{10}$.

It is obvious to see $\Im \in \Upsilon(\Omega, \alpha)$. Here $w(\zeta, \xi) \in (2, 4)$.

We will verify that $\Im$ satisfy (15.1) for $\zeta \neq \xi$ with $\Im(\zeta) \neq \Im(\xi)$ and $\alpha(\zeta, \xi) > 1$. For instance, we demonstrate that the (15.1) is valid in two cases:

Case 1. $\zeta = 0, \xi = 1$ (or vice versa if ζ, ξ swap the values). Then

$$\theta(w(\zeta,\xi)b_e(\Im\zeta,\Im\xi)) = 8.1581373$$

and

$$\left[\theta\begin{pmatrix}\lambda_1(\zeta)\, b_e(\zeta,\xi)+\lambda_2(\zeta)\, b_e(\zeta,\Im\zeta)+\lambda_3(\zeta)\, b_e(\xi,\Im\xi) \\ +\lambda_4(\zeta)\, \frac{b_e(\xi,\Im\xi)[1+b_e(\zeta,\Im\zeta)]}{w(\zeta,\xi)[1+b_e(\zeta,\xi)]}+\lambda_5(\zeta)\, \frac{b_e(\zeta,\Im\zeta).b_e(\xi,\Im\xi)}{w(\zeta,\xi).b_e(\zeta,\xi)}\end{pmatrix}\right]^\gamma$$

$$= 13.32.$$

Therefore, (15.1) implies that $8.1581373 < 13.32$, and (15.1) is holds true.

Case 2. $\zeta = 1, \xi = 0.9$ (or vice versa if ζ, ξ change places). Then

$$\theta(w(\zeta,\xi)b_e(\Im\zeta,\Im\xi)) = 1.204644$$

and

$$\left[\theta\begin{pmatrix}\lambda_1(\zeta)\, b_e(\zeta,\xi)+\lambda_2(\zeta)\, b_e(\zeta,\Im\zeta)+\lambda_3(\zeta)\, b_e(\xi,\Im\xi) \\ +\lambda_4(\zeta)\, \frac{b_e(\xi,\Im\xi)[1+b_e(\zeta,\Im\zeta)]}{w(\zeta,\xi)[1+b_e(\zeta,\xi)]}+\lambda_5(\zeta)\, \frac{b_e(\zeta,\Im\zeta).b_e(\xi,\Im\xi)}{w(\zeta,\xi).b_e(\zeta,\xi)}\end{pmatrix}\right]^\gamma$$

$$= 218.93$$

Therefore, (15.1) implies that $1.204644 < 218.93$, and (15.1) is holds true.

Similarly, it can be verified for any $\zeta \neq \xi \in \Omega$ with $\alpha(\zeta, \xi) \geq 1$. Consequently, all conditions of the theorem are satisfied, and $Fix(\Im) = \{0\}$ is a unique set.

15.5 Application

Consider the boundary value problem :

$$\begin{cases} \zeta''''(\rho) = Q(\rho, \zeta(\rho)), & 0 < \rho < 1; \\ \zeta(0) = \zeta'(0) = \zeta''(1) = \zeta'''(1) = 0, \end{cases} \tag{15.7}$$

with $I = [0, 1]$ and $Q \in C(I \times \mathbb{R}, \mathbb{R})$. This problem is equivalent to the integral equation

$$\zeta(\rho) = \int_0^1 H(\rho, \sigma) Q(\sigma, u(\sigma))\, d\sigma, \quad \text{for } \rho \in I, \tag{15.8}$$

where $H : I \times I \to [0, \infty)$ is the Green function given by

$$H(\rho, \sigma) = \frac{1}{6} \begin{cases} \sigma^2(3\rho - \sigma), & 0 \le \sigma \le \rho \le 1 \\ \rho^2(3\sigma - \rho), & 0 \le \rho \le \sigma \le 1. \end{cases}$$

Consider the set $\Omega = C(I, \mathbb{R}) := \{\zeta : I \to \mathbb{R} \mid \zeta \text{ is continuous on } I\}$ and define an EBbDS $b_e : \Omega \times \Omega \to [0, \infty)$ by

$$b_e(\zeta, \xi) = \max_{\rho \in I} |\zeta(\rho) - \xi(\rho)|^2 \ \text{ for } \ \zeta, \xi \in \Omega.$$

Then (Ω, b_e) is a complete EBbDS with $w(\zeta, \xi) = |u(\sigma)| + |v(\sigma)| + 2$.

Theorem 15.4 *Let $\Im : \Omega \times \Omega$ be the operator defined by*

$$\Im\zeta(\rho) = \int_0^1 H(\rho, \sigma) Q(\rho, \zeta(\sigma))\, d\sigma \ \text{ for } \ \sigma \in I.$$

Also, let $\hbar : \mathbb{R} \times \mathbb{R} \to \mathbb{R}$ be a given function. Suppose the following assertions hold:

(C1) $Q : I \times \mathbb{R} \to \mathbb{R}$ *is a continuous function, non-decreasing in the second variable;*
(C2) *there exists* $\zeta_0 \in \Omega$ *such that* $\hbar(\zeta_0(\rho), \Im\zeta_0(\rho)) \ge 0$ *for all* $\rho \in I$;
(C3) $\zeta \in \Omega$ *and* $\hbar(\zeta(\rho), \Im\zeta(\rho)) \ge 0$ *for all* $\rho \in I$ *implies that* $\hbar(\Im\zeta(\rho), \Im\Im\zeta(\rho)) \ge 0$ *for all* $\rho \in I$;
(C4) *for* $\zeta, \xi \in \Omega$ *with* $\hbar(\zeta, \xi) \ge 0$, *and* $\sigma \in I$, *we have*

$$(|\zeta(\sigma)| + |\xi(\sigma)| + 2)|Q(\sigma, \zeta(\sigma)) - Q(\sigma, \xi(\sigma))| \le [M(\zeta, \xi)(\sigma)]^{\frac{1}{2}}, \tag{15.9}$$

where

$$M(\zeta,\xi)(\sigma) = \left\{ \begin{array}{l} \lambda_1(\zeta(\sigma))|\zeta(\sigma)-\xi(\sigma)|^2 + \lambda_2(\zeta(\sigma))|\zeta(\sigma)-\Im\zeta(\sigma)|^2 \\ +\lambda_3(\zeta(\sigma))|\xi(\sigma)-\Im\xi(\sigma)|^2 \\ +\lambda_4(\zeta(\sigma))\frac{|\xi(\sigma)-\Im\zeta(\sigma)|^2[1+|\zeta(\sigma)-\Im\xi(\sigma)|^2]}{(|\zeta(\sigma)|+|\xi(\sigma)|+2)[1+|\zeta(\sigma)-\xi(\sigma)|^2]} \\ +\lambda_5(\zeta(\sigma))\frac{|\zeta(\sigma)-\Im\upsilon(\sigma)|^2|\xi(\sigma)-\Im\zeta(\sigma)|^2}{(|\zeta(\sigma)|+|\xi(\sigma)|+2)|\zeta(\sigma)-\xi(\sigma)|^2} \end{array} \right\}^{\frac{1}{2}};$$

(C7) $\sup_{\rho\in I}\int_0^1 (H(\rho,\sigma))^2 d\sigma \leq \frac{1}{2}$.

Then there exists a solution of the integral equation (15.8), *and hence, there exists a solution of the problem* (15.7).

Proof Define a function $\alpha : \Omega \times \Omega \to [0,\infty)$ by

$$\alpha(\zeta,\xi) = \begin{cases} 1, & \text{if } \hbar(\zeta(\rho),\xi(\rho)) \geq 0, \text{ for all } \rho \in I \\ \eta, & \text{otherwise,} \end{cases} \tag{15.10}$$

where $\eta \in (0,1)$. It is easy to see that the assumption (C2) readily implies the condition ($\mathscr{B}1$) of Theorem 15.1. The assumption (C3) easily implies that $\Im \in \Upsilon(\Omega,\alpha)$.

Next, we verify that $\Im \in \mathfrak{J}(\Omega,\alpha,\lambda)$. For this, let $\zeta,\xi \in \Omega$ be such that $\alpha(\zeta,\xi) \geq 1$, i.e., $\hbar(\zeta(\rho),\xi(\rho)) \geq 0$ for all $\rho \in I$. Using the fact $ln(\theta(\gamma\,\sigma)) \leq \gamma ln(\theta(\sigma))$ for each $\sigma \in I$, by the definition of operator $\Im$, we have

$$\begin{aligned} & ln[\theta(w(\zeta,\xi)b_e(\Im\zeta,\Im\xi))] \\ & = ln\left[\theta\left(\begin{array}{l}\max_{\sigma\in I}(|\zeta(\sigma)|+|\xi(\sigma)|+2) \\ \times\left(\left|\int_0^1 H(\rho,\sigma)Q(\sigma,\zeta(\sigma))d\sigma - \int_0^1 H(\rho,\sigma)Q(\sigma,\xi(\sigma))d\sigma\right|\right)\end{array}\right)^2\right] \\ & \leq ln\left[\theta\left(\max_{\sigma\in I}\left(\int_0^1 H(\rho,\sigma)(|\zeta(\sigma)|+|\xi(\sigma)|+2)\,|Q(\sigma,\zeta(\sigma))-Q(\sigma,\xi(\sigma))|\,d\sigma\right)\right)^2\right] \\ & \leq ln\left[\theta\left(\left(\max_{\sigma\in I}\int_0^1 H(\rho,\sigma)\,[M(\zeta,\xi)(\sigma)]^{\frac{1}{2}}\,d\sigma\right)^2\right)\right] \\ & \leq ln\left[\theta\left(\left(\max_{\sigma\in I}\int_0^1 [H(\rho,\sigma)]^2 d\sigma\right)^2\left(\max_{\sigma\in I}\int_0^1 [M_\Im(\zeta,\xi)]^{\frac{1}{2}}\,d\sigma\right)^2\right)\right] \\ & \leq ln[\theta(\frac{1}{4}M_\Im(\zeta,\xi))] \\ & \leq \frac{1}{4}ln[\theta(M_\Im(\zeta,\xi))] \end{aligned}$$

which implies

$$\theta(\omega(\zeta,\xi))b_e(\Im\zeta,\Im\xi)) \leq [\theta(M_\Im(\zeta,\xi))]^{1/4}$$

Thus, all the hypotheses of Theorem 15.1 are fulfilled. Thus, there exists a fixed point of $\Im$, that is, a continuous function $\zeta^* \in C(I, \mathbb{R})$ such that $\Im\zeta^* = \zeta^*$, that is,

$$\zeta^*(\rho) = \Im\zeta^*(\rho) = \int_0^1 H(\rho, \sigma)Q(\sigma, \zeta^*(\sigma))\, d\sigma.$$

Consequently, ζ^* is a solution of the boundary value problem (15.7). □

References

1. Abdeljawad, T., Karapinar, E., Panda, S.K., Mlaiki, N.: Solutions of boundary value problems on extended-Branciari b-distance. J. Ineq. Appl. **2020**(103), 1–16 (2020)
2. Bakhtin, I.A.: The contraction mapping principle in quasi metric spaces. Funkc. Anal. Ulianowsk Gos. Ped. Inst. **30**, 243–253 (1999)
3. Czerwik, S.: Contraction mappings in b-metric spaces. Acta Math. Inform. Univ. Ostraviensis. **5**, 5–11 (1993)
4. Czerwik, S.: Nonlinear set-valued contraction mappings in b-metric spaces. Atti Sem. Mat. Fis. Univ. Modena **46**, 263–276 (1998)
5. Branciari, A.: A fixed point theorem for mappings satisfying a general contractive condition of integral type. Int. J. Math. Math. Sci. **29**(9), 531–536 (2002)
6. Chen, L., Huang, S., Li, C., Zhao, Y.: Several fixed-point theorems for F-contractions in complete Branciari b-metrics, and applications. J. Funct. Spaces **2020**(7963242), 1–10 (2020)
7. Baleanu, D., Rezapour, S., Mohammadi, M.: Some existence results on nonlinear fractional differential equations. Philos. Trans. A **371**(20120144) (2013)
8. Ding, S., Imdad, M., Radenović, S., Vujakovic, J.: On some fixed point results in b-metric, rectangular and b-rectangular metric spaces. Arab J. Math. Sci. **22**(2), 151–164 (2016)
9. Dukić, D., Kadelburg, Z., Radenović, S.: Fixed points of Geraghty-type mappings in various generalized metric spaces. Abstract Appl. Anal. **561245**, 1–13 (2011)
10. Jain, R.: Fixed point results in partially ordered b-metric spaces. Inter. J. Pure Appl. Math. **120**(8), 177–185 (2018)
11. Jain, R.: Fixed point theorems for generalized weak contraction mapping in generating space of b-dislocated metric spaces. Asian-European J. Math. (2020) https://doi.org/10.1142/S1793557121501205
12. Jain, R., Nashine, H.K., Kadelburg, Z.: Some fixed point results on relational quasi partial metric spaces and application to non-linear matrix equations. Symmetry **13**(6), 993 (2021)
13. Jain, R., Nashine, H.K., Lee, J.R., Park, C.: Unified relational-theoretic approach in metric-like spaces with an application. AIMS Math. **6**(8), 8959–8977 (2021)
14. Jain, R., Kalaiarasi, R., Bag, U.: Fixed points results on a partially ordered multiplicative metric spaces. TWMS J. Appl. Eng. Math. **12**(4), 1295–1300 (2022)
15. Jain, R.: Strong convergence for α-nonexpansive mapping using a partial order induced by a function. Inter. J. Nonlinear Anal. Appl. **13**(2), 2503–2508 (2022)
16. Jain, R., Nashine, H.K.: Manuel De la Sen, An implicit relation, relational theoretic approach under w-distance and application to nonlinear matrix equations. J. Inequ. Appl. **2021**, 169 (2021)
17. Jain, R., Nashine, H.K., Kumar, S.: An implicit relation approach in metric spaces under w-distance and application to fractional differential equation. J. Funct. Spaces **9928881**, 12 (2021)
18. Jain, R., Nashine, H.K., Kadelburg, Z.: Positive solutions of nonlinear matrix equations via fixed point results in relational metric spaces with w-distance. Filomat **36**(14), 4811–4829 (2022)

19. Jain, R., Nashine, H.K., Kim, J.K.: Positive solutions for a nonlinear matrix equation using fixed point results in extended Branciari b-distance spaces. Nonlinear Funct. Anal. Appl. **27**(4), 709–730 (2022)
20. Nashine, H.K., Jain, R., Parvaneh, V.: A relational-theoretic approach to get solution of nonlinear matrix equations. J. Inequ. Appl. **2022**(79), 1–19 (2022)
21. Jeong, G.S., Rhoades, B.E.: Maps for which $F(T) = F(T^n)$. Fixed Point Theory Appl. **6**, 87–131 (2005)
22. Jleli, M., Samet, B.: A new generalization of the Banach contraction principle. J. Inequ. Appl. **38**, 1–8 (2014)
23. Kamran, T., Samreen, M., Ain, O.U.: A generalization of b-metric space and some fixed point theorems. Mathematics **5**(2), 19 (2017)
24. Mitrović, Z.D., Isık, H., Radenović, S.: The new results in extended b-metric spaces and applications. Int. J. Nonlinear Anal. Appl. **11**(1), 473–482 (2020)
25. Mustafa, Z., Parvaneh, V., Roshan, J.R., Kadelburg, Z.: b2-Metric spaces and some fixed point theorems. Fixed Point Theory Appl. **2014**(144), 1–23 (2014)
26. Rakić, D., Mukheimer, A., Došenović, T., Mitrović, Z.D., Radenović, S.: On some new fixed point results in fuzzy b-metric spaces. J. Ineq. Appl. **2020**(99), 1–14 (2020)
27. Samet, B., Vetro, C., Vetro, P.: Fixed point theorems for α-ψ-contractive type mappings. Nonlinear Anal. **75**, 2154–2165 (2012)
28. Sintunavarat, W.: A new approach to $\alpha - \psi$-contractive mappings and generalized Ulam-Hyers stability, well-posedness and limit shadowing results. Carpathian J. Math. **31**(3), 395–401 (2015)

Chapter 16
Reliability Analysis of a Batch Arrival Poisson Input Queue with Two Kinds of Heterogeneous Service and a General Start-Up Time Under $N - policy$

Anjana Begum and Gautam Choudhury

Abstract This article considers a single server vacation queueing model that requires some additional time to start a busy period (start-up time) under *N-policy* subject to unpredictable server breakdown. A single server provides two different services to its customers arriving in batches where the customers can facilitate optional repeated service once if desired. The server is allowed to take a single vacation after each busy period following a Bernoulli trial. The start-up, service, repeated service, and vacation times are generally distributed. For this model, the system reliability function (RF) in terms of Laplace Stieltjes Transform (LST) and mean time to first failure (MTTFF) of the server are analyzed analytically. Finally, the underlying model has been discussed in the context of functional analysis.

Keywords Reliability funtion · MTTFF · Breakdown · Re-service · Setup time · N-policy

16.1 Introduction

Queueing models having two kinds of service with an optional repeated service facility is characterized by the feature that the server renders two different kinds of services to its customers, and a customer on arrival can choose either kind of service. However, if the customers are not convinced with the quality of the service received, then they have an advantage of repeating the same service once. Madan et al. [18] discussed such type of model for batch arrival queue. For further discussion on this kind of model, we mention the works of Al-khedhari and Tadj [2], Baruah et al. [20], Choudhury and Kalita [8].

Now, adding more flexibility in the system server vacation is one of the important phenomena that can be considered. Queueing systems with server absence or vacation

A. Begum (✉) · G. Choudhury
Mathematical Sciences Division, Institute of Advanced Study in Science and Technology, Paschim Boragaon, Guwahati 781035, Assam, India
e-mail: ab.stats.ghy@gmail.com

B. Hazarika et al. (eds.), *Advances in Functional Analysis and Fixed-Point Theory*, Industrial and Applied Mathematics,
https://doi.org/10.1007/978-981-99-9207-2_16

have earned a lot of consideration in the last few decades owing to their wide range of applications in various problems like power plants, hospitals, call centers, banks, telecommunications, etc. For a detailed discussion on vacation queues, one can refer to the excellent works of Doshi [4], Takagi [11], and Tian and Zhang [21]. There are different vacation models available in the literature, out of which, we have considered the Bernoulli schedule under a single vacation policy. The concept of the classical vacation system under the Bernoulli policy was first introduced by Keilson and Servi [16]. Kella [22] suggested a generalized Bernoulli schedule in which if the queue is empty on his return, the server can take k consecutive vacations with probability p_k. A fundamental difference in the Bernoulli vacation policy between the earlier referred work and the present study is that in the articles cited above Bernoulli vacation commences only in customer's absence. However, the current investigation is more generic because it allows a vacation even in the presence of a customer. As such the vacation mechanism is characterized by the fact that, after each service commencement, the server can either undergo a vacation of arbitrary length or stay in the system and continue serving. To refer the readers to some recent supports on Bernoulli vacation, we cite the works by Madan et al. [17], Choudhury and Deka [10], Begum and Choudhury [1] among others.

Vacation queueing models with $N - policy$ is also an essential topic of discussion to handle the vacation period and queue size. According to this policy, the server becomes active when the queue size reaches a particular threshold value, say $N \geq 1$ and deactivates itself when the system is empty and waits in the system till the queue size builds up to N again. Lee and Srinivasan [12] were the first to study a batch arrival $N - policy$ queue. Later, Lee et al. [13] made an in-depth investigation on this model through different approaches. $N - policy$ queue with server start-up time where the server before activating itself for services, performs some preparatory job can be seen in Takagi [11], Medhi and Templeton [15] for an $M/G/1$ queueing system. Hur and Ahn [24] have analyzed an $M^X/G/1$ queue with server vacations and start-up period. Among some similar works in this area, we mention few works by Arumuganathan and Jeyakumar [23], Wang et al. [19], Choudhury et al. [9], and Kalita and Choudhury [6].

Though various aspects of queueing features under $N - policy$ with server start-up period has been discussed by many researchers separately; however, to the best of authors knowledge, no work on reliability analysis can be found in queueing literature on two kinds of service with optional repeated service under Bernoulli vacation schedule, N-policy and server start-up period for an unreliable server. Thus, through this investigation, we have analyzed the RF and MTTFF of an unreliable single server queue with two heterogeneous services under repeated service approach, Bernoulli vacation schedule, N-policy, and server start-up period. To cite a few studies available in the literature conducting the reliability analysis for an unreliable server, we cite the works of Li et al. [25], Tang [11], Wang [14], and Kalita and Choudhury [5].

16.2 Application of the Proposed Model

The applicability of this kind of queueing model from a practical point of view can be observed in the production process, manufacturing system, inventory system, communication networks, shopping malls, cafeterias, and many more.

For example, in a production-to-order system (POS), if two kinds of commodities are manufactured, the model under study may find an application in such systems. POS is a production scheme in production planning and management. Since the customer's demand for a commodity is highly random in such systems, we can model it as a Poisson Process. Let us assume that the arrival of raw materials for both the kinds of commodities are in bulk following the compound Poisson process and the production house does not start its production until some specific number of raw materials say $N \geq 1$ accumulates in an idle period. We also make the assumption that the machine undergoes some extra preventive operations before processing the raw materials indicating a start-up period. However, reproduction of some commodities may be required if the customers are not happy with the level of quality received. After each order completion, the production facility can take a break/perform some maintenance activity, or produce the next order in line, implying a Bernoulli vacation. In practice, it may so happen that, while producing a commodity of any kind, the progress of the production may be interrupted due to abrupt failure of machinery, thereby signifying a breakdown which can be fixed at a repair house. Since the performance of a POS is immensely affected by machine failures, it is worthy of studying such a system from the reliability point of view.

16.3 Mathematical Model

We consider an $M^X/G/1/N - policy$ queue with start-up time. Here, the server is assumed to become operative only when the number of customers builds up to a threshold value, i.e., $N \geq 1$ (build-up period). After the queue size becomes equal to or exceeds the value N, it does not become operative instantly but rather takes up some gear-up time, also known as start-up time, before starting its actual service. These two periods constitute an idle period, i.e., idle period = build-up period + start-up period, on completion of which busy period begins. Once the server becomes operative, it provides two heterogeneous services to its customers with an option of repeating the same service once if they are not contented with the quality of the service provided. After each busy period, the server can perform some maintenance work, do some secondary job, take a break of random duration with some probability, or continue serving with its complementary probability.

The server considered here is assumed to be unreliable, i.e., the server may be subject to unpredictable failures during which the server is incapable of rendering service to the customers. A sample path of the model discussed above is shown in Fig. 16.1.

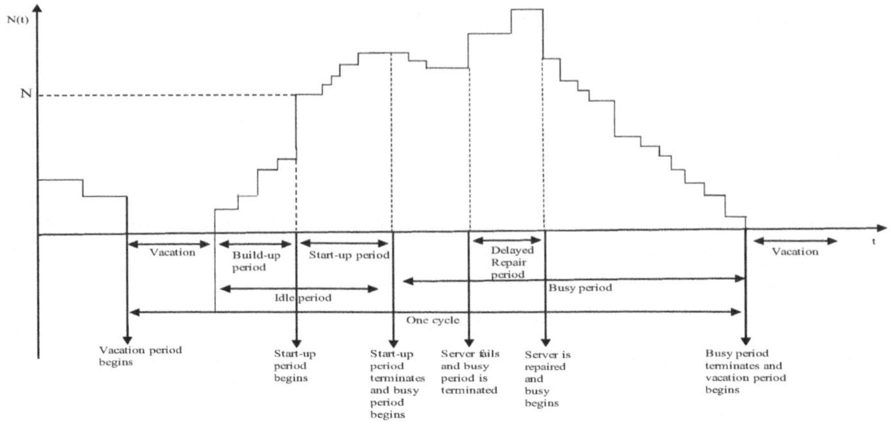

Fig. 16.1 Sample Path of the proposed model

Considering general service, repeated service, vacation, and start-up time distributions, we develop an unreliable non-Markovian single server model under the assumption that customers join the system in bulk. The necessary notations, assumptions, random variables (r.v), and probabilities used to illustrate the model under consideration are defined in Table 16.1.

Now, let us define a r.v $\sigma(t)$ at time t, which corresponds to the various status of the server as

$$\sigma(t) = \begin{cases} 0, & \text{idle} \\ 1, & \text{start-up period} \\ 2, & \text{rendering the first kind of service} \\ 3, & \text{rendering the second kind of service} \\ 4, & \text{rendering the first kind of repeated service} \\ 5, & \text{rendering the second kind of repeated service} \\ 6, & \text{vacation} \end{cases}$$

Then a bi-variate Markov process $\{N(t), A(t); t \geq 0\}$ is considered where $A(t) = 0, G^0(t), S_1^0(t), S_2^0(t), R_1^0(t), R_2^0(t), V^0(t)$; when $\sigma(t) = 0, 1, 2, 3, 4, 5, 6$, respectively.

Now, the limiting probabilities for the system states provided y as the elapsed time under the steady-state environment are defined as

$$Q_n(t) = Pr\{N(t) = n, \sigma(t) = 0\}; \qquad n = 0, 1, \cdots, N-1$$

$$f_{G_n}(y; t)dy = Pr\{N(t) = n, \sigma(t) = 1; y < G^0(t) \leq y + dy\}; \; y > 0; \; n = N, N+1, \cdots.$$

Table 16.1 Notations and parameters

λ	Arrival rate of customers
X	Size of a batch
w_l	Pr[X=l]; l=1,2,...
$P_X(s)$	Probability generating function of X
$\mu_X^{(l)}$	l^{th} finite factorial moments of X ($l = 1, 2, \cdots$)
$N(t)$	System size at time t
q_1/q_2	Probability to opt the first/second kind of service (FKS/SKS)
z_1/z_2	Probability to opt the first/second kind of repeated service (FKRS/SKRS)
$(1-z_1)/(1-z_2)$	Probability of leaving the system without repeating first/second kind of service
p_1/p_2	Probability to opting/not opting for a vacation
$G/S_i/R_i/V$	Start-up/Service/Repeated Service/ Vacation time r.v i=1(FKS/FKRS),2(SKS/SKRS)
$F_G(y)/F_{S_i}(y)/$ $F_{R_i}(y)/F_V(y)$	Cumulative distribution function of start-up/service/ repeated service/vacation times following general distributions i=1(FKS/FKRS),2(SKS/SKRS)
$F_G^*(\varphi)/F_{S_i}^*(\varphi)/$ $F_{R_i}^*(\varphi)/F_V^*(\varphi)$	Laplace Stieltjes Transform (LST) of $F_G(y)/F_{S_i}(y)/F_{R_i}(y)/F_V(y)$ i=1(FKS/FKRS),2(SKS/SKRS)
$\mu_G^{(l)}/\mu_{S_i}^{(l)}/$ $\mu_{R_i}^{(l)}/\mu_V^{(l)}$	l^{th} finite moments of Start-up/Service/Repeated Service/Vacation time distributions i=1(FKS/FKRS),2(SKS/SKRS)
α_1/α_2	Failures rate following exponential distribution; FKS(FKRS)/SKS(SKRS)
$G^0(t)/S_i^0(t)/$ $R_i^0(t)/V^0(t)$	Elapsed Start-up/Service/ Repeated Service/Vacation time at time t i=1(FKS/FKRS),2(SKS/SKRS)

$$f_{S_{1,n}}(y;t)dy = Pr\{N(t) = n, \sigma(t) = 2;\ y < S_1^0(t) \leq y + dy\};\ \ y > 0;\ n = 1, 2, \cdots$$

$$f_{S_{2,n}}(y;t)dy = Pr\{N(t) = n, \sigma(t) = 3;\ y < S_2^0(t) \leq y + dy\};\ \ y > 0;\ n = 1, 2, \cdots$$

$$f_{R_{1,n}}(y,t)dy = Pr\{N(t) = n, \sigma(t) = 4;\ y < R_1^0(t) \leq y + dy\};\ \ y > 0;\ n = 1, 2, \cdots$$

$$f_{R_{2,n}}(y,t)dy = Pr\{N(t) = n, \sigma(t) = 5;\ y < R_2^0(t) \leq y + dy\};\ \ y > 0;\ n = 1, 2, \cdots$$

$$f_{V_n}(y;t)dy = Pr\{N(t) = n, \sigma(t) = 6;\ y < V^0(t) \leq y + dy\};\ \ y > 0;\ n = 0, 1, \cdots.$$

Since $F_G(y)/F_{S_i}(y)/F_{R_i}(y)/F_V(y)(i = 1, 2)$ denotes the CDFs, therefore we have

$F_G(0) = 0$, $F_G(\infty) = 1$, $F_{S_i}(0) = 0$, $F_{S_i}(\infty) = 1$, $F_{R_i}(0) = 0$, $F_{R_i}(\infty) = 1$, $F_V(0) = 0$, $F_V(\infty) = 1$ $(i = 1, 2)$.

Further, we assume that $F_G(y)$, $F_{S_i}(y)$, $F_{R_i}(y)$, $F_V(y)$ are continuous at $y = 0$ $(i = 1, 2)$.

Subsequently, the hazard rate functions of $G/S_i/R_i/V$ $(i = 1, 2)$ are defined as follows:

$$\beta(y)dy = \frac{dF_G(y)}{1 - F_G(y)},\ \zeta_i(y)dy = \frac{dF_{S_i}(y)}{1 - F_{S_i}(y)},\ \kappa_i(y)dy = \frac{dF_{R_i}(y)}{1 - F_{R_i}(y)},\ \vartheta(y)dy = \frac{dF_V(y)}{1 - F_V(y)}.$$

16.4 Reliability Analysis

Let the server's time to the first failure be denoted by γ, then the RF of the server can be defined as $R(t) = Pr(\gamma > t)$.

The RF of the server is attained by assuming the failure state of the server to be an absorbing state, which indeed generates a new system. Therefore, to achieve the LST of the RF, the Kolmogorov forward equations (Cox [7]) for $i = 1, 2$ can be set as follows:

$$\frac{\partial Q_0(t)}{\partial t} + \lambda Q_0(t) = \left[\sum_{i=1}^{2} p_2 \left\{(1 - z_i) \int_0^\infty f_{S_{1,i}}(y;t)\zeta_i(y)dy + \int_0^\infty f_{R_{1,i}}(y;t)\kappa_i(y)dy\right\}\right] + \left[\int_0^\infty f_{V_0}(y;t)\vartheta(y)dy\right] \tag{16.1}$$

$$\frac{\partial Q_n(t)}{\partial t} + \lambda Q_n(t) = \lambda \sum_{l=1}^{n} w_l Q_{n-l}(t);\ n = 1, 2, \cdots, N-1 \tag{16.2}$$

$$\left(\frac{\partial}{\partial t} + \frac{\partial}{\partial y}\right) f_{G_n}(y;t) + [\lambda + \beta(y)] f_{G_n}(y;t) = \lambda \sum_{l=1}^{n-N+1} w_l f_{G_{n-l+1}}(y;t);\ n = N, N+1, \cdots \tag{16.3}$$

$$\left(\frac{\partial}{\partial t} + \frac{\partial}{\partial y}\right) f_{S_{n,i}}(y;t) + [\lambda + \alpha_i + \zeta_i(y)] f_{S_{n,i}}(y;t) = \lambda \sum_{l=1}^{n} w_l f_{S_{n-l,i}}(y;t);\ n = 1, 2, \cdots \tag{16.4}$$

$$\left(\frac{\partial}{\partial t} + \frac{\partial}{\partial y}\right) f_{R_{n,i}}(y;t) + [\lambda + \alpha_i + \kappa_i(y)] f_{R_{n,i}}(y;t) = \lambda \sum_{l=1}^{n} w_l f_{R_{n-l,i}}(y;t);\ n = 1, 2, \cdots \tag{16.5}$$

$$\left(\frac{\partial}{\partial t}+\frac{\partial}{\partial y}\right) f_{V_n}(y;t)+[\lambda+\vartheta(y)]f_{V_n}(y;t)=\lambda\sum_{l=1}^{n} w_l f_{V_{n-l}}(y;t);\ n=0,1,\cdots \quad (16.6)$$

The forward equations set above are solved against some boundary constraints set below for $i = 1, 2$:

at $y = 0$**:**

$$f_{G_n}(0;t)=\lambda\sum_{l=0}^{N-1} w_{n-l}Q_l(t);\ n=N,N+1,\cdots \quad (16.7)$$

$$f_{S_{i,n}}(0;t)=p_2\left[q_i\sum_{i=1}^{2}\left\{(1-z_i)\int_0^\infty f_{S_{n+1,i}}(y;t)\zeta_i(y)dy+\int_0^\infty f_{R_{n+1,i}}(y;t)\kappa_i(y)dy\right\}\right]$$
$$+q_i\int_0^\infty f_{V_n}(y;t)\vartheta(y)dy;\ n=1,2,\cdots,N-1 \quad (16.8)$$

$$f_{S_{i,n}}(0;t)=p_2\left[q_i\sum_{i=1}^{2}\left\{(1-z_i)\int_0^\infty f_{S_{n+1,i}}(y;t)\zeta_i(y)dy+\int_0^\infty f_{R_{n+1,i}}(y;t)\kappa_i(y)dy\right\}\right]$$
$$+q_i\int_0^\infty f_{V_n}(y;t)\vartheta(y)dy+q_i\int_0^\infty f_{G_n}(y;t)\beta(y)dy;\ n=N,N+1,\cdots \quad (16.9)$$

$$f_{R_{n,i}}(0;t)=z_i\int_0^\infty f_{S_{n,i}}(y;t)\zeta_i(y)dy \quad n=1,2,\cdots \quad (16.10)$$

$$f_{V_n}(0;t)=p_1\left[\sum_{i=1}^{2}\left\{(1-z_i)\int_0^\infty f_{S_{n+1,i}}(y;t)\zeta_i(y)dy+\int_0^\infty f_{R_{n+1,i}}(y;t)\kappa_i(y)dy\right\}\right];$$
$$n=0,1,2,\cdots \quad (16.11)$$

with initial condition:
$Q_n(0)=\delta_{n,0};\quad n=0,1,\cdots$
where $\delta_{j,k}$ represents the Konecker's delta function.

Now, to solve the system of equations defined above, we introduce the LST of the probability generating function for $|s|\le 1$ ($i=1,2$) as given below:

$$P_G^*(y,\varphi;s)=\sum_{n=N}^{\infty}s^n f_{G_n}^*(y;\varphi); \qquad P_G^*(0,\varphi;s)=\sum_{n=N}^{\infty}s^n f_{G_n}^*(0;\varphi)$$

$$P_{S_i}^*(y,\varphi;s)=\sum_{n=1}^{\infty}s^n f_{S_{n,i}}^*(y;\varphi); \qquad P_{S_i}^*(0,\varphi;s)=\sum_{n=1}^{\infty}s^n f_{S_{n,i}}^*(0;\varphi)$$

$$P_{R_i}^*(y,\varphi;s)=\sum_{n=1}^{\infty}s^n f_{R_{n,i}}^*(y;\varphi); \qquad P_{R_i}^*(0,\varphi;s)=\sum_{n=1}^{\infty}s^n f_{R_{n,i}}^*(0;\varphi)$$

$$P_V^*(y,\varphi;s)=\sum_{n=0}^{\infty}s^n f_{V_n}^*(y;\varphi); \qquad P_V^*(0,\varphi;s)=\sum_{n=0}^{\infty}s^n f_{V_n}^*(0;\varphi)$$

$$Q^*(\varphi; s) = \sum_{n=0}^{N-1} s^n Q_n^*(\varphi) \quad \text{and} \quad Q_n^*(\varphi) = \int_0^\infty e^{-\varphi t} dQ_n(t);$$
$$n = 0, 1, \cdots, N-1$$

Now, we apply LST from (16.1) to (16.11) which gives the followings ($i = 1, 2$):

$$(\lambda + \varphi)Q_0^*(\varphi) - Q_0(0) = \left[\sum_{i=1}^{2} p_2 \left\{(1 - z_i)\int_0^\infty f_{S_{1,i}}^*(y; \varphi)\zeta_i(y)dy + \int_0^\infty f_{R_{1,i}}^*(y; \varphi)\kappa_i(y)dy\right\}\right] + \int_0^\infty f_{V_0}^*(y; \varphi)\vartheta(y)dy \tag{16.12}$$

$$(\lambda + \varphi)Q_n^*(\varphi) - Q_n(0) = \lambda \sum_{l=1}^{n} w_l Q_{n-l}^*(\varphi);\ n = 1, 2, \cdots, N-1 \tag{16.13}$$

$$(\varphi + \lambda + \beta(y)) f_{G_n}^*(y; \varphi) + \frac{\partial}{\partial y} f_{G_n}^*(y; \varphi) = \lambda \sum_{l=1}^{n-N+1} w_l f_{G_{n-l+1}}^*(y; \varphi);\ n = N+1, N+2, \cdots \tag{16.14}$$

$$(\varphi + \lambda + \alpha_i + \zeta_i(y)) f_{S_{n,i}}^*(y; \varphi) + \frac{\partial}{\partial y} f_{S_{n,i}}^*(y; \varphi) = \lambda \sum_{l=1}^{n} w_l f_{S_{n-l,i}}^*(y; \varphi);\ n = 1, 2, \cdots \tag{16.15}$$

$$(\varphi + \lambda + \alpha_i + \kappa_i(y)) f_{R_{n,i}}^*(y; \varphi) + \frac{\partial}{\partial y} f_{R_{n,i}}^*(y; \varphi) = \lambda \sum_{l=1}^{n} w_l f_{R_{n-l,i}}^*(y; \varphi);\ n = 1, 2, \cdots \tag{16.16}$$

$$(\varphi + \lambda + \vartheta(y)) f_{V_n}^*(y; \varphi) + \frac{\partial}{\partial y} f_{V_n}^*(y; \varphi) = \lambda \sum_{l=1}^{n} w_l f_{V_{n-l}}^*(y; \varphi);\ n = 0, 1, \cdots \tag{16.17}$$

$$f_{G_n}^*(0; \varphi) = \lambda \sum_{l=0}^{N-1} w_{n-l} Q_l^*(\varphi);\ n = N, N+1, \cdots \tag{16.18}$$

$$\begin{aligned} f_{S_{n,i}}^*(0; \varphi) &= q_i p_2 \left[(1 - z_1)\int_0^\infty f_{S_{n+1,1}}^*(y; \varphi)\zeta_1(y)dy + (1 - z_2)\int_0^\infty f_{S_{n+1,2}}^*(y; \varphi)\zeta_2(y)dy\right] \\ &+ q_i p_2 \left[\int_0^\infty f_{R_{n+1,1}}^*(y; \varphi)\kappa_1(y)dy + \int_0^\infty f_{R_{n+1,2}}^*(y; \varphi)\kappa_2(y)dy\right] \\ &+ q_i \int_0^\infty f_{V_n}^*(y; \varphi)\vartheta(y)dy;\ n = 1, 2, \cdots, N-1 \end{aligned} \tag{16.19}$$

$$f^*_{S_{n,i}}(0;\varphi) = q_i p_2\left[(1-z_1)\int_0^\infty f^*_{S_{n+1,1}}(y;\varphi)\zeta_1(y)dy + (1-z_2)\int_0^\infty f^*_{S_{n+1,2}}(y;\varphi)\zeta_2(y)dy\right]$$
$$+ q_i p_2\left[\int_0^\infty f^*_{R_{n+1,1}}(y;\varphi)\kappa_1(y)dy + \int_0^\infty f^*_{R_{n+1,2}}(y;\varphi)\kappa_2(y)dy\right]$$
$$+ q_i\int_0^\infty f^*_{V_n}(y;\varphi)\vartheta(y)dy + q_i\int_0^\infty f^*_{G_n}(y;\varphi)\beta(y)dy;\ n = N, N+1,\cdots \tag{16.20}$$

$$f^*_{R_{n,i}}(0;\varphi) = z_i\int_0^\infty f^*_{S_{n,i}}(y;\varphi)\zeta_i(y)dy;\ n = 1,2,\cdots \tag{16.21}$$

$$f^*_{V_n}(0;\varphi) = p_1\left[\sum_{i=1}^{2}\left\{(1-z_i)\int_0^\infty f^*_{S_{n+1,i}}(y;\varphi)\zeta_i(y)dy + \int_0^\infty f_{R_{n+1,i}}(y;\varphi)\kappa_i(y)dy\right\}\right] \tag{16.22}$$

Now, by multiplying Eqs. (16.15) and (16.16), respectively, by appropriate powers of s and after that taking summation over $n = 1, 2, \cdots, (i = 1, 2)$, we get two set of differential equations of Lagrangian type whose solutions are given by

$$P^*_{S_i}(y,\varphi;s) = P^*_{S_i}(0,\varphi;s)[1 - F_{S_i}(y)]exp\{-(\varphi+\alpha_i+\varepsilon(s))y\};\ y > 0 \tag{16.23}$$

$$P^*_{R_i}(y,\varphi;s) = P^*_{R_i}(0,\varphi;s)[1 - F_{R_i}(y)]exp\{-(\varphi+\alpha_i+\varepsilon(s))y\};\ y > 0 \tag{16.24}$$

where $\varepsilon(s) = \lambda(1 - P_X(s))$.
On a similar note, from Eq. (16.17), we have

$$P^*_V(y,\varphi;s) = P^*_V(0,\varphi;s)[1 - F_V(y)]exp\{-(\varphi+\varepsilon(s))y\};\ y > 0 \tag{16.25}$$

Next, we multiply (16.22) by suitable powers of s and take summation over $n = 0, 1, \cdots$ to derive the following equation:

$$sP^*_V(0,\varphi;s) = p_1\left[(1-z_1)P^*_{S_1}(0,\varphi;s)F^*_{S_1}(\varphi+\alpha_1+\varepsilon(s)) + P^*_{R_1}(0,\varphi;s)F^*_{R_1}(\varphi+\alpha_1+\varepsilon(s))\right]$$
$$+ p_1\left[(1-z_2)P^*_{S_2}(0,\varphi;s)F^*_{S_2}(\varphi+\alpha_2+\varepsilon(s)) + P^*_{R_2}(0,\varphi;s)F^*_{R_2}(\varphi+\alpha_2+\varepsilon(s))\right] \tag{16.26}$$

Similarly, from (16.21), we have

$$P^*_{R_i}(0,\varphi;s) = z_i P^*_{S_i}(0,\varphi;s)F^*_{S_i}(\varphi+\alpha_i+\varepsilon(s));\ i = 1,2 \tag{16.27}$$

Now, we simplify Eq. (16.13), which yields the following:

$$\lambda\sum_{n=N}^{\infty}\left(\sum_{l=0}^{N-1} w_{n-l}Q^*_l(\varphi)\right)s^n = \lambda Q^*(\varphi;s)P_X(s) - (\lambda+\varphi)(Q^*(\varphi;s) - Q^*_0(\varphi)) \tag{16.28}$$

Now, we multiply Eq. (16.14) by suitable powers of s and then take summation over all possible values of n gives the following result:

$$P_G^*(y,\varphi;s) = P_G^*(0,\varphi;s)[1 - F_G(y)]exp\{-(\varphi+\varepsilon(s))y\};\ y > 0 \tag{16.29}$$

Likewise, proceeding with Eq. (16.18) and thereafter using Eq. (16.28) we get the following result:

$$P_G^*(0,\varphi;s) = \lambda Q^*(\varphi;s)P_X(s) - (\lambda+\varphi)(Q^*(\varphi;s) - Q_0^*(\varphi)) \tag{16.30}$$

Putting (16.30) in (16.29), we have

$$\begin{aligned} P_G^*(y,\varphi;s) = {} & [\lambda Q^*(\varphi;s)P_X(s) - (\lambda+\varphi)(Q^*(\varphi;s) - Q_0^*(\varphi))][1 - F_G(y)] \\ & exp\{-(\varphi+\varepsilon(s))y\};\ y > 0 \end{aligned} \tag{16.31}$$

Now, we multiply (16.19) and (16.20) by suitable powers of s and take summation summing all possible values of n and after that, Eqs. (16.12), (16.26), (16.27) and (16.31) give the followings on simplification:

$$\begin{aligned} & [s - \{p_2 + p_1F_V^*(\varphi+\varepsilon(s))\}\{(1-z_i) + z_iF_{R_i}^*(\eta_i(s))\}q_iF_{S_i}^*(\eta_i(s))]P_{S_i}(0,\varphi;s) = sq_i \\ & +[\{p_2 + p_1F_V^*(\varphi+\varepsilon(s))\}\{(1-z_{i'}) + z_{i'}F_{R_{i'}}^*(\eta_{i'}(s))\}q_iF_{S_{i'}}(\eta_{i'}(s))]P_{S_{i'}}^*(0,\varphi;s) \\ & +q_is[\lambda Q^*(\varphi;s)P_X(s) - (\lambda+\varphi)(Q^*(\varphi;s) - Q_0^*(\varphi))]F_G^*(\varphi+\varepsilon(s)) - q_is(\varphi+\lambda)Q_0^*(\varphi) \end{aligned} \tag{16.32}$$

where $\eta_i(s) = (\varphi + \alpha_i + \varepsilon(s))$, $i = 1, 2$, and $i' = 2, 1$.

Now, we solve Eq. (16.32) ($i = 1, 2$) which yields the following:

$$P_{S_i}^*(0,\varphi;s) = \frac{\begin{aligned} & sq_i[1 - \{(\varphi+\lambda) \\ & Q_0^*(\varphi)\{1 - F_G^*(\varphi+\varepsilon(s))\} + (\varphi+\varepsilon(s))Q^*(\varphi;s)F_G^*(\varphi+\varepsilon(s))\}] \end{aligned}}{\begin{aligned} & [s - \{p_2 + p_1F_V^*(\varphi+\varepsilon(s))\}\{\{(1-z_1) + z_1F_{R_1}^*(\eta_1(s))\}q_1F_{S_1}^*(\eta_1(s)) \\ & + \{(1-z_2) + z_2F_{R_2}^*(\eta_2(s))\}q_2F_{S_2}^*(\eta_2(s))\}] \end{aligned}} \tag{16.33}$$

Putting Eq. (16.33) in Eq. (16.27) ($i = 1, 2$), we get the following result:

$$P_{R_i}^*(0,\varphi;s) = \frac{\begin{aligned} & sq_iz_i[1 - \{(\varphi+\lambda) \\ & Q_0^*(\varphi)\{1 - F_G^*(\varphi+\varepsilon(s))\} + (\varphi+\varepsilon(s))Q^*(\varphi;s)F_G^*(\varphi+\varepsilon(s))\}]F_{S_i}^*(\eta_i(s)) \end{aligned}}{\begin{aligned} & [s - \{p_2 + p_1F_V^*(\varphi+\varepsilon(s))\}\{\{(1-z_1) + z_1F_{R_1}^*(\eta_1(s))\}q_1F_{S_1}^*(\eta_1(s)) \\ & + \{(1-z_2) + z_2F_{R_2}^*(\eta_2(s))\}q_2F_{S_2}^*(\eta_2(s))\}] \end{aligned}} \tag{16.34}$$

Finally, by substituting (16.33) and (16.34) in (16.26), we have

$$P_V^*(0,\varphi;s)=\frac{\begin{array}{l}p_1[1-\{(\varphi+\lambda)Q_0^*(\varphi)\{1-F_G^*(\varphi+\varepsilon(s))\}+(\varphi+\varepsilon(s))Q^*(\varphi;s)F_G^*(\varphi+\varepsilon(s))\}]\\ [\{(1-z_1)+z_1F_{R_1}^*(\eta_1(s))\}q_1F_{S_1}^*(\eta_1(s))+\{(1-z_2)+z_2F_{R_2}^*(\eta_2(s))\}q_2F_{S_2}^*(\eta_2(s))]\end{array}}{\begin{array}{r}[s-\{p_2+p_1F_V^*(\varphi+\varepsilon(s))\}\{\{(1-z_1)+z_1F_{R_1}^*(\eta_1(s))\}q_1F_{S_1}^*(\eta_1(s))\\ +\{(1-z_2)+z_2F_{R_2}^*(\eta_2(s))\}q_2F_{S_2}^*(\eta_2(s))\}]\end{array}} \tag{16.35}$$

Again, from Eq. (16.23) ($i=1,2$), we have

$$\begin{aligned}P_{S_i}^*(\varphi;s)&=\int_0^\infty P_{S_i}^*(y,\varphi;s)dy\\ &=\frac{\begin{array}{l}sq_i[1-F_{S_i}^*(\eta_i(s))]\\ [1-\{(\varphi+\lambda)Q_0^*(\varphi)\{1-F_G^*(\varphi+\varepsilon(s))\}+(\varphi+\varepsilon(s))Q^*(\varphi;s)F_G^*(\varphi+\varepsilon(s))\}]\end{array}}{\begin{array}{r}\eta_i(s)[s-\{p_2+p_1F_V^*(\varphi+\varepsilon(s))\}\{\{(1-z_1)+z_1F_{R_1}^*(\eta_1(s))\}q_1F_{S_1}^*(\eta_1(s))\\ +\{(1-z_2)+z_2F_{R_2}^*(\eta_2(s))\}q_2F_{S_2}^*(\eta_2(s))\}]\end{array}}\end{aligned} \tag{16.36}$$

On a similar note, from Eq. (16.24) ($i=1,2$), we have the following result:

$$\begin{aligned}P_{R_i}^*(\varphi;s)&=\int_0^\infty P_{R_i}^*(y,\varphi;s)dy\\ &=\frac{\begin{array}{l}sq_iz_i[1-F_{R_i}^*(\eta_i(s))][1-\{(\varphi+\lambda)\\ Q_0^*(\varphi)\{1-F_G^*(\varphi+\varepsilon(s))\}+(\varphi+\varepsilon(s))Q^*(\varphi;s)F_G^*(\varphi+\varepsilon(s))\}]F_{S_i}^*(\eta_i(s))\end{array}}{\begin{array}{r}\eta_i(s)[s-\{p_2+p_1F_V^*(\varphi+\varepsilon(s))\}\{\{(1-z_1)+z_1F_{R_1}^*(\eta_1(s))\}q_1F_{S_1}^*(\eta_1(s))\\ +\{(1-z_2)+z_2F_{R_2}^*(\eta_2(s))\}q_2F_{S_2}^*(\eta_2(s))\}]\end{array}}\end{aligned} \tag{16.37}$$

Likewise, from Eq. (16.25), we attain the following:

$$\begin{aligned}P_V^*(\varphi;s)&=\int_0^\infty P_V^*(y,\varphi;s)dy\\ &=\frac{\begin{array}{l}p_1[1-F_V^*(\varphi+\varepsilon(s))][1-\{(\varphi+\lambda)Q_0^*(\varphi)\{1-F_G^*(\varphi+\varepsilon(s))\}+(\varphi+\varepsilon(s))\\ Q^*(\varphi;s)F_G^*(\varphi+\varepsilon(s))\}][\{(1-z_1)+z_1F_{R_1}^*(\eta_1(s)\}q_1F_{S_1}^*(\eta_1(s)+\{(1-z_2)\\ +z_2F_{R_2}^*(\eta_2(s)\}q_2F_{S_2}^*(\eta_2(s)]\end{array}}{\begin{array}{r}(\varphi+\varepsilon(s))[s-\{p_2+p_1F_V^*(\varphi+\varepsilon(s))\}\{\{(1-z_1)+z_1F_{R_1}^*(\eta_1(s))\}\\ q_1F_{S_1}^*(\eta_1(s))+\{(1-z_2)+z_2F_{R_2}^*(\eta_2(s)\}q_2F_{S_2}^*(\eta_2(s))\}]\end{array}}\end{aligned} \tag{16.38}$$

Similarly, from (16.31), we have

$$\begin{aligned}P_G^*(\varphi;s)&=\int_0^\infty P_G^*(y,\varphi;s)dy\\ &=\frac{\lambda Q^*(\varphi;s)P_X(s)-(\lambda+\varphi)(Q^*(\varphi;s)-Q_0^*(\varphi))[1-F_G^*(\varphi+\epsilon(s))]}{(\varphi+\epsilon(s))}\end{aligned} \tag{16.39}$$

Now, let us consider a convex function as given below:

$$f(s) = [p_2 + p_1 F_V^*(\varphi + \varepsilon(s))][\{(1 - z_1) + z_1 F_{R_1}^*(\eta_1(s))\}q_1 F_{S_1}^*(\eta_1(s)) + \{(1 - z_2) + z_2 F_{R_2}^*(\eta_2(s))\}q_2 F_{S_2}^*(\eta_2(s))] - s$$

We know that, if $f(s)$ happens to be a convex function, then, by Rouche's theorem, it will have strictly one root within the unit circle $| s |=1$ for $Re(\varphi)$. Accordingly,

$$Q^*(\varphi; s) = \sum_{n=0}^{N-1}\left[\frac{1}{\lambda + \varphi - \lambda[h_0(\varphi)]^{n+1}}\right]s^n$$

and

$$Q_0^*(\varphi) = \frac{1}{\lambda + \varphi - \lambda h_0(\varphi)}$$

where $h_o(\varphi)$ is the root of the following equation:

$$s = [p_2 + p_1 F_V^*(\varphi + \varepsilon(s))][\{(1 - z_1) + z_1 F_{R_1}^*(\eta_1(s))\}q_1 F_{S_1}^*(\eta_1(s)) + \{(1 - z_2) + z_2 F_{R_2}^*(\eta_2(s))\}q_2 F_{S_2}^*(\eta_2(s))]$$

Thus, from Eqs. (16.36), (16.37), (16.38), and (16.39), we have the LST of the system reliability function as given below:

$$\begin{aligned} \mathrm{R}^*(\varphi) &= Q^*(\varphi) + P_G^*(\varphi) + P_V^*(\varphi) + \sum_{i=1}^{2}[P_{S_i}^*(\varphi) + P_{R_i}^*(\varphi)] \\ &= Q^*(\varphi; 1) + P_G^*(\varphi; 1) + P_V^*(\varphi; 1) + \sum_{i=1}^{2}[P_{S_i}^*(\varphi; 1) + P_{R_i}^*(\varphi; 1)] \end{aligned} \tag{16.40}$$

where

$$Q^*(\varphi; 1) = \sum_{n=0}^{N-1}\left[\frac{1}{\lambda + \varphi - \lambda[h_0(\varphi)]^{n+1}}\right] \tag{16.41}$$

$$P_{S_i}^*(\varphi; 1) = \frac{q_i[1 - F_{S_i}^*(\varphi + \alpha_i)] \, [1 - \{(\varphi + \lambda)Q_0^*(\varphi)\{1 - F_G^*(\varphi)\} + \varphi Q^*(\varphi)F_G^*(\varphi)\}]}{(\varphi + \alpha_i)[1 - \{p_2 + p_1 F_V^*(\varphi)\}\{\{(1 - z_1) + z_1 F_{R_1}^*(\varphi + \alpha_1)\}q_1 F_{S_1}^*(\varphi + \alpha_1) + \{(1 - z_2) + z_2 F_{R_2}^*(\varphi + \alpha_2)\}q_2 F_{S_2}^*(\varphi + \alpha_2)\}]};$$

$$i = 1, 2 \tag{16.42}$$

$$P^*_{R_i}(\varphi;1)=\frac{\begin{array}{l}q_i z_i[1-F^*_{R_i}(\varphi+\alpha_i)]\\ \quad[1-\{(\varphi+\lambda)Q^*_0(\varphi)\{1-F^*_G(\varphi)\}+\varphi Q^*(\varphi)F^*_G(\varphi)\}]F^*_{S_i}(\varphi+\alpha_i)\end{array}}{\begin{array}{l}(\varphi+\alpha_i)[1-\{p_2+p_1F^*_V(\varphi)\}\{\{(1-z_1)+z_1F^*_{R_1}(\varphi+\alpha_1)\}q_1F^*_{S_1}(\varphi+\alpha_1)\\ \quad+\{(1-z_2)+z_2F^*_{R_2}(\varphi+\alpha_2)\}q_2F^*_{S_2}(\varphi+\alpha_2)\}]\end{array}};$$

$$i=1,2 \tag{16.43}$$

$$P^*_V(\varphi;1)=\frac{\begin{array}{l}p_1[1-F^*_V(\varphi)][1-\{(\varphi+\lambda)Q^*_0(\varphi)\{1-F^*_G(\varphi)\}+\varphi Q^*(\varphi)F^*_G(\varphi)\}]\\ \quad[\{(1-z_1)+z_1F^*_{R_1}(\varphi+\alpha_1)\}q_1F^*_{S_1}(\varphi+\alpha_1)+\{(1-z_2)+z_2F^*_{R_2}(\varphi+\alpha_2)\}\\ \quad q_2F^*_{S_2}(\varphi+\alpha_2)]\end{array}}{\begin{array}{l}\varphi[1-\{p_2+p_1F^*_V(\varphi)\}\{\{(1-z_1)+z_1F^*_{R_1}(\varphi+\alpha_1)\}q_1F^*_{S_1}(\varphi+\alpha_1)\\ \quad+\{(1-z_2)+z_2F^*_{R_2}(\varphi+\alpha_2)\}q_2F^*_{S_2}(\varphi+\alpha_2)\}]\end{array}} \tag{16.44}$$

$$P^*_G(\varphi;1)=\frac{\lambda Q^*(\varphi)-(\lambda+\varphi)(Q^*(\varphi)-Q^*_0(\varphi))[1-F^*_G(\varphi)]}{\varphi} \tag{16.45}$$

Finally, we achieve the LST of the system reliability function $R^*(\varphi)$ by putting (16.41), (16.42), (16.43), (16.44), and (16.45) in (16.40),

$$\begin{aligned}R^*(\varphi)&=\frac{\varphi Q^*(\varphi)+[\lambda Q^*(\varphi)-(\lambda+\varphi)(Q^*(\varphi)-Q^*_0(\varphi))][1-F^*_G(\varphi)]}{\varphi}\\
&+\frac{\begin{array}{l}[1-\{(\varphi+\lambda)Q^*_0(\varphi)\{1-F^*_G(\varphi)\}+\varphi Q^*(\varphi)F^*_G(\varphi)\}][q_1\{(1-F^*_{S_1}(\varphi+\alpha_1))\\ \quad+z_1(1-F^*_{R_1}(\varphi+\alpha_1))F^*_{S_1}(\varphi+\alpha_1)\}(\varphi+\alpha_2)+q_2\{(1-F^*_{S_2}(\varphi+\alpha_2))\\ \quad+z_2(1-F^*_{R_2}(\varphi+\alpha_2))F^*_{S_2}(\varphi+\alpha_2)\}(\varphi+\alpha_1)]\end{array}}{\begin{array}{l}(\varphi+\alpha_1)(\varphi+\alpha_2)[1-\{p_2+p_1F^*_V(\varphi)\}\{\{(1-z_1)+z_1F^*_{R_1}(\varphi+\alpha_1)\}\\ \quad q_1F^*_{S_1}(\varphi+\alpha_1)+\{(1-z_2)+z_2F^*_{R_2}(\varphi+\alpha_2)\}q_2F^*_{S_2}(\varphi+\alpha_2)\}]\end{array}}\\
&+\frac{\begin{array}{l}p_1[1-\{(\varphi+\lambda)Q^*_0(\varphi)\{1-F^*_G(\varphi)\}+\varphi Q^*(\varphi)F^*_G(\varphi)\}][1-F^*_V(\varphi)][\{(1-z_1)\\ \quad+z_1F^*_{R_1}(\varphi+\alpha_1)\}q_1F^*_{S_1}(\varphi+\alpha_1)+\{(1-z_2)+z_2F^*_{R_2}(\varphi+\alpha_2)\}\\ \quad q_2F^*_{S_2}(\varphi+\alpha_2)\}]\end{array}}{\begin{array}{l}\varphi[1-\{p_2+p_1F^*_V(\varphi)\}\{\{(1-z_1)+z_1F^*_{R_1}(\varphi+\alpha_1)\}\\ \quad q_1F^*_{S_1}(\varphi+\alpha_1)+\{(1-z_2)+z_2F^*_{R_2}(\varphi+\alpha_2)\}q_2F^*_{S_2}(\varphi+\alpha_2)\}]\end{array}}\end{aligned}$$

16.5 Mean Time to First Failure of the Server

We know that the MTTFF of a server (Wang [14]) is given by

$$MTTFF=\int_0^\infty R(t)dt=R^*(\varphi)|_{\varphi=0} \tag{16.46}$$

Also by the Tauberian theorem of LST, we have the following result:

$$\lim_{\varphi \to 0} \varphi Q^*(\varphi) = \lim_{t \to \infty} Q(t) = Q_0$$

Let, us now define the following expression,

$$Q_n = Q_0 \Gamma_n \; ; \quad n = 0, 1, \cdots, (N-1) \quad \text{and} \, \Gamma_0 = 1 \tag{16.47}$$

where
$Q_n = \lim_{t \to \infty} Q_n(t)$
Γ_n is the probability of arrival of a batch of 'n'units in an idle period
Q_0 is a normalizing constant.

Now, let us consider the normalizing condition under the steady-state regime:

$$\sum_{n=0}^{N-1} Q_n + \sum_{n=N}^{\infty} \left[\int_0^{\infty} f_G(y) dy \right] + + \sum_{n=0}^{\infty} \left[\int_0^{\infty} f_{V_n}(y) dy \right] + \sum_{n=1}^{\infty} \left[\sum_{i=1}^{2} \left\{ \int_0^{\infty} f_{S_{n,i}}(y) dy + \int_0^{\infty} f_{R_{n,i}}(y) dy \right\} \right] = 1$$

Solving this normalizing condition, we get the value of Q_0 as

$$Q_0 = \frac{(1 - \rho_u)}{\sum_{n=0}^{N-1} \Gamma_n + \lambda \mu_G^{(1)}} \tag{16.48}$$

where $\rho_u = p_1 \rho_V + \{q_1(\rho_{S_1} + z_1 \rho_{R_1}) + q_2(\rho_{S_2} + z_2 \rho_{R_2})\}$ such that $\rho_V = \lambda \mu_X^{(1)} \mu_V^{(1)}$, $\rho_{S_i} = \lambda \mu_X^{(1)} \mu_{S_i}^{(1)}$ and $\rho_{R_i} = \lambda \mu_X^{(1)} \mu_{R_i}^{(1)}$ for $i = 1, 2$.

Finally, by putting the value of Q_0 from Eq. (16.48) in Eq. (16.46) gives the MTTFF of the server as

$$MTTFF = [Q^*(0) + \lambda \mu_G^{(1)} Q_0^*(0)] \left[1 - \rho_u - \left(\sum_{n=0}^{N-1} \Gamma_n + \lambda \mu_G^{(1)} \right) \right] + \frac{\begin{bmatrix} q_1 \alpha_2 \{(1 - F_{S_1}^*(\alpha_1)) + z_1(1 - F_{R_1}^*(\alpha_1)) F_{S_1}^*(\alpha_1)\} \\ + q_2 \alpha_1 \{(1 - F_{S_2}^*(\alpha_2)) + z_2(1 - F_{R_2}^*(\alpha_2)) F_{S_2}^*(\alpha_2)\} \end{bmatrix}}{\alpha_1 \alpha_2 (\sum_{n=0}^{N-1} \Gamma_n + \lambda \mu_G^{(1)}) \, [[\{(1 - z_1) + z_1 F_{R_1}^*(\alpha_1)\} q_1 F_{S_1}^*(\alpha_1) + \{(1 - z_2) + z_2 F_{R_2}^*(\alpha_2)\} q_2 F_{S_2}^*(\alpha_2)] - 1]}$$

where

$$Q^*(0) = \sum_{n=0}^{N-1}\left[\frac{1}{\lambda - \lambda[h_0(0)]^{n+1}}\right]$$

and

$$Q_0^*(0) = \frac{1}{\lambda - \lambda h_0(0)}$$

$h_0(0)$ being the root of the equation given below:

$$s = [p_2 + p_1 F_V^*(\varepsilon(s))][\{(1 - z_1) + z_1 F_{R_1}^*(\alpha_1 + \varepsilon(s))\}q_1 F_{S_1}^*(\alpha_1 + \varepsilon(s)) + \{(1 - z_2) + z_2 F_{R_2}^*(\alpha_2 + \varepsilon(s))\}q_2 F_{S_2}^*(\alpha_2 + \varepsilon(s))]$$

16.6 Proposed Model in the Light of Functional Analysis

Functional analysis is that branch of Mathematics, some parts of which help in better understanding of ideas from probability and stochastic processes. In functional analysis, rather than the properties of a certain probability measure, the property of the whole space or a certain subspace of such measures is more important. Different transformations used to derive the RF of the underlying model, namely the LST, PGF, etc., can also be viewed in terms of functional analysis.

PGF and LST are particular cases of the Gelfrad transform, which falls under the umbrella of functional analysis. The Gelfrad transform is a multiplicative function on a Banach algebra, satisfying the following bounded linear function:

$$(F, xy) = (F, x)(F, y);\, \forall x, y \in \mathbb{A} \tag{16.49}$$

where F is fixed, while x and y vary, and $\mathbb{A}$ is a compact subset of a vector space. For a detailed discussion about the topic, the reader may refer to Chap. 6, pages 201–233 Bobrowski [3].

References

1. Begum, A., Choudhury, G.: Analysis of an unreliable single server batch arrival queue with two types of services under Bernoulli vacation policy. Commun. Stat. Theory Methods **50**(9), 2136–2160 (2021)
2. Al-khedhairi, A., Tadj, L.: A bulk service queue with a chose pf service and re-service under Bernoulli schedule. Int. J. Contemp. Math. Sci. **2**(23), 1107–1120 (2007)
3. Bobrowski, A.: Functional Analysis for Probability and Stochastic Processes: An Introduction. Cambridge University Press (2005)
4. Doshi, B.T.: Queueing systems with vacations a survey. Queue. Syst. **1**(1), 29–66 (1986)

5. Kalita, C.R., Choudhury, G.: A note on reliability analysis of an N-policy unreliable $M^X/\binom{G_1}{G_2}/1$ queue with optional repeated service. RAIRO Oper. Res. **52**, 713–724 (2018)
6. Kalita, C.R., Choudhury, G.: An $M^X/\binom{G_1}{G_2}/1$ repeated service queue with N-policy and setup time subject to server's breakdown and delayed repair. Commun. Stat. Theory Methods **49**(23), 5709–5737 (2020)
7. Cox, D.R.: The analysis of non-Markovian stochastic processes by the inclusion of supplementary variables. Math. Proceed. Cambridge Philos. Soc. **51**, 433–441 (1955)
8. Choudhury, G., Kalita, C.R.: An $M/G/1$ queue with two types of general heterogeneous service and optional repeated service subject to server's breakdown and delayed repair. Qual. Tech. Quant. Manag. **15**(5), 622–654 (2017)
9. Choudhury, G.G., Tadj, L., Ke, J.C.: A two-phase service system with bernoulli vacation schedule, setup time and N-policy for an unreliable server with delaying repair. Qual. Technol. Quant. Manag. **8**(3), 271–284 (2011)
10. Choudhury, G., Deka, M.: A single server queueing system with two phases of service subject to server breakdown and Bernoulli vacation. Appl. Math. Modell. **36**, 6050–6060 (2012)
11. Takagi, H.: Queueing Analysis-A Foundation of Performance Evaluation, vol. 1. Elsevier, Amsterdam (1991)
12. Lee, H.S., Srinivasan, M.M.: Control policies for the $M^X/G/1$ queueing system. Manage. Sci. **35**, 708–721 (1989)
13. Lee, H.W., Lee, S.S., Chae, K.C.: Operating characteristic of $M^X/G/1$ queue with N-policy. Queue. Syst. **15**, 387–399 (1994)
14. Wang, J.: An $M/G/1$ queue with second optional service and server breakdowns. Comput. Math. Appl. **47**, 1713–1723 (2004)
15. Medhi, J., Temptleton, J.G.C.: A Poisson input queue under N-policy and with a general start up time. Comput. Oper. Res. **19**(1), 35–41 (1992)
16. Keilson, J., Servi, L.D.: Oscillating random walk models for $GI/G/1$ vacation system with Bernoulli schedules. J. Appl. Probab. **23**, 790–802 (1986)
17. Madan, K.C., Al-Rawi, Z.R., Al- Nasser, A.D.: On $M^X/\binom{G_1}{G_2}/1/G(BS)/V_s$ vacation queue with two types of general heterogeneous service. J. Appl. Math. Decis. Sci. **3**, 123–135 (2005)
18. Madan, K.C., Al- Nasser, A.D., Al-Masri, A.Q.: On $M^X/\binom{G_1}{G_2}/1$ queue with optional re-service. Appl. Math. Comput. **152**, 71–88 (2004)
19. Wang, K.H., Wang, T.Y., Pearn, W.L.: Optimal control of the N policy $M/G/1$ queueing system with server breakdowns and general startup times. Appl. Math. Model. **31**, 2199–2212 (2007)
20. Baruah, M., Madan, K.C., Eldabi, T.: Balking and re-service in a vacation queue with batch arrival and two types of heterogeneous service. J. Math. Res. **4** (2012)
21. Tian, N., Zhang, Z.G.: Vacation Queueing Models: Theory and Applications. Springer, New York (2006)
22. Kella, O.: Optimal control of vacation scheme in an M/G/1 queue. Oper. Res. **38**, 724–728 (1990)
23. Arumuganathan, R., Jeyakumar, S.: Steady state analysis of a bulk queue with multiple vacations, setup times with N-policy and closedown times. Appl. Math. Model. **29**, 972–986 (2005)
24. Hur, S., Ahn, S.: Batch arrival queues with vacations and server setup. Appl. Math. Model. **29**, 1164–1181 (2005)
25. Li, W., Shi, D., Chao, X.: Reliability analysis of $M/G/1$ queueing systems with server breakdowns and vacations. J. Appl. Probab. **34**, 546–555 (1997)
26. Tang, Y.H.: A single server $M/G/1$ queueing system subject to breakdowns-some reliability and queueing problems. Microelectron. Relab. **37**, 315–321 (1997)

Chapter 17
Altering JS-Metric Space

Paulraj Gnanachandra, Jeffin Varunnya, Santanu Acharjee, and Saeid Jafari

Abstract In this paper, we introduce Altering JS-metric space and prove some results. We show that metric space, b-metric space, dislocated metric space, and JS-metric space are Altering JS-metric spaces. Later, we prove some fixed point theorems in Altering JS-metric space. We also prove a coupled fixed point theorem in a partial ordered set using Altering JS-metric space.

Keywords Altering JS metric · Metric space · Fixed point theory

17.1 Introduction

The investigation of metric spaces was started in the year 1905 by Frechet [10]. It is a mathematical area that gives a structure to the study of continuity, convergence, and other related ideas. The metric is a function that defines the concept of distance between any two members of a non-empty set which are usually called points.

Let S be a non-empty set and let $d : S \times S \longrightarrow [0, \infty)$ be a function called a metric [10] or a distance function if for all $r, s, t \in S$,
(d1) $d(s, s) = 0$,
(d2) $d(s, t) = 0 \Leftrightarrow s = t$,
(d3)$d(s, t) = d(t, s)$,
(d4) $d(s, t) \leq d(s, r) + d(r, t)$.

P. Gnanachandra · J. Varunnya
Centre for Research and Post Graduate Studies in Mathematics, Ayya Nadar Janaki Ammal College, Sivakasi, Tamil Nadu, India

S. Acharjee (✉)
Department of Mathematics, Gauhati University, Guwahati 781014, Assam, India
e-mail: sacharjee326@gmail.com

S. Jafari
Mathematical and Physical Science Foundation, 4200 Slagelse, Denmark
e-mail: saeidjafari@topositus.com

B. Hazarika et al. (eds.), *Advances in Functional Analysis and Fixed-Point Theory*, Industrial and Applied Mathematics,
https://doi.org/10.1007/978-981-99-9207-2_17

The non-empty set S together with the metric d is called a metric space and is denoted by the pair (S, d). Since then metric space has attracted the attention of researchers. On the other hand, fixed point theory is a fine fusion of analysis, topology, and geometry. During the past few decades, fixed point theory was a prime part in the study of non-linear phenomena. It has immense applications in various areas of mathematics like differential equations, integral equations, numerical analysis, etc. Fixed point theorem also has its applications in the problems of mathematical economics [25] and game theory [26].

For the non-empty set S, and the self-map $T : S \to S$, an element $s \in S$ is called a fixed point of T if $s = Ts$. A self-map may or may not have a fixed point. Also, if a function has a fixed point, then it may or may not be unique. For example, consider a translation map, $T : \mathbb{R} \to \mathbb{R}$, defined by $Ts = s + 1$. Then, T has no point in $\mathbb{R}$ such that $s = Ts$. So, T has no fixed point in $\mathbb{R}$ whereas, in case of an identity map, every point of $\mathbb{R}$ turns out to be a fixed point.

Several concepts of metric spaces are playing significant roles in various disciplines of science and social science. Most importantly they play a key role in fixed point theory. We have found numerous extensions of the metric spaces. One such generalization was given by Jleli and Samet [1] in the year 2015. This generalized form of metric space retrieves a huge category of prevailing metrics in the literature. In recent years, a lot of researchers have studied these generalized forms of metric space. Various contractive and expansive conditions have been used to prove the existence of fixed points for self-maps in these spaces [7–9, 11]. In 1984, Wang et al. [12] studied some fixed point theorems using expansive type mappings, which inspired many researchers to inquire the expansive mappings [13, 21–24].

The altering distance function [3, 4] is a function that alters the distance between two points. Delbosco [3] and Scof [4] initiated this concept, and proved some fixed point theorems for self-maps in complete metric spaces. Khan [2] used the notion of altering distance function and proved some contractive fixed point theorems. Following them many others were interested in altering the distances, and established a fair number of results [14–17]. Many contractive conditions were formed with the use of these new notions.

Hence, we extend the JS-metric due to Jleli and Samet [1] by altering the distances with the help of altering distance function and define the Altering JS-metric. Also, we prove some fixed point results for mappings, not necessarily continuous in these generalized spaces. We give an application on coupled fixed point.

17.2 Preliminaries

In this section, we discuss some preliminary results, which are important for this paper. Consider a non-empty set S and a mapping $J_\phi : S \times S \to [0, \infty]$. For every $s \in S$, let us define the set

$$C(J_\phi, S, s) = \left\{ \{s_n\} \subset S : \lim_{n\to\infty} J_\phi(s_n, s) = 0 \right\}$$

Definition 17.1 [1] Let S be a non-empty set and $J_\phi : S \times S \to [0, \infty]$ be a function satisfying the conditions:
(D1) $J_\phi(s,t) = 0 \Rightarrow s = t, \forall (s,t) \in S \times S$,
(D2) $J_\phi(s,t) = J_\phi(t,s), \forall (s,t) \in S \times S$,
(D3) there exists $k > 0$ such that

$$(s,t) \in S \times S, \{s_n\} \in C(J_\phi, S, s) \Rightarrow J_\phi(s,t) \leq k \lim_{n\to\infty} \sup J_\phi(s_n, t)$$

Then, J_ϕ is called a JS-metric and (S, J_ϕ) is called JS-metric space.

Definition 17.2 [3, 4] A non-negative function $\psi : [0, \infty) \to [0, \infty)$ is an altering distance function if
(1) ψ is continuous and monotonically non-decreasing,
(2) $\psi(t) = 0 \Leftrightarrow t = 0$,
(3) $h.t^r \leq \psi(t)$; for all $t > 0$ and $h, r > 0$ are constants.

Example 17.1 We have the trivial example, $\psi : [0, \infty) \to [0, \infty)$ defined by $\psi(t) = t$.

Example 17.2 Consider the function $\psi : [0, \infty) \to [0, \infty)$ defined by $\psi(t) = te^t$. Then, ψ is an altering distance function.
Now, if we consider Ψ as the set of all altering distance functions, then $\Psi \neq \emptyset$.

Dislocated metric space is a restriction of metric space. Here, the self-distances need not be zero. Dislocated metrics assume a significant role in topology as well as in different areas of science, particularly in logic programming and electronic engineering [20]. Hitzler and Seda [20] commenced the study of dislocated metric spaces and validated a fixed point theorem that generalized the Banach contraction principle. The b-metric space is a generalization of metric space. This space was introduced by Bakhtin [18] and Czerwik [19]. One may refer to An et al. [26] for topological aspects of b-metric space, and Karapınar and Salimi [27] and Pasicki [28] for fixed point theorems on dislocated metric space.

Definition 17.3 [20] Consider a non-empty set S. The function $d : S \times S \to [0, \infty]$ is said to be a dislocated metric if it satisfies the conditions:
(i) $d(s,t) = 0 \Rightarrow s = t, \forall\ (s,t) \in S \times S$,
(ii) $d(s,t) = d(t,s), \forall\ (s,t) \in S \times S$,
(iii) $d(s,t) \leq d(s,r) + d(r,t), \forall\ s,t,r \in S$.
The ordered pair (S, d) is known as dislocated metric space.

Definition 17.4 [18, 19] A b-metric on a non-empty set S is a function $d : S \times S \to [0, \infty]$ satisfying the conditions:
(i) $d(s,t) = 0 \Rightarrow s = t, \forall\ (s,t) \in S \times S$,
(ii) $d(s,t) = d(t,s), \forall\ (s,t) \in S \times S$,
(iii) There exists $b \geq 1$, such that, $d(s,t) \leq b\,[d(s,r) + d(r,t)], \forall\ s,t,r \in S$.
Then (S, d) is the b-metric space.

17.3 Altering JS-metric Spaces

Now, we introduce the Altering JS-metric space. It generalizes the JS-metric space by utilizing the altering distance function.

Definition 17.5 Let S be a non-empty set and $J_\phi : S \times S \to [0, \infty]$ be a given mapping. We say that J_ϕ is an Altering JS-metric on S if it satisfies the following conditions:
(A1) $J_\phi(s, t) = 0 \Rightarrow s = t$, for all $s, t \in S$,
(A2)$J_\phi(s, t) = J_\phi(t, s)$, for all $s, t \in S$,
(A3) there exists a function $\phi \in \Psi$ and a constant $k > 0$ such that

$$s, t \in S, \{s_n\} \in C(J_\phi, S, s) \Rightarrow \phi\left(J_\phi(s, t)\right) \leq \lim_{n\to\infty} \sup\left[\phi(k.J_\phi(s_n, t))\right]$$

Then, the pair (S, J_ϕ) is said to be an Altering JS-metric space.

If $\phi(t) = t$, then it can be observed that the Altering JS-metric space becomes a JS-metric space. Suppose the set $C(J_\phi, S, s)$ is empty for every element of S, then (S, J_ϕ) becomes a JS-metric space iff the conditions (A1) and (A2) are satisfied.

Theorem 17.1 *Every metric space is an Altering JS-metric space.*

Proof Let (S, d) be a metric space. Then,
(i) $d(s, t) = 0 \Rightarrow s = t \ \forall\, s, t \in S$ (Since (S, d) is a metric space)
(ii) $d(s, t) = d(t, s) \ \forall\, s, t \in S$
(iii) Let $\{s_n\} \in C(d, S, s)$ and we consider, $\phi(t) = t$.
So,

$$d(s, t) \leq d(s_n, s) + d(s_n, t) \text{ for all } s_n, s, t \in S$$

As $n \to \infty$, $d(s_n, s) \to 0$.
Thus,

$$d(s, t) \leq \lim_{n\to\infty} \sup[d(s_n, t)]$$

$$\Rightarrow \phi\left(d(s, t)\right) \leq \lim_{n\to\infty} \sup[\phi\left(d(s_n, t)\right)]$$

Hence, (S, d) is an Altering JS-metric space. □

Example 17.3 The metric defined by usual distance between two points is an Altering JS-metric space. Consider the non-empty set $S = [0, \infty)$ and let $J_\phi : S \times S \to [0, \infty]$ be defined by $J_\phi(s, t) =\mid s - t \mid$.
Obviously the axioms (A1) and (A2) are satisfied. It is enough to check the axiom (A3).
Let $s, t \in S$, $\{s_n\} \in C(J_\phi, S, s)$ and take $\phi(t) = t$.
Now,

$$\begin{aligned} J_\phi(s, t) &=\mid s - t \mid \\ &=\mid s - s_n + s_n - t \mid \end{aligned}$$

$$\leq \mid s_n - s \mid + \mid s_n - t \mid$$
$$= J_\phi(s_n, s) + J_\phi(s_n, t)$$

As $n \to \infty$ we get $J_\phi(s,t) \leq \lim_{n\to\infty} \sup\left[J_\phi(s_n, s) + J_\phi(s_n, t)\right]$

$$\Rightarrow J_\phi(s,t) \leq \lim_{n\to\infty} \sup[J_\phi(s_n, t)]$$

$$\Rightarrow \phi\left(J_\phi(s,t)\right) \leq \lim_{n\to\infty} \sup\left[\phi(J_\phi(s_n, t))\right]$$

Hence, (S, J_ϕ) is an Altering JS-metric space.

Example 17.4 Let $S = \mathbb{R}$ and let $J_\phi : S \times S \to [0, \infty]$ be defined by $J_\phi(s,t) = (s-t)^2$ and let $\phi(t) = t^2$.
If $J_\phi(s,t) = 0$, then $s = t$. Thus, axiom (A1) is true.
And also we have $(s-t)^2 = (t-s)^2$. Thus, axiom (A2) is satisfied.
Let there exist a sequence $\{s_n\}$ such that

$$\lim_{n\to\infty} J_\phi(s_n, s) = 0$$

i.e.

$$\lim_{n\to\infty} (s_n - s)^2 = 0$$

It can be noticed that for any $t \in S$, $J_\phi(t,t) = 0 \Rightarrow \lim_{n\to\infty} J_\phi(t,t) = 0$.
Hence, $s_n \longrightarrow s$ and $t \longrightarrow t$ implies $(s_n - t)^2 \longrightarrow (s-t)^2$ for $n \to \infty$.
That is,

$$J_\phi(s,t) = \lim_{n\to\infty} J_\phi(s_n, t)$$
$$= \lim_{n\to\infty} \sup[J_\phi(s_n, t)]$$

For any $k > 0$, we get

$$J_\phi(s,t) \leq k. \limsup_{n\to\infty} J_\phi(s_n, t)$$
$$= \lim_{n\to\infty} \sup[k.J_\phi(s_n, t)]$$

Since $\phi(t)$ is a non-decreasing function,

$$\phi\left(J_\phi(s,t)\right) \leq \phi\left[\lim_{n\to\infty} \sup(k.J_\phi(s_n, t))\right]$$

$$\text{So, } \phi\left(J_\phi(s,t)\right) \leq \lim_{n\to\infty} \sup[\phi\left(k.J_\phi(s_n, t)\right)]$$

Thus, (S, J_ϕ) is an Altering JS-metric space. But, (S, J_ϕ) fails to satisfy the triangle inequality. Hence, it is not a metric.

Remark 17.1 Suppose that (S, J_ϕ) is an Altering JS-metric space. Then by axiom (A3),

$$\phi\left(J_{\phi}(s,t)\right) \leq \lim_{n\to\infty} \sup\left[\phi(k.J_{\phi}(s_n,t))\right]$$

$$J_{\phi}(s,t) \leq \lim_{n\to\infty} \sup(k.J_{\phi}(s_n,t))$$

$$\leq \sup(k.J_{\phi}(s_n,t))$$

Let $\exists\, r \in S$ be such that $\sup(k.J_{\phi}(s_n,t)) = J_{\phi}(r,t)$. Then, we get

$$J_{\phi}(s,t) \leq J_{\phi}(r,t)$$

$$\leq J_{\phi}(r,t) + J_{\phi}(r,s)$$

$$\Rightarrow J_{\phi}(s,t) \leq J_{\phi}(s,r) + J_{\phi}(r,t)$$

This inequality is not true for all $s, t, r \in S$. Thus, the triangle inequality is not satisfied for all possible elements of the set. Hence, an Altering JS-metric does not need not be a metric.

Proposition 17.1 *If* $J_{\phi} : S \times S \to [0,\infty]$ *is an Altering JS-metric, then* $n.J_{\phi} : S \times S \to [0,\infty]$ *defined by* $(n.J_{\phi})(s,t) = n.J_{\phi}(s,t)$ *is also an Altering JS-metric for* $n \in (0,\infty)$.

Proof Let $n.J_{\phi}(s,t) = 0$ which implies $J_{\phi}(s,t) = 0 \Rightarrow s = t$. Hence, the axiom (A1) is true.
Also, $J_{\phi}(s,t) = J_{\phi}(t,s) \Rightarrow n.J_{\phi}(s,t) = n.J_{\phi}(t,s)$. Thus, (A2) is true.
From axiom (A3),

$$\phi\left(J_{\phi}(s,t)\right) \leq \lim_{n\to\infty} \sup\left[\phi(k.J_{\phi}(s_n,t))\right]$$

Multiplying both sides by n where $n \in (0,\infty)$, we get

$$n.\phi\left(J_{\phi}(s,t)\right) \leq n.\lim_{n\to\infty} \sup\left[\phi(k.J_{\phi}(s_n,t))\right]$$

$$\phi\left(n.J_{\phi}(s,t)\right) \leq \lim_{n\to\infty} \sup\left[\phi(k.n.J_{\phi}(s_n,t))\right]$$

Hence, $n.J_{\phi}$ is an Altering JS-metric. □

Proposition 17.2 *Let* $J_{\phi} : S \times S \to [0,\infty]$ *be an Altering JS-metric. Then,* $\sqrt[n]{J_{\phi}} : S \times S \to [0,\infty]$ *defined by* $\sqrt[n]{J_{\phi}}(s,t) = \sqrt[n]{J_{\phi}(s,t)}$ *is also an Altering JS-metric for a positive integer* n.

Proof Let us consider

$$\sqrt[n]{J_{\phi}(s,t)} = 0$$

Then,

$$J_\phi(s,t)=0$$

Since J_ϕ is an Altering JS-metric, thus, by axiom (A1), we get $s=t$.
Again,

$$\sqrt[n]{J_\phi(s,t)}=\sqrt[n]{J_\phi(t,s)}$$

Since J_ϕ is an Altering JS-metric, thus by axiom (A3) we have

$$\phi\left(J_\phi(s,t)\right)\leq \lim_{n\to\infty}\sup\left[\phi(k.J_\phi(s_n,t))\right]$$

$$\phi\left(J_\phi(s,t)\right)\leq \phi\left(\lim_{n\to\infty}\sup\left[(k.J_\phi(s_n,t))\right]\right)$$

Since ϕ is monotonically non-decreasing,

$$J_\phi(s,t)\leq \lim_{n\to\infty}\sup\left[(k.J_\phi(s_n,t))\right]$$

$$\Rightarrow \sqrt[n]{(J_\phi(s,t))}\leq \sqrt[n]{\left(\lim_{n\to\infty}\sup\left[(k.J_\phi(s_n,t))\right]\right)}$$

$$\Rightarrow \left(J_\phi(s,t)\right)^{\frac{1}{n}}\leq \left(\lim_{n\to\infty}\sup\left[(k.J_\phi(s_n,t))\right]\right)^{\frac{1}{n}}$$

$$\leq \lim_{n\to\infty}\sup\left(k.J_\phi(s_n,t)\right)^{\frac{1}{n}}$$

$$\leq \lim_{n\to\infty}\sup[k^{\frac{1}{n}}.\left(J_\phi(s_n,t)\right)^{\frac{1}{n}}]$$

$$\Rightarrow \sqrt[n]{(J_\phi(s,t))}\leq \lim_{n\to\infty}\sup[k^{\frac{1}{n}}\sqrt[n]{J_\phi(s_n,t)}]$$

Let $k'=k^{\frac{1}{n}}$

$$\Rightarrow \phi\left(\sqrt[n]{(J_\phi(s,t))}\right)\leq \phi\left(\lim_{n\to\infty}\sup[k'\sqrt[n]{.J_\phi(s_n,t)}]\right)$$

$$\Rightarrow \phi\left(\sqrt[n]{(J_\phi(s,t))}\right)\leq \lim_{n\to\infty}\sup[\phi\left(k'\sqrt[n]{.J_\phi(s_n,t)}\right)]$$

Thus, $\sqrt[n]{J_\phi}$ is an Altering JS-metric. □

Proposition 17.3 *Let (S,J_ϕ) be an Altering JS-metric space. Then, (S,J_ϕ) satisfies the following condition:*

$$\sqrt[n]{J_\phi(s,t)}\leq \lim_{n\to\infty}\sup[\phi\left(k.\frac{J_\phi(s_{n+1},t)}{J_\phi(s_n,t)}\right)]$$

Proof Since (S, J_ϕ) is an Altering JS-metric space, $(S, \sqrt[n]{J_\phi})$ is also an Altering JS-metric space. Thus, by axiom (A3), we have

$$\phi\left(\sqrt[n]{(J_\phi(s,t))}\right) \leq \lim_{n\to\infty} \sup[\phi\left(k.\sqrt[n]{J_\phi(s_n,t)}\right)]$$

Since ϕ is an altering distance function,

$$\Rightarrow \sqrt[n]{(J_\phi(s,t))} \leq \phi\left(\sqrt[n]{(J_\phi(s,t))}\right) \leq \lim_{n\to\infty} \sup[\phi\left(k.\sqrt[n]{.J_\phi(s_n,t)}\right)]$$

Since for any sequence $\{s_n\}$ of positive numbers, $\lim_{n\to\infty} \sup\sqrt[n]{s_n} \leq \lim_{n\to\infty} \sup\left(\frac{s_{n+1}}{s_n}\right)$ holds for all n, thus from the above inequality we obtain

$$\sqrt[n]{(J_\phi(s,t))} \leq \lim_{n\to\infty} \sup[\phi\left(k.\frac{J_\phi(s_{n+1},t)}{J_\phi(s_n,t)}\right)]$$

Proposition 17.4 *If $J_\phi^{'}$ and $J_\phi^{''}$ are Altering JS-metrics, then $max\left\{J_\phi^{'}, J_\phi^{''}\right\}$ is also an Altering JS-metric.*

Proof Let (S, J_ϕ) be an Altering JS-metric space and let $J_\phi(s,t) = max\left\{J_\phi^{'}(s,t), J_\phi^{''}(s,t)\right\}$.
(i) $J_\phi(s,t) = 0 \Rightarrow max\left\{J_\phi^{'}(s,t), J_\phi^{''}(s,t)\right\} = 0$
$\Rightarrow J_\phi^{'}(s,t) = J_\phi^{''}(s,t) = 0$
Since $J_\phi^{'}$ and $J_\phi^{''}$ are Altering JS-metrics, thus we get $s = t$
(ii) $J_\phi(s,t) = max\left\{J_\phi^{'}(s,t), J_\phi^{''}(s,t)\right\}$
$= max\left\{J_\phi^{'}(t,s), J_\phi^{''}(t,s)\right\} = J_\phi(t,s)$
(iii) Let $\{s_n\}$ be a sequence in S converging to s.
Now, $\phi\left[J_\phi(s,t)\right] = \phi\left[max\left\{J_\phi^{'}(s,t), J_\phi^{''}(s,t)\right\}\right]$
Without loss of generality, let $max\left\{J_\phi^{'}(s,t), J_\phi^{''}(s,t)\right\} = J_\phi^{'}(s,t)$.
Hence, we have $\phi\left[J_\phi(s,t)\right] = \phi\left[J_\phi^{'}(s,t)\right]$.
Since $J_\phi^{'}$ is an Altering JS-metric, by axiom (A3), there exists $\phi \in \Psi$ and $k > 0$ such that
$\phi\left[J_\phi^{'}(s,t)\right] \leq \lim_{n\to\infty} \sup\left[\phi(k.J_\phi^{'}(s_n,t))\right]$
Thus, we get

$$\phi\left[J_\phi(s,t)\right] \leq \lim_{n\to\infty} \sup\left[\phi(k.J_\phi^{'}(s_n,t))\right]$$
$$= \lim_{n\to\infty} \sup\left[\phi(k.max\left\{J_\phi^{'}(s_n,t), J_\phi^{''}(s_n,t)\right\})\right]$$
$$\phi\left[J_\phi(s,t)\right] \leq \lim_{n\to\infty} \sup\left[\phi(k.J_\phi(s_n,t))\right]$$

Hence, $max\left\{J_\phi^{'}, J_\phi^{''}\right\}$ is an Altering JS-metric. □

Proposition 17.5 *If $J_\phi^{'}$ and $J_\phi^{''}$ are Altering JS-metrics then $min\left\{J_\phi^{'}, J_\phi^{''}\right\}$ is also an Altering JS-metric.*

Proposition 17.6 *Every JS-metric space is an Altering JS-metric space.*

Proof Let (S, J_ϕ) be a JS-metric space. To prove that (S, J_ϕ) is an Altering JS-metric space, it is enough to prove that J_ϕ satisfies the axiom (A3).
By the property (D3), there exists $k > 0$

$$J_\phi(s,t) \leq k \lim_{n\to\infty} \sup[J_\phi(s_n,t)]$$

$$= \lim_{n\to\infty} \sup[k.J_\phi(s_n,t)]$$

Let $\phi \in \Psi$, then

$$\phi\left(J_\phi(s,t)\right) \leq \phi\left(\lim_{n\to\infty} \sup[k.J_\phi(s_n,t)]\right)$$

Since ϕ is continuous and monotonically non-decreasing,

$$\phi\left(J_\phi(s,t)\right) \leq \lim_{n\to\infty} \sup[\phi\left(k.J_\phi(s_n,t)\right)]$$

Thus, (S, J_ϕ) is an Altering JS-metric space. □

Jleli and Samet gave the following results.

Proposition 17.7 *[1] (i) Every b-metric space is a JS-metric space.*
(ii) Every dislocated metric space is a JS-metric space.

Thus, we can conclude that b-metric space and dislocated metric space are Altering JS-metric spaces.

Proposition 17.8 *Let (S, J_ϕ) be an Altering JS-metric space. If the axiom (A3) holds for the Altering functions ϕ_1 and ϕ_2 then it so does for $\phi_1 + \phi_2$ and $\phi_1.\phi_2$.*

Proof Let $\{s_n\}$ be a sequence in (S, J_ϕ) such that $\{s_n\}$ converges to s.
By axiom (A3), let there exist $\phi_1 \in \Psi$ and $k_1 > 0$ such that

$$\phi_1\left(J_{\phi_1}(s,t)\right) \leq \lim_{n\to\infty} \sup[\phi_1(k_1.J_{\phi_1}(s_n,t))] \tag{17.1}$$

Again by axiom (A3), let there exist $\phi_2 \in \Psi$ and $k_2 > 0$ such that

$$\phi_2\left(J_{\phi_2}(s,t)\right) \leq \lim_{n\to\infty} \sup[\phi_2(k_2.J_{\phi_2}(s_n,t))] \tag{17.2}$$

Adding the above two inequalities, we get

$$\phi_1\left(J_{\phi_1}(s,t)\right) + \phi_2\left(J_{\phi_2}(s,t)\right) \leq \lim_{n\to\infty} \sup[\phi_1(k_1.J_{\phi_1}(s_n,t))] + \lim_{n\to\infty} \sup[\phi_2(k_2.J_{\phi_2}(s_n,t))]$$

$$= \phi_1[\lim_{n\to\infty} \sup(k_1.J_{\phi_1}(s_n,t))] + \phi_2[\lim_{n\to\infty} \sup(k_2.J_{\phi_2}(s_n,t))]$$
$$= k_1\phi_1[\lim_{n\to\infty} \sup(J_{\phi_1}(s_n,t))] + k_2\phi_2[\lim_{n\to\infty} \sup(J_{\phi_2}(s_n,t))]$$

Let $k = max\{k_1, k_2\}$. Then,

$$\phi_1\left(J_{\phi_1}(s,t)\right) + \phi_2\left(J_{\phi_2}(s,t)\right) \leq k\phi_1[\lim_{n\to\infty} \sup(J_{\phi_1}(s_n,t))] + k\phi_2[\lim_{n\to\infty} \sup(J_{\phi_2}(s_n,t))]$$

$$(\phi_1 + \phi_2)[J_{\phi_1+\phi_2}(s,t)] \leq k(\phi_1 + \phi_2)[\lim_{n\to\infty} \sup(J_{\phi_1+\phi_2}(s,t))]$$

Multiplying the inequalities (1) and (2),

$$\phi_1\left(J_{\phi_1}(s,t)\right).\phi_2\left(J_{\phi_2}(s,t)\right) \leq \lim_{n\to\infty} \sup[\phi_1(k_1.J_{\phi_1}(s_n,t))].\lim_{n\to\infty} \sup[\phi_2(k_2.J_{\phi_2}(s_n,t))]$$

$$= \phi_1[\lim_{n\to\infty} \sup(k_1.J_{\phi_1}(s_n,t))].\phi_2[\lim_{n\to\infty} \sup(k_2.J_{\phi_2}(s_n,t))]$$
$$= k_1\phi_1[\lim_{n\to\infty} \sup(J_{\phi_1}(s_n,t))].k_2\phi_2[\lim_{n\to\infty} \sup(J_{\phi_2}(s_n,t))]$$

Taking $k' = k_1.k_2$ implies

$$\phi_1\left(J_{\phi_1}(s,t)\right).\phi_2\left(J_{\phi_2}(s,t)\right) \leq k'\phi_1[\lim_{n\to\infty} \sup(J_{\phi_1}(s_n,t))].\phi_2[\lim_{n\to\infty} \sup(J_{\phi_2}(s_n,t))]$$

$$(\phi_1.\phi_2)[J_{\phi_1.\phi_2}(s,t)] \leq k'(\phi_1.\phi_2)[\lim_{n\to\infty} \sup(J_{\phi_1.\phi_2}(s,t))]$$

Thus, axiom (A3) holds for $\phi_1 + \phi_2$ and $\phi_1.\phi_2$. □

Definition 17.6 Let (S, J_ϕ) be an Altering JS-metric space. Let $\{s_n\}$ be a sequence in S and $s \in S$. We say that the sequence $\{s_n\}$ is J_ϕ-convergent to s if

$$\lim_{n\to\infty} J_\phi(s_n, s) = 0$$

$$\text{i.e. } \{s_n\} \in C(J_\phi, S, s)$$

In this case, s is called the J_ϕ-limit of $\{s_n\}$ and we write $s_n \xrightarrow{J_\phi} s$.

Example 17.5 Let $S = [0, 1]$ and $J_\phi : S \times S \to [0, \infty]$ be defined by $J_\phi(s,t) = \mid s - t \mid$. Then, (S, J_ϕ) defines an Altering JS-metric space. Consider the sequence $\left\{\frac{n}{n+1}\right\} \in S$.

Taking limit as $n \to \infty$, we get

$$\lim_{n\to\infty} \mid \frac{n}{n+1} - 1 \mid = 0$$

Thus, $\left\{\frac{n}{n+1}\right\} \in C(J_\phi, S, 1)$ and hence 1 is the J_ϕ-limit of the sequence $\left\{\frac{n}{n+1}\right\}$.

The result that any constant sequence is convergent does not hold in case of an Altering JS-metric space. The following example shows that a constant sequence may not converge in an Altering JS-metric space.

Example 17.6 Let $S = [0, 1]$ and $J_\phi : S \times S \to [0, \infty]$ be defined by

$$J_\phi(s,t) = \begin{cases} 1, \text{ if } s, t \text{ are both rational or both irrational} \\ \infty, \text{otherwise} \end{cases}$$

The constant sequence does not converge in (S, J_ϕ).

Proposition 17.9 *The J_ϕ-limit of a sequence is unique, i.e. if (S, J_ϕ) is an Altering JS-metric space and $\{s_n\}$ is a sequence in S such that $\{s_n\} \in C(J_\phi, S, s) \bigcap C(J_\phi, S, t)$, then $s = t$.*

Proof Given $\{s_n\} \in C(J_\phi, S, s) \bigcap C(J_\phi, S, t)$, i.e. $\{s_n\}$ J_ϕ-converges to both s and t. Then, $\lim_{n\to\infty} J_\phi(s_n, s) = 0$ and $\lim_{n\to\infty} J_\phi(s_n, t) = 0$. By the axiom (A3), there exists a $\phi \in \Psi$ and $k > 0$ such that

$$s, t \in S, \{s_n\} \in C(J_\phi, S, s) \Rightarrow \phi(J_\phi(s,t)) \leq \lim_{n\to\infty} \sup \left[\phi(k.J_\phi(s_n, t))\right]$$

Since $\{s_n\} \in C(J_\phi, S, t)$, then $\phi\left(J_\phi(s,t)\right) = 0$ where $n \to \infty$. Thus, $J_\phi(s,t) = 0 \Rightarrow s = t$. □

Definition 17.7 A sequence $\{s_n\}$ in an Altering JS-metric space (S, J_ϕ) is said to be J_ϕ-Cauchy if for given $\varepsilon > 0$, $\exists\, n_0 \in N$ such that $J_\phi(s_n, s_m) < \varepsilon$ for all $m, n \geq n_0$.

Definition 17.8 The Altering JS-metric space (S, J_ϕ) is said to be J_ϕ-complete if every J_ϕ-Cauchy sequence in S is J_ϕ-convergent in it.

17.4 Some Fixed Point Theorems in Altering JS-metric Spaces

Here, we discuss some fixed point theorems of self-maps with certain contractive conditions under the frame of Altering JS-metric space.

Theorem 17.2 *Let (S, J_ϕ) be a complete Altering JS-metric space and let $T : S \to S$ be a continuous mapping such that $\lambda(J_\phi(Ts, Tt) \leq \delta\lambda(J_\phi(s, Ts)) - (1-\delta)\mu$ $(J_\phi(t, Tt))$, where $\lambda, \mu \in \Psi$ and $\delta \in (\frac{1}{2}, 1)$. Then, T has a fixed point in S.*

Proof Let $s_0 \in S$ and let $s_1 = Ts_0$. If $s_0 = s_1 = Ts_0$, then s_0 is a fixed point. Suppose not, then $J_\phi(s_0, s_1) > 0$. By the hypothesis, we have

$$\lambda(J_\phi(Ts_0, Ts_1) \leq \delta\lambda(J_\phi(s_0, Ts_0)) - (1-\delta)\mu(J_\phi(s_1, Ts_1))$$

Let $Ts_1 = s_2$. If $s_1 = s_2 = Ts_1$, then s_1 is a fixed point. Suppose $J_\phi(s_1, s_2) > 0$. Now, using the above inequality,

$$\begin{aligned}\lambda(J_\phi(s_1, s_2) &= \lambda(J_\phi(Ts_0, Ts_1)\\ &\leq \delta\lambda(J_\phi(s_0, Ts_0)) - (1-\delta)\mu(J_\phi(s_1, Ts_1))\\ &= \delta\lambda(J_\phi(s_0, s_1)) - (1-\delta)\mu(J_\phi(s_1, s_2))\\ &\leq \delta\lambda(J_\phi(s_0, s_1))\\ &\leq \lambda(J_\phi(s_0, s_1))\end{aligned}$$

Since λ is non-decreasing, we get

$$J_\phi(s_1, s_2) \leq J_\phi(s_0, s_1)$$

Again let $Ts_2 = s_3$ and let $J_\phi(s_2, s_3) > 0$:

$$\begin{aligned}\lambda(J_\phi(s_2, s_3) &= \lambda(J_\phi(Ts_1, Ts_2)\\ &\leq \delta\lambda(J_\phi(s_1, Ts_1)) - (1-\delta)\mu(J_\phi(s_2, Ts_2))\\ &= \delta\lambda(J_\phi(s_1, s_2)) - (1-\delta)\mu(J_\phi(s_2, s_3))\\ &\leq \delta\lambda(J_\phi(s_1, s_2))\\ &\leq \lambda(J_\phi(s_1, s_2))\end{aligned}$$

Since λ is non-decreasing, we get

$$J_\phi(s_2, s_3) \leq J_\phi(s_1, s_2)$$

By induction, we acquire an iterative sequence $\{s_n\} \in S$, such that $J_\phi(s_n, s_{n+1}) \leq J_\phi(s_{n-1}, s_n) \forall\ n$. It is clear that $\left\{J_\phi(s_n, s_{n+1})\right\}$ is a non-increasing sequence and $J_\phi(s_n, s_{n+1}) \geq 0\ \forall\ n$. Hence, there exists some $s^* \geq 0$ such that $J_\phi(s_n, s_{n+1}) \to s^*$ as $n \to \infty$.

By the hyposthesis,

$$\begin{aligned}\lambda(J_\phi(Ts_n, Ts_{n+1}) &\leq \delta\lambda(J_\phi(s_n, Ts_n)) - (1-\delta)\mu(J_\phi(s_{n+1}, Ts_{n+1}))\\ &= \delta\lambda(J_\phi(s_n, s_{n+1})) - (1-\delta)\mu(J_\phi(s_{n+1}, s_{n+2}))\end{aligned}$$

Letting $n \to \infty$, we have $\lambda(s^*) \leq \delta\lambda(s^*) - (1-\delta)\mu(s^*)$

$$\Rightarrow \lambda(s^*) \leq -\mu(s^*)$$

Since $\lambda, \mu \in \Psi$, we get $\lambda(s^*) = \mu(s^*) = 0$, thus giving $s^* = 0$. Hence, $J_\phi(s_n, s_{n+1}) \to 0$. Now, let us prove that the sequence $\{s_n\}$ is a $J_\phi-$Cauchy sequence in S:

$$\begin{aligned}\lambda(J_\phi(s_n, s_m)) &= \lambda(J_\phi(Ts_{n-1}, Ts_{m-1}))\\ &\leq \delta\lambda(J_\phi(s_{n-1}, Ts_{n-1})) - (1-\delta)\mu(J_\phi(s_{m-1}, Ts_{m-1}))\\ &= \delta\lambda(J_\phi(s_{n-1}, s_n)) - (1-\delta)\mu(J_\phi(s_{m-1}, s_m))\end{aligned}$$

As $n, m \to \infty$, thus $J_\phi(s_n, s_m) \to 0$

Thus, $\{s_n\}$ is a $J_\phi-$Cauchy sequence in S. Since S is $J_\phi-$complete, thus $\{s_n\}$ is $J_\phi-$convergent to $s \in S$.

$$\begin{aligned}\text{Now, } \phi((J_\phi(s, Ts)) &\leq \lim_{n\to\infty} \sup\left[\phi(k.(J_\phi(s_n, Ts))\right]\\ &\leq \lim_{n\to\infty} \sup\left[\phi(k.\lambda(J_\phi(s_n, Ts))\right]\end{aligned}$$

$$\leq \lim_{n\to\infty} \sup\left[\phi(k.\lambda(J_\phi(Ts_{n-1}, Ts))\right]$$
$$\leq \lim_{n\to\infty} \sup k.\phi[\delta\lambda[J_\phi(s_{n-1}, Ts_{n-1}) - (1-\delta)\mu(J_\phi(s, Ts))$$
$$= \lim_{n\to\infty} \sup[k.\phi[\delta\lambda[J_\phi(s_{n-1}, s_n) - (1-\delta)\mu(J_\phi(s, Ts))]$$

$\Rightarrow \phi((J_\phi(s, Ts)) \leq -k(1-\delta).\phi[\mu(J_\phi(s, Ts))]$. This is possible if and only if $\phi((J_\phi\ (s, Ts)) = 0 \Rightarrow J_\phi(s, Ts) = 0$. Thus, s is the fixed point of T. □

Corollary 17.1 *Let (S, J_ϕ) be a complete b-metric space and let $T : S \to S$ be a continuous mapping such that $\lambda(J_\phi(Ts, Tt) \leq \delta\lambda(J_\phi(s, Ts)) - (1-\delta)\mu(J_\phi(t, Tt))$; where $\lambda, \mu \in \Psi$ and $\delta \in (\frac{1}{2}, 1)$. Then, T has a fixed point in S.*

Proof The proof follows from Proposition 17.7 and Theorem 17.2. □

Corollary 17.2 *If (S, J_ϕ) is a complete dislocated metric space and $T : S \to S$ is a continuous mapping satisfying the condition $\lambda(J_\phi(Ts, Tt) \leq \delta\lambda(J_\phi(s, Ts)) - (1-\delta)\mu(J_\phi(t, Tt))$, where $\lambda, \mu \in \Psi$ and $\delta \in (\frac{1}{2}, 1)$. Then, T has a fixed point in S.*

Proof The proof is an immediate outcome of Proposition 17.7 and Theorem 17.2. □

Corollary 17.3 *Let (S, J_ϕ) be a complete JS-metric space and let $T : S \to S$ be a continuous mapping satisfying the contractive condition $\lambda(J_\phi(Ts, Tt) \leq \delta\lambda(J_\phi(s, Ts)) - (1-\delta)\mu(J_\phi(t, Tt))$, where $\lambda, \mu \in \Psi$ and $\delta \in (\frac{1}{2}, 1)$. Then, T has a fixed point in S.*

Proof Proof follows from Proposition 17.6 and Theorem 17.2. □

17.5 Coupled Fixed Point Theorems and Partial Ordered Set

The idea of the coupled fixed point theorem was initiated by Bhaskar and Lakshmikantham [5]. In 2012, Choudhury and Kundu [6] proved a Kannan type coupled fixed point result. In this section we apply our results to obtain the coupled fixed point for the mapping $G : S \times S \to S$, where S is an Altering JS-metric space.

Definition 17.9 [5] Let $(S, \preceq)$ be a partially ordered set and $G : S \times S \to S$ be a single-valued mapping. Then, the mapping G is said to have mixed monotone property if G is monotonically non-decreasing with respect to its first entry and is monotonically non-increasing with respect to its second entry. That is, for any $s, t \in S$,

$$s_1, s_2 \in S,\ s_1 \preceq s_2 \Rightarrow G(s_1, t) \preceq G(s_2, t)$$

and

$$t_1, t_2 \in S,\ t_1 \preceq t_2 \Rightarrow G(s, t_1) \succeq G(s, t_2).$$

An element $(s, t) \in S \times S$ is called a coupled fixed point of the mapping G if $G(s, t) = s$ and $G(t, s) = t$.

Theorem 17.3 *[6] Let $(S, \preceq)$ be a partially ordered set and suppose there is a metric d on S such that (S, d) is a complete metric space. Let $G : S \times S \to S$ be a mapping such that G has the mixed monotone property and satisfies $d(G(s,t), G(u,v)) \leq \lambda(M(s,t,u,v)).(M(s,t,u,v))$,*
where $M(s,t,u,v) = \dfrac{d(s,G(s,t)) + d(t,G(t,s)) + d(u,G(u,v)) + d(v,G(v,u))}{4}$
for all $s,t,u,v \in S$ with $s \preceq u$ and $t \succeq v$ and $\lambda : \mathbb{R}^+ \to [0,1)$ be a function such that $\lambda(s_n) \to 1 \Rightarrow s_n \to 0$. Also suppose that
(a) G is continuous or
(b) S has the properties,
(i) if a non-decreasing sequence $\{s_n\} \to s$, then $s_n \preceq s$, for all n,
(ii) if a non-increasing sequence $\{t_n\} \to t$, then $t \preceq t_n$, for all n.
If there exist $s_0, t_0 \in S$ such that $s_0 \preceq G(s_0, t_0)$ and $t_0 \succeq G(t_0, s_0)$, then G has a coupled fixed point in $S \times S$.

Now, we will prove the following theorem that connects the Altering JS-metric space with the coupled fixed point. The following theorem is the consequence of Theorem 17.2.

Theorem 17.4 *Let $(S, \preceq)$ be a partially ordered set and suppose there is a metric J_ϕ on S such that (S, J_ϕ) is a complete Altering JS-metric space. Let $G : S \times S \to S$ be a continuous mapping such that G has the mixed monotone property and satisfies*

$$\lambda\left[max\left\{J_\phi(G(s,t), G(u,v)), J_\phi(G(u,v), G(s,t))\right\}\right] \leq \delta\lambda\left[max\left\{J_\phi(s, G(s,t)),\right.\right.$$
$$\left.\left.J_\phi(t, G(t,s))\right\}\right] - (1-\delta)\mu\left[max\left\{J_\phi(u, G(u,v)), J_\phi(v, G(v,u))\right\}\right]$$

where $\lambda, \mu \in \Psi$ and $\delta \in (\frac{1}{2}, 1)$. Also suppose that S has the properties,
(i) if a non-decreasing sequence $\{s_n\} \to s$, then $s_n \preceq s$, for all n,
(ii) if a non-increasing sequence $\{t_n\} \to t$, then $t \preceq t_n$, for all n.
If there exist $s_0, t_0 \in S$ such that $s_0 \preceq G(s_0, t_0)$ and $t_0 \succeq G(t_0, s_0)$, then G has a coupled fixed point in $S \times S$.

Proof Let $s, t, u, v, s^*, t^* \in S$ and $max\left\{J_\phi(s,t), J_\phi(u,v)\right\} = J'_\phi(s^*, t^*)$. By the Proposition 4, J'_ϕ is an Altering JS-metric. Since (S, J_ϕ) is a complete Altering JS-metric space, (S, J'_ϕ) is also a complete Altering JS-metric space. Thus, all the conditions of Theorem 17.2 are satisfied. Thus, we get the required result. □

17.6 Conclusion

In this paper, we introduced Altering JS-metric space and proved some results. We showed that metric space, b-metric space, dislocated metric space, and JS-metric space are Altering JS-metric spaces. Later, we proved some fixed point theorems in Altering JS-metric space. We also proved a coupled fixed point theorem in a partial ordered set using Altering JS-metric space. It has been proved that a metric space, *b*-metric space, dislocated metric space are Altering JS-metric spaces. The topological properties of metric space, b-metric space [26], dislocated metric space [27, 28] have been studied in the literature, it is important to note that topological properties of Altering JS-metric space may be studied in future. One may refer to [30] for connections of Altering JS-metric space with graph theory via fixed point theory. At the end, it must be noted that richer structure like Altering JS-metric space may open several new dimensions of research in several areas of mathematics.

References

1. Jleli, M., Samet, B.: A generalized metric space and related fixed point theorems. Fixed Point Theo. Appl. **2015**, 61 (2015)
2. Khan, M.S., Swaleh, M., Sessa, S.: Fixed point theorems by altering distances between the points. Bull. Austral. Math. Soc. **30**(1), 1–9 (1984)
3. Delbosco, D.: Un'estensione di un teorema sul punto fissi di S.Reich. Rend. Sem. Mat. Univers. Politecn. Torino **35**, 233–238 (1976–1977)
4. Skof, F.: Teorema di punti fisso per applicazioni negli spazi metrici. Atti. Accad. Sci. Torino **111**, 323–329 (1977)
5. Bhaskar, T.G., Lakshmikantham, V.: Fixed point theorems in partially ordered metric spaces and applications. Nonlinear Anal. TMA **65**(7), 1379–1393 (2006)
6. Choudhury, B.S., Kundu, A.: On coupled generalised Banach and Kannan type contractions. J. Nonlinear Sci. Appl. **5**, 259–270 (2012)
7. Shoaib, A., Mehmood, Q.: Fixed point results in Js multiplicative metric spaces. Turkish J. Anal. Number Theo. **6**, 159–163 (2018)
8. Pariya, A., Sindersiya, A., Badshah, V.H., Gupta, N.: Fixed point theorem for quasi-js-metric spaces. Asian J. Math. Stat. **10**(1), 13–17 (2017)
9. Gajić, L., Ralević, N.: A Common Fixed Point Result in JS-metric space, Bulletin T. CLI de l'Académie serbe des sciences et des arts-2018, Classe des Sciences mathématiques et naturelles, Sciences mathématiques, vol. 43 (2018)
10. Fréchet, M.: Sur Quelques points du calcul fonctionnel. Rend. Circ. Matem, Palermo XXII (1906)
11. Turkoglu, D., Manav, N.: Fixed point theorems in a new type of modular metric spaces. Fixed Point Theo. Appl. **2018**, 25 (2018)
12. Wang, S.Z., Li, B.Y., Gao, Z.M., Iseki, K.: Some fixed point theorems on expansion mappings. Math. Japon. **29**, 631–636 (1984)
13. Khan, M.A., Khan, M.S., Sessa, S.: Some theorems on expansion mappings and their fixed points. Demonstratio Math. **3**, 673–683 (1986)
14. Naidu, S.V.R.: Some fixed point theorems in metric spaces by altering distances. Czechoslovak Math. J. **53**, 205–212 (2003)
15. Sastry, K.P.R., Babu, G.V.R.: Some fixed point theorems by altering distances between the points. Indian J. Pure Appl. Math. **30**, 641–647 (1999)
16. Rhoades, B.E.: Some theorems on weakly contractive maps. Nonlinear Anal. **47**, 2683–2693 (2001)
17. Dhutta, P.N., Choudhury, B.S.: Ageneralization of contraction principle in metric spaces. Fixed Point Theo. Appl. **87**, 109–116 (2008)
18. Bakhtin, I.A.: The contraction principle in quasimetric spaces. Funct. Anal. **30**, 26–37 (1989)
19. Czerwik, S.: Contraction mappings in b-metric spaces. Acta Math. Inf. Univ. Ostrav. **1**, 5–11 (1993)
20. Hitzler, P., Seda, A.K.: Dislocated topologies. J. Electr. Eng. **51**(12), 3–7 (2000)

21. Park, S., Rhoades, B.E.: Some fixed point theorems for expansion mappings. Math. Japonica **33**, 129–132 (1988)
22. Pathak, H.K., Kang, S.M., Ryu, J.W.: Some fixed points of expansion mappings. Int. J. Math. Math. Sci. **19**(1), 97–102 (1996)
23. Kang, S.M.: Fixed points for expansion mappings. Math. Japonica **38**, 713–717 (1993)
24. Taniguchi, T.: Common fixed point theorems on expansion mappings on complete metric spaces. Math. Japonica **34**, 139–142 (1989)
25. Kakutani, S.: A generalization of Brouwer's fixed point theorem. Duke Math. J. **7**, 457–459 (1941)
26. Nash, J.F., Jr.: Equilibrium points in N-person games. Proc. Nat. Acad. Sci. U.S.A. **36**, 48–49 (1950)
27. An, T.V., Tuyen, L.Q., Dung, N.V.: Stone-type theorem on b-metric spaces and applications. Top. Appl. **50–64**, 185–186 (2015)
28. Karapınar, E., Salimi, P.: Dislocated metric space to metric spaces with some fixed point theorems. Fixed Point Theo. Appl. **2013**, 222 (2013)
29. Pasicki, L.: Dislocated metric and fixed point theorems. Fixed Point Theo. Appl. **2015**, 82 (2015)
30. Acharjee, S., Gnanachandra, P., Jafari, S., Varunnya, X.M.J.: Fixed point theorems for generalised contractions in Altering JS-metric space with graph (submitted)

SPRINGER NATURE

GPSR Compliance

The European Union's (EU) General Product Safety Regulation (GPSR) is a set of rules that requires consumer products to be safe and our obligations to ensure this.

If you have any concerns about our products, you can contact us on ProductSafety@springernature.com

In case Publisher is established outside the EU, the EU authorized representative is:

Springer Nature Customer Service Center GmbH
Europaplatz 3
69115 Heidelberg, Germany

The manufacturer's authorised representative in the EU is Springer Nature Customer Service Centre GmbH, Europaplatz 3, 69115 Heidelberg, Germany. If you have any concerns regarding our products, please contact ProductSafety@springernature.com

Printed and bound by CPI Group (UK) Ltd, Croydon, CR0 4YY
01/10/2026
02225896-0009